Environmentally Conscious Transportation

Environmentally Conscious Transportation

Edited by
Myer Kutz

John Wiley & Sons, Inc.

This book is printed on acid-free paper. ⊗

Copyright © 2008 by John Wiley & Sons, Inc. All rights reserved

Published by John Wiley & Sons, Inc., Hoboken, New Jersey
Published simultaneously in Canada

No part of this publication may be reproduced, stored in a retrieval system, or transmitted in any form or by any means, electronic, mechanical, photocopying, recording, scanning, or otherwise, except as permitted under Section 107 or 108 of the 1976 United States Copyright Act, without either the prior written permission of the Publisher, or authorization through payment of the appropriate per-copy fee to the Copyright Clearance Center, 222 Rosewood Drive, Danvers, MA 01923, (978) 750-8400, fax (978) 646-8600, or on the Web at www.copyright.com. Requests to the Publisher for permission should be addressed to the Permissions Department, John Wiley & Sons, Inc., 111 River Street, Hoboken, NJ 07030, (201) 748-6011, fax (201) 748-6008, or online at www.wiley.com/go/permissions.

Limit of Liability/Disclaimer of Warranty: While the publisher and the author have used their best efforts in preparing this book, they make no representations or warranties with respect to the accuracy or completeness of the contents of this book and specifically disclaim any implied warranties of merchantability or fitness for a particular purpose. No warranty may be created or extended by sales representatives or written sales materials. The advice and strategies contained herein may not be suitable for your situation. You should consult with a professional where appropriate. Neither the publisher nor the author shall be liable for any loss of profit or any other commercial damages, including but not limited to special, incidental, consequential, or other damages.

For general information about our other products and services, please contact our Customer Care Department within the United States at (800) 762-2974, outside the United States at (317) 572-3993 or fax (317) 572-4002.

Wiley publishes in a variety of print and electronic formats and by print-on-demand. Some material included with standard print versions of this book may not be included in e-books or in print-on-demand. If this book refers to media such as a CD or DVD that is not included in the version you purchased, you may download this material at http://booksupport.wiley.com. For more information about Wiley products, visit www.wiley.com.

Library of Congress Cataloging-in-Publication Data:

Environmentally conscious transportation / edited by Myer Kutz.
    p. cm.
 Includes bibliographical references and index.
 ISBN 978-0-471-79369-4 (cloth)
 1. Transportation—Environmental aspects. 2. Transportation—Social aspects. 3. Sustainable engineering. I. Kutz, Myer.
 HE147.65.E585 2008
 388'.049—dc22

                                                    2007029089

To Noreen, Diane and Chris, and Liz and Richard

Contents

Contributors   ix
Preface   xi

1. The Economic and Environmental Footprints of Transportation   1
   Lester B. Lave and W. Michael Griffin

2. Public Transportation and the Environment   15
   Michael D. Meyer

3. Transportation and Air Quality   47
   Mohan M. Venigalla

4. The Social Cost of Motor Vehicle Use in the United States   57
   Mark A. Delucchi

5. Traffic Congestion Management   97
   Nagui M. Rouphail

6. Electric and Hybrid Vehicle Design and Performance   129
   Andrew Burke

7. Hydraulic Hybrid Vehicles   191
   Amin Mohaghegh Motlagh, Mohammad Abuhaiba, Mohammad H. Elahinia, and Walter W. Olson

8. Biofuels for Transportation   213
   Aaron Smith, Cesar Granda, and Mark Holtzapple

9. Life-Cycle Assessment as a Tool for Sustainable Transportation Infrastructure Management   257
   Gerardo W. Flintsch

10. Pavement and Bridge Management and Maintenance   283
    Sue McNeil

11. Impacts of the Aviation Sector on the Environment   301
    Victoria Williams

Index   331

Contributors

Mohammad Abuhaiba
The University of Toledo
Toledo, Ohio

Andrew Burke
University of California, Davis
Davis, California

Mark A. Delucchi
University of California, Davis
Davis, California

Mohammad H. Elahinia
The University of Toledo
Toledo, Ohio

Gerardo W. Flintsch
Virginia Polytechnic Institute
and State University
Blacksburg, Virginia

Cesar Granda
Texas A&M University
College Station, Texas

W. Michael Griffin
Carnegie Mellon University
Pittsburgh, Pennsylvania

Mark Holtzapple
Texas A&M University
College Station, Texas

Lester B. Lave
Carnegie Mellon University
Pittsburgh, Pennsylvania

Sue McNeil
University of Delaware
Newark, Delaware

Michael D. Meyer
Georgia Institute of Technology
Atlanta, Georgia

Amin Mohaghegh Motlagh
The University of Toledo
Toledo, Ohio

Walter W. Olson
The University of Toledo
Toledo, Ohio

Nagui M. Rouphail
North Carolina State University
Raleigh, North Carolina

Aaron Smith
Texas A&M University
College Station, Texas

Mohan M. Venigalla
George Mason University
Fairfax, Virginia

Victoria Williams
Imperial College
London, United Kingdom

Preface

Many readers will approach the volumes in the **Wiley Series in Environmentally Conscious Engineering** with some degree of familiarity with, knowledge about, or even expertise in, one or more of a range of environmental issues, such as climate change, pollution, and waste. Such capabilities may be useful for readers of this series, but they aren't strictly necessary, for the purpose of this series is not to help engineering practitioners and managers deal with the *effects* of man-induced environmental change. Nor is it to argue about whether such effects degrade the environment only marginally or to such an extent that civilization as we know it is in peril, or that any effects are nothing more than a scientific-establishment-and-media-driven hoax and can be safely ignored. (Authors of a plethora of books, even including fiction, and an endless list of articles in scientific and technical journals, have weighed in on these matters, of course.) On the other hand, this series of engineering books does take as a given that the overwhelming majority in the scientific community is correct, and that the future of civilization depends on minimizing environmental damage from industrial, as well as personal, activities. At the same time, the series does not advocate solutions that emphasize only curtailing or cutting back on these activities. Instead, its purpose is to exhort and enable engineering practitioners and managers to reduce environmental impacts—to engage, in other words, in *environmentally conscious engineering*, a catalog of practical technologies and techniques that can improve or modify just about anything engineers do, whether they are involved in designing something, making something, obtaining or manufacturing materials and chemicals with which to make something, generating power, or transporting people and freight.

Increasingly, engineering practitioners and managers need to know how to respond to challenges of integrating environmentally conscious technologies, techniques, strategies, and objectives into their daily work, and, thereby, find opportunities to lower costs and increase profits while managing to limit environmental impacts. Engineering practitioners and managers also increasingly face challenges in complying with changing environmental laws. So companies seeking a competitive advantage and better bottom lines are employing environmentally responsible practices to meet the demands of their stakeholders, who now include not only owners and stockholders, but also customers, regulators, employees, and the larger, even worldwide community.

Engineering professionals need references that go far beyond traditional primers that cover only regulatory compliance. They need integrated approaches centered on innovative methods and trends in design, manufacturing, power generation, transportation, and materials handling that help them focus on using environmentally friendly processes. They need resources that help them participate in strategies for designing environmentally responsible products, methods, and processes—resources that provide a foundation for understanding and implementing principles of environmentally conscious engineering.

To help engineering practitioners and managers meet these needs, I envisioned a flexibly connected series of edited books, each devoted to a broad topic under the umbrella of environmentally conscious engineering. The series started with three volumes that are closely linked—environmentally conscious mechanical design, environmentally conscious manufacturing, and environmentally conscious materials and chemicals processing. The series has continued with another set of linked volumes—a volume on environmentally conscious alternative energy production and this volume on environmentally conscious transportation. Two other volumes in this second linked set are in the works—environmentally conscious materials handling and environmentally conscious conventional energy production.

The intended audience for the series is practicing engineers and upper-level students in a number of areas—mechanical, chemical, industrial, manufacturing, plant, environmental, civil, and transportation—as well as engineering managers. This audience is broad and multidisciplinary. Some of the practitioners who make up this audience are concerned with design, some with manufacturing and processing, some with the everyday aspects of energy production and moving people and goods, and others with economics and setting and implementing corporate and public policies. These practitioners work in a wide variety of organizations, including institutions of higher learning, design, manufacturing, and consulting firms, as well as federal, state and local government agencies. So what made sense in my mind was a series of relatively short books, rather than a single, enormous book, even though the topics in each of the smaller volumes have linkages and some of the topics might be suitably contained in more than one freestanding volume. In this way, each volume is targeted at a particular segment of the broader audience. At the same time, a linked series is appropriate because every practitioner, researcher, and bureaucrat can't be an expert on every topic, especially in so broad and multidisciplinary a field, and may need to read an authoritative summary on a professional level of a subject that he or she is not intimately familiar with but may need to know about for a number of different reasons.

The **Wiley Series in Environmentally Conscious Engineering** is composed of practical references for engineers who are seeking to answer a question, solve a problem, reduce a cost, or improve a system or facility. These books are not research monographs. The purpose is to show readers what options are available in a particular situation and which option they might choose to solve problems at hand. I want these books to serve as a source of practical advice to readers.

I would like them to be the first information resource a practicing engineer reaches for when faced with a new problem or opportunity—a place to turn to even before turning to other print sources, even any officially sanctioned ones, or to sites on the Internet. So the books have to be more than references or collections of background readings. In each chapter, readers should feel that they are in the hands of an experienced consultant who is providing sensible advice that can lead to beneficial action and results.

The fifth volume in the series, *Environmentally Conscious Transportation*, is an important reference for environmental, civil, transportation, mechanical, and industrial engineers, as well as public policy planners and officials. This book examines the societal costs of transportation in their broadest sense, both environmental and economic. The book's contributors discuss a wide range of transportation modes, from private automobiles, with a separate chapter on biofuels, to heavy trucks and buses, to rail and public transportation systems, to aircraft. This book also focuses on pollution from both ground vehicles and aircraft, traffic congestion management, and transportation infrastructure management, with special attention paid to life-cycle assessment (LCA).

I asked the contributors, located in the United States and the United Kingdom, to provide short statements about the contents of their chapters and why the chapters are important. Here are their responses:

Lester B. Lave (Carnegie Mellon University in Pittsburgh, Pennsylvania), who, together with Michael Griffin, contributed Chapter 1, **The Economic and Environmental Footprints of Transportation**, writes, "This chapter explores the marvelous achievement of modern passenger and freight transport systems. We are reliably transported halfway around the world in less than a day for hundreds of dollars and enjoy fresh fruit delivered from half a world away for a few cents. Technology advances and market competition have produced a $500 billion per year industry (5% of GDP) that employs three million workers. The market competition does not deal with the externalities of these systems, such as the social costs of imported oil for fuel, associated carbon dioxide emissions, air and water pollution, and injuries and deaths. Transportation consumes 7 percent of the 100 quadrillion Btus used by the U.S. economy, consuming a much larger proportion of total petroleum. We quantify the direct resource use and effects, as well as the full life cycle (extraction to disposal) of the transport systems. The total externalities of the sector were $118 billion in 2000, 23 percent of the total revenue of the sector. On a life-cycle basis, externalities are only 13 percent of total revenue, since transportation has larger externalities than the systems that support it."

Michael D. Meyer (Georgia Institute of Technology in Atlanta), who contributed Chapter 2, **Public Transportation and the Environment**, writes, "Public transit systems are often viewed as one of the most important strategies for minimizing environmental impacts from transportation systems. By providing an alternative to the use of the automobile, transit services can reduce the overall level of vehicle-related emissions in a metropolitan area. Over the long term, effective transit services can also influence where businesses and households locate in a metropolitan area. Thus, it is not surprising that in many developed countries, land-use policies go hand-in-hand with transit investment. At the individual station level, incentives to encourage transit-oriented development have become a prerequisite to successful land developments.

"This chapter examines the linkage between public transportation systems and environmental impacts. This relationship relates not only to the potential substitution of transit mobility for that based on private auto use, but also to the environmental impacts that occur when transit infrastructure is built. In addition, transit operations can create environmental impacts as well, such as, noise, air emissions, water quality impacts, and vibrations. Mitigating the impacts of guideway or station construction and transit operations is necessary, but not the sole component of a true environmentally conscious transit program. As described in this chapter, environmentally conscious transit should be linked strongly to urban design, community development, and management of the daily operations of the transit agency. By so doing, public transit can become a major contributor to the much broader goal of developing sustainable communities."

Mohan M. Venigalla (George Mason University in Fairfax, Virginia), who contributed Chapter 3, **Transportation and Air Quality**, writes, "Transportation sources contribute heavily to the poor quality of ambient air. Presented in this chapter are discussions on National Ambient Air Quality Standards (NAAQS), air quality designations and the basis for those designations. This chapter also presents an overview of the air quality management, with specific emphasis on managing the contribution from the transportation sources. Strategies outlined include emission control at the source (vehicle), fuel and transportation planning levels."

Mark A. Delucchi (University of California, Davis), who contributed Chapter 4, **The Social Cost of Motor Vehicle Use in the United States**, writes, "The social cost of motor vehicle use includes all of the public, private, environmental, and energy costs of using motor vehicles. Analyses of the social cost of motor vehicle use usually excite considerable interest, if only because nearly all of us use motor vehicles. Researchers have performed social-cost analyses for a variety reasons, and have used them in a variety of ways, to support a wide range of policy positions. Some researchers have used social-cost analyses to argue that motor vehicles and gasoline are terrifically underpriced, while others have used them to downplay the need for drastic policy intervention in the transportation sector.

"My colleagues at the University of California and I have done a detailed analysis of some of the costs of motor vehicle use in the United States. We explain the purpose of estimating the total social cost of motor vehicle use, briefly review some of the pertinent research, explain the conceptual framework and cost classification, and present and discuss their cost estimates. We emphasize that while social-cost analysis can help analyze the costs and benefits of transportation projects, set efficient prices for fuels and vehicles, or prioritize research funding, it cannot tell us precisely what should be done to improve transportation systems."

Nagui M. Rouphail (North Carolina State University in Raleigh), who contributed Chapter 5, **Traffic Congestion Management**, writes, "This chapter presents fundamental information on measuring vehicle activity and emissions, highlights the effects of traffic congestion on emissions and air quality, discusses transportation demand and supply-oriented methods that can be implemented to mitigate those effects, and finally discusses methods for assessing the effect of the mitigation measures. With surface transportation directly accounting for 40 percent of nitrogen oxides, 56 percent of carbon monoxide and 28 percent of volatile organic compounds in the national emission inventory, issues of traffic congestion and mitigation should be high on the list of priorities for scientists, engineers, and policy makers who promulgate the need for an environmentally friendly transportation system."

Andrew Burke (University of California, Davis), who contributed Chapter 6, **Electric and Hybrid Vehicle Design and Performance**, writes, "Electric and hybrid vehicles, including fuel cell vehicles, are expected to be key new technologies for the design and commercialization of ultra-clean and energy efficient vehicles in the future. Whether these new technologies are successfully commercialized and when that commercialization is likely to occur depends both on technical progress and the resultant vehicle performance and on the cost of the various advanced vehicles compared to conventional ICE vehicles. In this chapter, both performance and cost of the various technologies are assessed and cost estimates are given from which the relative attractiveness of vehicles using the various technologies can be judged. Relationships are presented from which the greenhouse gas emissions of the advanced vehicles using petroleum, electricity, hydrogen, and biomass-based fuels can be calculated."

Walter W. Olson (The University of Toledo in Toledo, Ohio), who contributed Chapter 7, **Hydraulic Hybrid Vehicles**, together with Amin Mohaghegh Motlagh, Mohammad Abuhaiba, and Mohammad H. Elahinia, writes, "A major contributor to today's environmental footprint of transportation is heavy vehicles including trucks and buses. The needs of mass transportation of people and goods require the use of large vehicles with high power densities. To achieve these power densities, these vehicles are almost uniformly powered by large diesel engines. While much work has been performed on reducing emissions and improving fuel economy of the diesel engines through better control of the combustion process, further achievements will be made through hybridization of the powertrain. But today's research in the electric hybridization of light vehicles, mainly passenger cars, cannot be directly applied to heavy vehicles because of the large differences in power density needed. This requires an alternative path to heavy vehicle hybridization, a path provided by the use of hydraulics and fluid power.

"With the exception of engineers directly in the design and maintenance areas of work vehicles and agriculture, this technology is little understood by transportation and environmental engineers in general. This chapter provides access to the concepts and devices of fluid power to transportation and environmental engineers. In addition, this chapter allows practitioners and problem solvers in environmentally conscious transportation to see how to make choices to improve the environment while considering large vehicle needs."

Aaron Smith (Texas A&M University in College Station), who contributed Chapter 8, **Biofuels for Transportation**, together with Cesar Granda and Mark Holtzapple, writes, "The dominate biofuels known today (ethanol, biodiesel and hydrogen), as well as other less well-known biofuels, are reviewed objectively with regard to key topics such as biomass source, manufacturing methods, quality standards, vehicle use, performance, emissions, distribution, safety, subsidy, and fuel availability. The focus of the chapter is to present the "big picture" surrounding each biofuel with quantitative and qualitative data presented to compare biofuels to petroleum. An environmentally conscious future will need biofuels to become independent from fossil fuels, address global warming, improve emissions, and help rural economies."

Gerardo W. Flintsch (Virginia Polytechnic Institute and State University in Blacksburg), who contributed Chapter 9, **Life-Cycle Assessment of Transportation Infrastructure**, writes, "This chapter discusses the use of life-cycle assessment (LCA) as a tool for sustainable transportation infrastructure management. The strategic, network, and project-level decisions supported by transportation infrastructure asset management have many technical, economic, social, and environmental impacts over the life-cycle of the transportation infrastructure.

"The assessment of these impacts over the whole life of infrastructure systems is necessary for making informed decisions on how to define policies, allocate resources, select projects, and/or design and construct these projects in a sustainable manner. Therefore, to support sustainable development, transportation asset management decisions should be constructed over three pillars: economic development, ecological sustainability, and social desirability. Current practice typically includes the consideration of technical and economic issues through life-cycle cost analysis (LCCA).

"Environmental aspects are increasingly being considered though LCA, and although the inclusion of social aspects is in its infancy, they should not be neglected. The consideration of environmental and social goals in conjunction with economic considerations promises to enhance transportation infrastructure asset management while promoting sustainable transportation infrastructure systems."

Sue McNeil (University of Delaware in Newark), who contributed Chapter 10, **Pavement and Bridge Management and Maintenance**, writes, "This chapter introduces the concept of asset management and then reviews opportunities for integrating environmental issues into the asset management framework. The chapter concludes with a brief case study based in New Zealand. The relationship between asset management and the environment is particularly important in the context of sustainable practices and the full life-cycle cost of various strategies."

Victoria Williams (Imperial College in London, United Kingdom), who contributed Chapter 11, **Impacts of the Aviation Sector on the Environment**, writes, "Air transport has transformed the way the world interacts, but not without environmental consequences. Chapter 11 discusses these impacts, focusing on aviation's share in climate change, noise nuisance, and poor air quality. For each of these, the decisions and practices that contribute to the impact and to its mitigation are explored. Other environmental issues relating to aircraft and airport systems are also discussed. The chapter highlights the challenge of comparing and prioritizing the environmental impacts of aviation and the difficult balance between social benefits and environmental costs for this rapidly growing industry."

That ends the contributors' comments. I would like to express my heartfelt thanks to all of them for having taken the opportunity to work on this book. Their lives are terribly busy, and it is wonderful that they found the time to write thoughtful and complex chapters. I developed the book because I believed it could have a meaningful impact on the way many engineers approach their daily work, and I am gratified that the contributors thought enough of the idea that they were willing to participate in the project. Thanks also to my editor, Bob Argentieri, for his faith in the project from the outset. And a special note of thanks to my wife, Arlene, whose constant support keeps me going.

Myer Kutz
Delmar, New York

CHAPTER 1

THE ECONOMIC AND ENVIRONMENTAL FOOTPRINTS OF TRANSPORTATION

Lester B. Lave and W. Michael Griffin
Green Design Institute
Carnegie Mellon University
Pittsburgh, Pennsylvania

1	INTRODUCTION	1	6 LIFE CYCLE VERSUS DIRECT EFFECTS	11
2	ECONOMIC ACTIVITY	2	7 TRANSPORTATION EXTERNALITIES	11
3	ENERGY USE	4		
4	POLLUTION EMISSIONS	5	8 SUMMARY AND CONCLUSION	11
5	INJURIES AND DEATHS	9		

1 INTRODUCTION

This chapter explores both passenger and freight transport in the United States for a recent year, and over time where available. Our focus is on commercial transport:

1. Sales, ton-miles or passenger-miles, employment for each industry
2. Fuel use and CO_2 emissions for each industry
3. Air pollution emissions for each industry
4. Injuries and deaths by industry
5. Direct resource use and environmental discharges and the full life cycle
6. Social cost in dollars of the externalities, compared to the industry revenue

Passenger transportation and freight transportation, from commuting to oil pipelines, play an essential role in our economy and lifestyles. Mass-transit strikes all but shutter major cities; winter storms interrupt truck and air traffic, reducing our food choices. We Americans are mobile people who prefer to choose where we want to live, work, and shop, assuming that inexpensive transportation will be available to get us from one place to another. We choose to purchase petroleum from Kuwait and Nigeria, apples from New Zealand and Chile, and electronics from China and Japan, assuming that transportation of these goods

2 The Economic and Environmental Footprints of Transportation

will be reliable and inexpensive. We make our purchasing decisions at the local store or on the Internet with little concern for transportation prices, since they are generally low. Few Americans regard the fuel economy of a new automobile or light truck as of central concern relative to the appearance, size, and power of the alternatives. Even for a vacation across the continent or around the world, the cost of transportation is generally a small fraction of the total price of a vacation.

In short, we take transportation of passengers and freight for granted, unless there is a strike or storm that disrupts the flow. Even the doubling of the prices for gasoline and jet fuel caused little more than a momentary ripple of concern for most Americans.

Modern transportation is a remarkable achievement, compared to most of human existence. For transporting passengers on foot or by horse, 20 miles is a day's journey; an aircraft could transport a passenger 500 times faster. If freight is carried by a porter or yak, as is still done in the mountains of Nepal, transportation becomes a major fraction of the price of a good. In contrast, petroleum is transported in a supertanker from the Persian Gulf to the United States for about $1 per barrel, about 7/100 of a cent per ton-mile.

The remarkable achievements use large quantities of fuel and raw materials, emit large amounts of air pollution and greenhouse gases, and use large amounts of land and labor. Here we review and estimate the economic and environmental costs of passenger and freight transportation for the United States. We estimate the direct effects of transportation, as well as the full life-cycle implications of the activities.

2 ECONOMIC ACTIVITY

Table 1 shows that the transportation modes had a total revenue of $541 billion in 2000, just over 5 percent of GDP (2000 was the last year not distorted by 9-11 that is available).[1] Trucking was the dominant mode, with $223 billion in revenue, follow by air transport at $130 billion, then gas pipeline at $72 billion, rail at $36 billion, water at $33 billion, transit at $24 billion, oil pipeline at $9 billion, and bus at $4 billion. These figures are for commercial transportation and so don't include cars and light trucks driver by consumers, which are about 10 percent of GDP (consisting of the purchase of new vehicles, service and maintenance, fuel and insurance, and licensing).

The sector employs 5 million workers, with 40 percent employed in trucking. Air employs 24 percent, transit 12 percent, rail and water each employ 6 percent, and the pipelines employ 5 percent of transportation workers.

The sector produces 933 billion passenger-miles, with air having 77 percent of the share, followed by bus at 17 percent, transit at 5 percent, and rail at 1 percent. The freight sector in 2000 had 4 trillion ton-miles. Truck and rail have the lion's share with 28 and 34 percent, respectively, followed by water at 15 percent, oil with 14 percent, and gas pipelines with 8 percent. Air freight is slightly less than 1 percent of the total.

Table 1 Select Transportation Statistics for 1990 and 2000

Sector	Revenue ($ Millions in 2000 $)		Employment (Thousands)		Passenger-Miles (Millions)		Ton-Miles (Millions)	
	1990	2000	1990	2000	1990	2000	1990	2000
Air	93,085	130,299	589	732	472,567	708,926	16,514[a]	30,863[a]
Truck	156,037	223,197	1,247	1,406	—	—	854,000	1,203,000
Transit	19,675	24,243	262	348	41,143	47,666	—	—
Bus	10,999	13,237	1,649	2,136	121,398	160,919	—	—
Rail	36,374	36,213	240	194	6,057	5,498	1,033,969	1,465,960
Water	26,365	33,333	364	361	—	—	833,544[b]	645,799[b]
Oil pipeline	10,425	8,958	19	13	—	—	584,100	577,300
Gas pipeline	80,923	72,075	192	125	—	—	280,692[c]	350,889[c]
Total	433,881	541,555	4,562	5,315	641,165	923,009	3,602,829	4,273,811

Note: From Ref. 1. and Ref. 2.
[a] Freight-related ton-miles only.
[b] Domestic only.
[c] Data found in Table 1-46b update of Ref. 1.

The importance of the sector is far greater than its 5 percent share of GDP. The transportation sector has gotten highly efficient at moving goods and people, so the costs are relatively small. The low costs have led to rapid expansion from 1990 to 2000, particularly rail and trucking for freight traffic (31% and 29%, respectively) and air for passenger traffic (46%). Revenue expanded more rapidly: 29 percent for air, 30 percent for truck, 19 percent for transit, 20 percent for water, and 17 percent for bus. Pipeline and rail revenue fell when viewed in constant 2000 $.

Efficiency is also measured by the revenue per passenger-mile or per ton-mile and the number of workers by passenger-mile or per ton-mile. Revenue per passenger-mile increased from 25 to 27 cents from 1990 to 2000, and air fell by the same amount from 20 to 18 cents. Interestingly, if you include government subsidies for transit, the "revenue" per passanger-mile was 48 cents in 1990 and increased to 51 cents in 2000. The bus sector's revenue per passanger-mile dropped from just over 9 cents to 8.2 cents from 1990 to 2000.

Revenue per ton-mile increased from 18.3 to 18.6 cents for truck, remained essentially flat for water at a cent, and fell from 3.5 to 2.5 cents for rail. Natural gas pipelines fell from 29 to 20 cents while oil pipelines dropped slightly from 1.8 to 1.6 cents per ton-mile. It should be noted that the water transport revenue shown in Table 1 includes revenue revenues generated from international freight and domestic and international passenger transport, and the ton-mile data are only freight related. Only domestic freight-related revenue was used in the calculated.

Employees per million passenger-miles increased only in the transit sector, going from 6.4 to 7.3. In contrast, employees per million ton-miles fell from 1.3 to 1.0 for air, 39.6 to 35.3 for rail, and remained essentially flat for bus varying from 13.6 to 13.3. Employees per million ton-miles for truck, rail, and oil and gas pipelines all dropped. Truck dropped from 1.5 to 1.2, rail from 0.23 to 0.13, oil pipelines from 0.03 to 0.02, and gas pipelines from 0.68 to 0.36. Employees per million ton-miles rose only for water transportation during this period from 0.4 to 0.6. Thus, the record is one of decreased productivity for passenger traffic and increased productivity for freight traffic.

3 ENERGY USE

The United States uses a great deal of energy, about 100 quadrillion Btu per year. Commercial transportation is an energy-intensive sector using about 7 percent of the total, while automobiles and light trucks for personal use contribute another 18 percent. Table 2 provides energy use by transportation mode in 2000. Trucks use 57 percent of total energy, followed by air with 32 percent, rail with 7 percent, and transit with 4 percent.[2] Surprisingly, trucking uses the most Btu per dollar of revenue, followed by air, rail, transit, and pipelines, in that order. Per passenger-mile, transit is less efficient than air. Rail is a factor of 10 more efficient than truck in Btu per ton-mile; pipelines are three times less efficient than rail.

Table 3 shows fuel use by transport mode. Cars and trucks use the lion's share of fuel, 163 billion gallons of gasoline or diesel. Air uses 16 billion gallons of jet fuel and aviation gasoline. Water transport uses 6 billion gallons of residual fuel oil, 2 billion gallons of distillate, and 1 billion gallons of gasoline. Rail uses 6 billion gallons of distillate and 350 million kilowatt-hours of electricity. Transit uses 5.5 billion kWh of electricity and almost 1 billion gallons of diesel and gasoline. Pipelines are the only major user of natural gas, consuming 640 billion cubic feet.[3]

Table 2 Transportation Energy Use for 2000

Sector	Energy Use		Energy Efficiency	
	(Trilion Btu)	(Btu/$)	(Btu/Passenger-Mile)	(Btu/Ton-Mile)
Air	2,374	18,220	531–822	
Truck	4,252	19,050	—	3,565
Transit	300	12,375	1,403	—
Rail (Freight)	516	15,131		334
Pipelines	911	11,241		981
Total	8,353	76,017		4,880

Note: From Ref. 3.

Table 3 Fuel Use by Sector

Sector	Fuel	1990	2000
Air	Jet fuel (million gallons)	12,323	14,845
	Aviation gasoline (million gallons)	353	333
	Jet fuel (million gallons)	663	972
Highway	Gasoline, diesel and other fuels (million gallons)	130,755	162,554
Transit	Electricity (million kWh)	4,837	5,510
	Gasoline, diesel and other fuels (million gallons)	685	889
Rail	Distillate/diesel fuel (million gallons)	3,197	3,776
	Electricity (million kWh)	330	350
Water	Residual fuel oil (million gallons)	6,326	6,410
	Distillate/diesel fuel oil (million gallons)	2,065	2,261
	Gasoline (million gallons)	1,300	1,124
Pipeline	Natural gas (million cubic feet)	659,816	642,210

Note: From Ref. 3.

4 POLLUTION EMISSIONS

Burning these huge quantities of fossil fuel generates millions of tons of air pollutants and billions of tons of carbon dioxide and other greenhouse gases. Extracting petroleum, transporting it, distilling it into products, and then transporting the products generate still more air pollution and greenhouse gases.

While transportation was the dominant source of carbon monoxide (CO), Oxides of nitrogen (NO_x), volatile organic compounds (VOC), and lead in 1970, eliminating lead from gasoline and installing catalytic converters on automobiles actually reduced the emissions of these pollutions by 2000, with the most spectacular decrease in lead emissions—from 173,000 tons to 560 tons. These declines testify to the success of major social efforts to clean the air by reducing emissions, especially for automobiles. Eliminating lead from gasoline was the key for reducing emissions of CO, NO_x, VOC, and lead because it eliminated the major source of lead emissions and allowed installation of the modern three-way catalyst that reduces CO, NO_x, and VOC by about 90 percent.

The amount of economic activity in the United States increased from $3.8 trillion in 1970 to $10 trillion in 2000 (in 2000 dollars). In the absence of air pollution control laws, emissions of each of these pollutants should have increased. The significant drop in all pollutants is a testimony to the will, enforcement, and ingenuity in finding ways to lower air pollutant emissions.

The improvements in air pollution emissions is in marked contrast to the increases in carbon dioxide emissions. Each mode expanded, using more fuel and emitting more CO_2. The energy efficiency of the U.S. economy increased from 1990 to 2000, going from 11,600 Btu/$ of GDP to 10,100 Btu/$ of GDP.[4] The transportation section increased its energy efficiency as well, going from 50,500 Btu/$ of sector output to 49,100 Btu/$ of sector output.[4] The transportation sector

Table 4 Transportation Direct and Life-Cycle Emission for 2000 for Selected Priority Pollutants

Sector	Direct (Tons)	Life Cycle (Tons)
CO	92,239	100,678
NO_x	12,560	13,738
SO_x	697	1,546
VOC	7,969	8,667
PM10	552	795

Note: From Ref. 4.

is five times more energy intensive than the U.S. economy as a whole, but both are learning to use energy more efficiently.

One indication of the energy intensity of transportation is that, while the sector uses almost 7 percent of total energy, it employs only about 2 percent of the total work force.

Shown in Tables 4, 5, and 6 are air pollution emissions from transportation. Table 4 estimates air pollution emissions for carbon monoxide (CO), oxides of nitrogen (NO_x), sulfur oxides (SO_x), volatile organic compounds (VOC) and suspended particulate matter less than 10 microns is diameter (PM_{10}) for the entire transportation sector using the Economic Input-Output Life-Cycle Analysis (EIOLCA) model.[5] This model brings together the input-output table of the United States for 1997 together with emissions data from EPA. Transportation is directly responsible for huge amounts of CO emissions, large amounts of NO_x and VOC emissions, and smaller amounts of SO_x and PM_{10} emissions. Life-cycle emissions for CO, NO_x, and VOC are about 9 percent higher than the direct emission, but SO_x and PM_{10} are dramatically increased: 120 percent for SO_x and 44 percent for PM_{10}. The air pollution emissions come mainly from the reliance of this sector on burning gasoline and distillates in internal-combustion engines. This link is illustrated in Table 5, albeit obliquely.

Table 5 shows a detailed breakdown of air pollution emissions by vehicle class, e.g., light- and heavy-duty gasoline or diesel engine vehicles. Unfortunately, there is no easy way to aggregate various categories of vehicles and allocate separate emissions to a specific transportation sector.[6] Aircraft are of course the dominant vehicle in air transportation, and it is likely that passenger aircraft are the dominant emitters. The data in Table 5 likely reflect this, but the others are not straightforward.

Table 6 shows estimates for air pollution emissions for each mode. The EPA estimates from Table 5 for air, railroad, and water are shown. They are supplemented with emissions estimates from the EIOLCA model.[5] The two sets of estimates are not identical since the EPA figures are for 1990 and 2000, while the EIOLCA estimates are for 1997. EPA sources do not provide estimates for the

Table 5 Priority Pollutant Emission from the Transportation Sector

Sector	CO 1990	CO 2000	NO$_x$ 1990	NO$_x$ 2000	SO$_x$ 1990	SO$_x$ 2000	VOC 1990	VOC 2000	PM10 1990	PM10 2000
					(1,000 short tons)					
TRANSPORTATION TOTAL	131,702	92,239	13,373	12,560	874	697	12,790	7,969	713	552
HIGHWAY VEHICLES	110,255	68,061	9,592	8,394	503	260	10,175	5,325	385	230
Light-duty gas vehicles & motorcycles	67,237	36,398	4,262	2,312	111	103	6,268	2,903	56	51
Light-duty gas vehicles	66,972	36,229	4,240	2,297	111	102	6,217	2,877	56	51
Motorcycles	265	169	22	15	1	0	51	25	1	0
Light-duty gas trucks	32,228	27,041	1,504	1,436	52	70	2,720	1,929	31	31
Light-duty gas trucks 1	19,739	18,138	962	999	31	47	1,618	1,251	17	22
Light-duty gas trucks 2	12,489	8,903	542	437	21	23	1,102	678	14	10
Heavy-duty gas vehicles	8,921	3,422	567	453	16	14	710	256	17	10
Diesels	1,869	1,200	3,259	4,192	324	73	476	238	281	137
Heavy-duty diesel vehicles	1,806	1,185	3,194	4,178			441	230	266	135
Light-duty diesel trucks	25	7	23	6			17	4	5	1
Light-duty diesel vehicles	38	8	43	7			19	3	11	1
OFF-HIGHWAY (Tier 1–12)	21,447	24,178	3,781	4,167	371	437	2,616	2,644	328	322
Nonroad gasoline	19,356	21,647	120	192	9	11	2,322	2,344	58	70
Recreational	1,876	1,892	7	10			409	467	5	5
Construction	805	711	4	6			79	53	2	2
Industrial	1,215	847	18	14			63	29	1	0
Lawn & garden	10,118	11,499	42	86			910	811	18	21
Farm	313	365	3	4			15	14	0	0
Light commercial	3,245	4,232	14	35			201	171	2	2
Logging	52	78	0	0			8	12	0	1
Airport service	11	9	0	0			1	0	0	0
Railway maintenance	6	7	0	0			0	0	0	0
Recreational marine vessels	1,716	2,008	32	35			637	787	30	38
Nonroad diesel	875	932	1,454	1,600	131	198	196	202	199	179
Recreational	2	3	1	2			0	1	0	0
Construction	396	444	702	762			82	90	82	74
Industrial	70	64	136	133			18	16	16	13
Lawn & garden	15	26	24	45			4	8	3	5
Farm	328	325	478	530			77	70	83	72
Light commercial	38	50	48	72			10	14	8	10
Logging	18	9	39	22			3	2	5	2
Airport service	3	4	7	9			1	1	1	1
Railway maintenance	2	3	2	3			0	1	0	1
Recreational marine vessels	3	4	16	22			1	1	0	1
Aircraft	239	270	70	88	7	9	29	27	3	4
Marine vessels	132	133	1,003	1,008	167	163	32	32	43	43
Diesel	132	133	1,003	1,008			25	32	43	43
Residual oil	NA	NA	NA	NA			7	NA	NA	NA
Other	NA	NA	NA	NA			NA	NA	23	25
Railroads	94	99	945	1,001	56	56	36	39	1	1
Other	752	1,097	189	278	0	0	1	1	1	1
Liquified petroleum gas	641	969	162	246			0	0	0	0
Compressed natural gas	110	128	27	32			0	0		

Note: From Ref. 6.

8 The Economic and Environmental Footprints of Transportation

Table 6 Emissions from Transportation Sectors

	CO				NO_x			
	EPA		EIO		EPA		EIO	
	1990	2000	1997		1990	2000	1997	
			Sector	Life Cycle			Sector	Life Cycle
Sector	(1,000 short tons)				(1,000 short tons)			
Air	250	279	331	1,046	70	88	103	209
Railroads	102	109	104	339	947	1,004	902	936
Water	132	133	158	605	1,003	1,008	1,217	1,242
Truck Transportation	—	—	26,777	27,602	—	—	1,930	2,125
Transit & Ground	—	—	118	196	—	—	854	867
Pipeline	—	—	499	761	—	—	26	126

	SO_x				VOC			
	EPA		EIO		EPA		EIO	
	1990	2000	1997		1990	2000	1997	
			Sector	Life Cycle			Sector	Life Cycle
Sector	(1,000 short tons)				(1,000 short tons)			
Air	7	9	7	133	30	27	45	115
Railroads	56	56	48	73	36	40	42	62
Water	167	163	230	256	32	32	1,170	1,186
Truck Transportation	—	—	81	233	—	—	1,992	2,101
Transit & Ground	—	—	22	35	—	—	60	71
Pipeline	—	—	3	140	—	—	13	80

	PM_{10}				CO_2	
	EPA		EIO		EIO	
	1990	2000	1997		1997	
			Sector	Life Cycle	Sector	Life Cycle
Sector	(1,000 short tons)					
Air	3	4	6	26	161,310	203,815
Railroads	1	2	20	27	28,889	36,968
Water	43	43	113	118	47,857	36,396
Truck Transportation	—	—	47	93	373,497	430,080
Transit & Ground	—	—	30	33	7,931	12,536
Pipeline	—	—	0	15	124,732	162,358

Note: From Refs. 2 and 7.

EIOLCA sectors Truck Transportation, Transit & Ground, and Pipeline. For the modes where comparison is possible, the emissions data are reasonably comparable, given the accuracy of the emissions estimates and the way they are tabulated. The EIOLCA estimate for VOC emissions from water transport has an error that we were not able to reconcile within the time limits of this paper.

Table 6 shows the dominant role of trucks in emissions of CO, since emissions are 50 times higher than pipelines, the next highest mode. CO emissions indicate inefficient combustion. For NO_x emissions, trucks are again the leading source, but are followed closely by water, railroads, and transit. Emissions from air and pipelines are small. Truck transport is again responsible for the larges emissions of VOC. Water transport is the largest source of SO_x emissions, since a great deal of residual fuel oil is used, which has a high sulfur content. PM_{10} emissions are smaller than other categories, with water having the most emissions.[8]

Table 6 also shows CO_2 emissions for each sector. CO_2 emissions are roughly proportional to fuel use, since almost all modes rely on petroleum products. Air and truck transportation are the two largest CO_2-emitting sectors, releasing over 200 and 400 million tons of CO_2, respectively.

5 INJURIES AND DEATHS

Table 7 shows the fatalities and injuries for each sector for 1990 and 2000. The fatality data are more accurate, since there is no problem with defining the outcome, as there is with injuries. The injury data tend to follow the same patterns as the mortality data.

The fatality and injury rates for air and the two pipelines are subject to major statistical variation from year to year. For example, in many years, there are no fatalities for domestic scheduled airline operation. The mortality rate rose slightly for trucks, but fell slightly for transit, and fell significantly for rail and water. Somewhat surprisingly, rail has the most fatalities, due to collisions at grade crossings. Trucks are somewhat safer, despite the much greater number of trucks and truck-miles and the greater possibility of crashes, since they operate on highways with other vehicles. Transit has the third-highest number of fatalities, which is not surprising, given that buses and taxicabs operate in crowded urban settings.

Per passenger-mile, air is safer than transit, with 0.3 deaths per billion passenger-miles compared to 4 for transit. Assuming that aircraft average 100 passengers on a flight, that is 30 deaths per billion aircraft miles, or 1 death every 33 million aircraft miles. Since some of the air fatalities are due to freight operations, dividing all fatalities by passenger-miles will overstate the fatalities for passenger operation. Thus, air is 10 times safer than transit operations, including city buses and taxis. This safety is testimony to the vast care to make air transport safe.

Per ton-mile, all modes are safe, although water and oil pipelines are much safer than truck or rail. Trucks are responsible for 0.6 deaths per billion ton-miles. For an 80,000 pound truck, that would be 2.4 fatalities every million

Table 7 Fatalities and Injuries by Sector for 1990 and 2000

	1990		2000	
Sector	Fatalities	Injuries	Fatalities	Injuries
Air	96	76	168	48
Truck	705	41,822	754	30,832
Transit	679	87,247	585	74,466
Rail	1,300	24,143	937	10,424
Water	85	175	53	150
Oil pipeline	3	7	1	4
Gas pipeline	6	69	37	77
Total	2,874	153,539	2,535	116,001

By Ton-Miles (Values $\times\ 10^{-6}$)

	1990		2000	
Sector	Fatalities/ Ton-Mile	Injuries/ Ton-Mile	Fatalities/ Ton-Mile	Injuries/ Ton-Mile
Air	0.0092	0.0073	0.0106	0.0030
Truck	0.0008	0.0493	0.0006	0.0258
Transit	—	—	—	—
Rail	0.0012	0.0227	0.0006	0.0067
Water	0.0001	0.0002	0.0001	0.0002
Oil pipeline	0.0000	0.0000	0.0000	0.0000
Gas pipeline	—	—	—	—

By Passenger-Miles (Values $\times\ 10^{-6}$)

	1990		2000	
Sector	Fatalities/ Passenger-Mile	Injuries/ Passenger-Mile	Fatalities/ Passenger-Mile	Injuries/ Passenger-Mile
Air	0.0003	0.0002	0.0003	0.0001
Truck	—	—	—	—
Transit	0.0040	0.5175	0.0027	0.3483
Rail	0.2146	3.9860	0.1704	1.8960
Water	—	—	—	—
Oil pipeline	—	—	—	—
Gas pipeline	—	—	—	—

truck-miles. For a 10,000 ton freight train, that would be 6 deaths every 10,000 train-miles.

6 LIFE CYCLE VERSUS DIRECT EFFECTS

The previous tabulations focus on the direct effects of the transportation sector. However, a more relevant analysis would examine the whole life cycle, from extraction of raw materials and energy through use and ultimate disposal. In this section, we estimate these full life-cycle effects using the EIOLCA model developed at Carnegie Mellon University.[5]

Table 4 shows the direct and total life-cycle emissions of air pollutants for the whole transportation sector. The transportation sector is so energy intensive and has such high direct emissions, that the life-cycle emissions are only 9.1 percent higher for CO, 9.7 percent higher for NO_x, and 8.8 percent higher for VOC. SO_x and PM_{10} are the major exceptions, with life-cycle SO_x emissions being 121.8 percent higher for SO_x and 44.0 percent higher for PM_{10}.[5] Transportation fuels, with the exception of residual fuel oil, have little sulfur. However, half of electricity is generated by coal, so the indirect SO_x emissions are high.

7 TRANSPORTATION EXTERNALITIES

Murphy and Delucchi estimates the externalities associated with motor vehicle use.[9] It is not possible to estimate the externalities for the whole transportation sector in the same detail. Nonetheless, it is possible to translate the air pollution emissions, as well as injuries and deaths, into dollar costs to society.

Table 8 shows the direct and life-cycle implications for each mode separately, accounting for the air pollution emissions, CO_2 emissions, and fatalities and injuries.[10] The air pollution emissions are valued using current allowance prices, where available. They are valued at the cost of abatement in other cases. CO_2 is valued at the price that allowances were recently trading in the European Union. Fatalities are valued at $5 million, the figure used by EPA in its "Cost of Clean Air" report.[11] The severity of injury is not specified; we assume that the reported injuries are serious and so value each at $50,000. Also shown is the revenue for each mode and the proportion of mode revenue needed to account for the externalities. Total external costs are about a quarter of the sector revenue costs, with CO_2 external cost being the largest single value.

8 SUMMARY AND CONCLUSION

The transportation sector is energy intensive. When the modes were largely uncontrolled, prior to 1970, they emitted large quantities of air pollution. Eliminating lead from gasoline permitted the use of catalytic converters, leading to reductions of 90 percent or more in emissions of CO, NO_x, and VOC. The

Table 8 Transportation Emissions and External Costs

Pollutant	Emissions		External Costs	
	Sector	Life Cycle	Sector	Life Cycle
	(1,000 Short Tons)		($ Million)	
CO	92,239	139,144	47,964	52,353
NO_x	12,560	13,331	35,168	38,465
SO_x	697	1,365	1,394	3,092
VOC	7,969	9,591	12,750	13,866
CO_2	744,216	882,153	9,674	11,467
PM10	552	761	2,374	3,419
	Number			
Fatalities	2,535	—	12,675	—
Injuries	116,001	—	5,800	—
Total external costs ($ millions)			118,125	129,670
Total sector revenue ($ millions)			503,000	970,790
Ratio external costs/Revenue			0.23	0.13

Data from other tables
Note: From Ref. 10.

achievement of motor vehicles in reducing air pollution emissions has been immense, and the manufacturers are to be commended.

While there are some indications of greater energy efficiency, CO_2 emissions are high and growing. The energy intensity and sheer size of the sector leads to large externalities. The sector also shows that where regulation or other pressures have focused on a goal, such as reducing pollution emissions from cars and light trucks, major advances can be accomplished. The European Union (EU) is pressing manufacturers to build vehicles with lower CO_2 emissions per mile and more efficient jet aircraft. The preliminary indication is that major improvements are likely.

The transportation sector is to be commended for its essential contributions to our lives, from food and transport of goods to personal mobility. While consumers will continue to demand this level of performance, society will demand major reductions in environmental externalities, from air pollution emissions to CO_2 emissions. We feel confident that the transportation sector will rise to meet this challenge.

REFERENCES

1. Bureau of Transportation Statistics, Department of Transportation, *National Transportation Statistics, 2006*. U.S. Government Printing Office, Washington, D.C., February 2007.
2. U.S. Census Bureau, *The 2007 Statistical Abstract*, www.census.gov/compendia/stab/. Accessed on June 13, 2007.
3. S. C. Davis and S. W. Diegel, *Transportation Energy Data Book*, 22nd ed., Oak Ridge National Laboratory, Oak Ridge, 2002.
4. Authors' calculation from data in Refs. 1 and 3.
5. C. T. Hendrickson, L. B. Lave, H. S. Matthews, J. Bergerson, G. Cicas, A. Horvath, S. Joshi, H. L. MacLean, D. Matthews, and F. C. McMichael, *Environmental Life Cycle Assessment of Goods and Services: An Input-Output Approach*, Resources for the Future, 2006.
6. U.S. Environmental Protection Agency, *National Emissions Trends*, U.S. Government Printing Office, Washington, D.C., 2007.
7. Carnegie Mellon University Green Design Institute, "Economic Input-Output Life Cycle Assessment (EIOLCA) Model", http://www.eiolca.net/. Accessed on November 29, 2007.
8. Energy Information Agency, Department of Energy, 2006. *Annual Energy Review 2005*. DOE/EIA-0384(2005).
9. J. J. Murphy and M. A. Delucchi, "A Review of the Literature on the Social Cost of Motor-Vehicle Use in the United States," *Journal of Transportation and Statistics*, **1**(1) (January 1998): 15–42.
10. H. S. Matthews and L. B. Lave, "Applications of Environmental Valuation for Determining Externality Costs," *Environmental Science and Technology*, **34** (2000): 1390–1395.
11. U.S. Environmental Protection Agency, "Benefits and Costs of the Clean Air Act," www.epa.gov/air/sect812/design.html. Assessed June 13, 2007.

CHAPTER 2

PUBLIC TRANSPORTATION AND THE ENVIRONMENT

Michael D. Meyer
School of Civil and Environmental Engineering
Georgia Institute of Technology
Atlanta, Georgia

1	INTRODUCTION	15	
2	PUBLIC TRANSPORTATION DEFINED	16	
3	ENVIRONMENTAL LINKAGES TO PUBLIC TRANSPORTATION SYSTEMS	18	
	3.1 Environmental Impacts of Transit Facility Design and Construction	20	
	3.2 Environmental Impacts of Transit Operations: Overview	25	
	3.3 Technology-based Environmental Impacts	26	
	3.4 Ridership-based Environmental Impacts	33	
	3.5 Mitigating the Environmental Impacts	34	
	3.6 Environmental Stewardship and Sustainable Development	36	
4	CONCLUSION	43	

1 INTRODUCTION

The efficient movement of people and goods enables the economic success and quality of life of metropolitan areas throughout the world. Although the current technology of transportation provides for unprecedented levels of urban mobility, it also significantly affects the natural environment and, in some cases, the human condition. By its very nature, at least in today's world, the vast majority of urban transportation systems and services consume petroleum resources at increasingly growing levels, due primarily to the rapid growth in urban population and automobile ownership worldwide. These resources are used to generate propulsion energy for the actual movement of vehicles, but in the process also produce waste products or pollutants.

It is not surprising that in most of the major cities of the world, the transportation sector is at or near the top of the list of major sources of air pollution. Transportation is also a major source of water pollutants, noise levels, and ecological disruption, both during construction as well as during subsequent operation.

And over the long term, the accessibility provided by the transportation system has a significant influence on how a metropolitan area or community develops. Improved access to land creates opportunities for new development to occur, whether for housing, office, or retail purposes.

The purpose of this chapter is to examine how public transportation systems can be viewed from an environmentally conscious perspective. The next section defines what is meant by a public transportation system. The following section then examines the relationship between environmental impacts and transit system planning and operations. This relationship not only includes the physical impacts that result from the construction and operation of such systems, but also the important symbiotic relationship between transportation accessibility and land use/urban form, which over the longer term can affect the overall sustainability of a metropolitan area. The final section concludes with important observations relating to a more environmentally conscious approach to public transportation planning and system design.

2 PUBLIC TRANSPORTATION DEFINED

One of the challenges in looking at the environmental impacts of public transit is that there are many different forms of public transportation, each having potentially different impacts on the environment. For purposes of this chapter, *public transportation* is broadly defined as consisting of the transit infrastructure, operations, and corresponding services provided to the general public for a fee or fare. A typical metropolitan area can include many different types of public transportation services, ranging from high-capacity, high-speed services such as rapid rail or bus rapid transit systems to more local services primarily serving communities or individual employment centers. These latter services often depend on buses or other forms of ridesharing vehicles.

Figure 1 illustrates the wide range of public transportation services that are usually available in a metropolitan area. As shown in the figure, high-capacity transit services link major population and employment centers in the region, which, in turn, are also served by local transit services and often by targeted collector/distributor systems, such as people movers or shuttles. In addition, Figure 1 indicates another phenomenon that is quite typical of metropolitan transportation systems. Private transit or employer-based employee services offer premium transit service for a higher price (although technically, these services are not public transit). All types of transit must be considered when discussing the relationship between public transportation and the environment.

Mode of travel is one of the most important characteristics of public transportation that relates to expected environmental impacts. Transportation planners refer to the mode of travel as the means or technology associated with the trip being made. For example, the automobile is considered a mode of travel, as is bus, subway, ferry, and walking. The importance of the mode of travel for

2 Public Transportation Defined 17

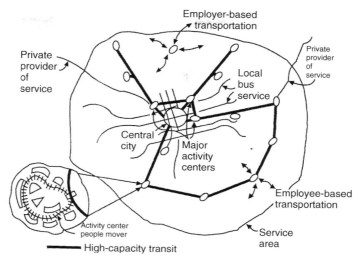

Figure 1 Typical metropolitan transit system.

environmental analysis is that each mode has different operating characteristics, fuel needs, infrastructure requirements, and potential influence on surrounding land-use patterns. Table 1 shows how the modal distinction could have important implications to environmental impacts, depending on which mode is considered.

The *land-use influence* characteristic shown in Table 1 is one of the most studied in the transportation planning literature (see, for example, Refs. 1 to 7). This reflects one of the fundamental relationships of transportation planning—that is, transportation investment influences land-use development, which, in turn, influences the performance of the transportation system, which over the long term influences land-use development, and so forth. For those interested in the sustainability of metropolitan areas and individual communities, one of the most important levers of influence thus becomes transportation system modal investment.

It is not surprising that many studies have shown that the interstate highway system has been the most significant influence on U.S. urban form over the past 50 years.[8] It is also not surprising that many view a combined public transportation/land-use investment strategy as the most effective way of limiting negative environmental impacts of growing urban communities. This interest has led many cities to adopt policies and strategies aimed at encouraging development around transit stations, known as transit-oriented development (TOD). The transportation focus of TOD approaches is to reduce the number of automobile trips by promoting a closer linkage between urban living, work locations and travel choices, thus enhancing environmental quality.[9–11]

In summary, the infrastructure, vehicle, and service characteristics of public transportation can have important implications to environmental impacts. In the short run, because of their need for infrastructure, many public transportation

Table 1 Modal Definition and Related Characteristics

Transportation Mode	Rapid Rail	Express Bus	Downtown Shuttle
Facility guideway	Fixed in place; new tracks can be laid, but will take time and substantial funding.	Can use existing road network and thus can flexibly respond to new service demands.	Can use existing road network and thus flexibly respond to new service demands, although limited to local service area.
Service provision	Limited to rail corridors; important premium placed on easy access to stations.	Flexible; can be restructured to meet new opportunities.	Flexible; can be restructured to meet new opportunities.
Ridership potential	Can handle the largest number of riders of all transit modes.	Generally limited by service strategies and vehicle size.	Very limited
Infrastructure impacts	Given high speeds, alignments need to be fairly straight; thus, potentially significant impacts on natural environment.	Service uses already provided road network; new bus rapid transit systems would have similar impacts as rail.	Very limited
Energy source	Almost always electric; third rail or overhead wires used.	Different fuel use found in practice; most fleets now using non-petroleum-based fuels.	Different fuel use found in practice; some fleets use diesel, others non-petroleum-based fuels.
Land-use influence	Over time, has shown to have important influence on land use development near stations.	Minimal influence on land-use patterns, although express bus combined with bus rapid transit facility could have more significant impact.	Very limited

modes have environmental impacts similar to any other form of surface transportation, such as energy consumption, air pollution, noise levels, erosion, habitat infringement, and so on. However, over the long run, public transportation investment also has a potentially significant role in shaping community development and urban form, thus requiring a broader spatial and temporal examination of the environmental consequences of such investment.

3 ENVIRONMENTAL LINKAGES TO PUBLIC TRANSPORTATION SYSTEMS

Much of the transportation literature as well as popular opinion often portray public transportation as a more environmentally friendly form of transportation than automobile use, and one that deserves greater priority in metropolitan transportation investments. Some jurisdictions have, in fact, viewed transportation investment as a means of achieving much broader sustainability goals. The

Oregon Department of Transportation (ODOT), for example, has adopted the following definition for a sustainable transportation system.[12]

"A sustainable transportation system strives to achieve objectives including, but not limited to, the following:

- Reinforce livable and economically strong communities,
- Encourage modal choice throughout the state,
- Support efficient land uses that reduce travel distances and increase travel options,
- Distribute system benefits and burdens equitably across society,
- Be affordable,
- Improve safety to reduce injuries and fatalities,
- Reduce emissions of greenhouse gases to reduce climate change,
- Protect air and water quality from pollutants,
- Operate with clean and fuel-efficient vehicles,
- Use maintenance and construction practices that are compatible with native habitats and species and which consider habitat fragmentation concerns,
- Minimize raw material use and disposal during construction and maintenance, and
- Apply life-cycle costs to transportation investments."

Public transportation certainly fits within this general concept of a sustainable transportation system, and indeed plays a primary role in the objective of encouraging modal choice. This definition of a sustainable transportation system also provides guidance on those elements of any transportation system that relate to environmentally conscious design. In particular, there are three key elements of an environmentally conscious public transportation system:

1. *How does the physical design and construction of the transit system infrastructure affect the surrounding natural environment and community?* The time frame of this element is often short term and deals with the more immediate physical effects associated with building a transit facility.
2. *What are the environmental impacts of operating transit systems?* Given that transit systems operate over many years, the time frame for this element can be many decades. Operations-oriented environmental considerations can include such things as air pollutant emissions, energy consumption, vibration, and noise.
3. *What are the longer-term, environmental stewardship factors that can be influenced by transit systems?* These factors, including such things as land-use impacts and overall management of a transit agency's operations, can have long-lasting effects, and generally position a transit system to be a positive contributor to environmental quality and sustainable development.

Figure 2 identifies the most important environmental impacts that are associated with both the design and operations of a transit service. It is important

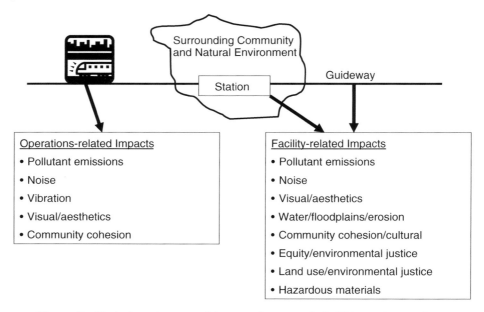

Figure 2 Typical environmental impacts from transit facilities and operations.

to note that, in many cases, a specific environmental impact might be relevant in more than one of these categories. For example, noise can be an issue not only during the construction of a facility, but also during the ongoing operations after a facility is built. The following section discusses each of these three questions, with emphasis given to the transit-related context for each environmental impact.

3.1 Environmental Impacts of Transit Facility Design and Construction

Environmental considerations during transit design and construction are very similar to those for any other transportation facility. In general, the types of impacts that are most associated with the design of a transit facility could relate to such issues as endangered species, erosion, visual presentation/aesthetics, floodplains, equity or environmental justice, hazardous materials/brownfields, historic/archeological/cultural resources, parklands, social and economic impacts, and water resources. Construction-related impacts associated with air quality and vibration could also be an important issue, especially for nearby communities (air quality and vibration impacts are discussed in the next section).

It is beyond the scope of this chapter to discuss in detail the specific characteristics and approaches used to analyze each type of environmental impact. In every case, numerous technical manuals and handbooks describe the approaches that are available to examine a specific impact. However, there are some design-related impacts that merit special attention because of special circumstances surrounding the design of a transit facility. The following sections discuss how transit facility design and construction relate to these specific environmental impacts.

Endangered Species/Sensitive Habitats

Issue: Urban encroachment upon the natural habitats of species that are in danger of extinction is an issue that has become more important in recent years. With respect to transit, this impact is not often a major issue for facility design or operations. However, there are some special circumstances where potential habitat disruption needs to be addressed because of the unique nature of the environment in which a facility is being constructed (e.g., salmon spawning waterways in the U.S. Pacific Northwest).

Approach: The major approach toward assessing potential impacts to biological resources is through field surveys and biological assessments. These assessments examine a broad range of factors relating to ecological functions, and the potential disruption associated with the construction of structures in such an environment. Because of the nature of the expertise needed to conduct a biological assessment, much of this work is done by very specialized companies and experts.

Mitigation: Possible mitigation strategies could include organizing construction staging away from sensitive ecological areas; using construction strategies that reduce impacts on the ecology, such as using pile-driving techniques that have lower vibration effects; designing drainage and stormwater discharge systems that minimize the amount of effluent entering local streams; and redesigning the structure or guideway to avoid the sensitive habitat area altogether.

Equity/Environmental Justice Issues

Issue: In the United States, a 1994 presidential executive order directed every federal agency to consider *environmental justice* as part of its mission. This was defined primarily as identifying and addressing the effects of all programs, policies, and activities on minority populations and low-income populations. For transportation, this was further defined as meeting the following three objectives:

1. Avoid, minimize, or mitigate disproportionately high and adverse human health and environmental effects, including social and economic effects, on minority populations and low-income populations.
2. Ensure the full and fair participation by all potentially affected communities in the transportation decision-making process.
3. Prevent the denial of, reduction in, or significant delay in the receipt of benefits by minority and low-income populations.

Clearly, the location of transit facilities and services in a metropolitan area represents conscious decisions of providing one part of the area with improved accessibility and mobility, thus potentially resulting in equity implications on the relative sharing of the benefits associated with this investment. At a higher decision-making level, some of the more contentious environmental justice issues relate to the relative investment in different transit modes. For example, substantially funding regional commuter rail services that primarily serve higher-income suburbs and not investing in local city bus service, which is often a major mode of transportation for low income households, would be viewed as an environmental justice issue.

Approach: According to the U.S. Council on Environmental Quality, the following steps should be included in any environmental justice assessment of transit projects:[13]

1. *Identification of minority or low-income populations:* Agencies should consider the composition of the affected area to determine whether minority populations, low-income populations, or Indian tribes are present, and if so, whether there may be disproportionately high and adverse human health or environmental effects on these populations. This identification should occur as early as possible during the environmental impact statement (EIS) process.
2. *Public participation:* Agencies should develop effective public participation strategies that assure meaningful community representation in the EIS process.
3. *Numeric analysis:* Where a disproportionate and adverse environmental impact is identified, agencies should consider relevant demographic, public health and industry data concerning the potential for exposure to human health or environmental hazards in the affected population, to the extent that such information is reasonably available.
4. *Alternatives and mitigation:* The relative impact of alternatives should be considered, and measures to avoid, minimize, and mitigate impacts should be evaluated as part of the EIS.

Mitigation: Reducing the consequences of transit investment on minority and low-income populations could include relocating guideway facilities and stations; providing mitigation actions such as noise walls, landscaping, or other forms of buffering; implementing new services targeted at the impacted communities; and relocating households and businesses if the impact is severe.

Hazardous Materials/Brownfields

Issue: Fast-paced growth in many metropolitan areas is increasing the need to provide better transit service not only to newly populated areas, but also to growing communities in the existing service area. With land prices rising due to development pressures, the best locations for transit guideways and transit stations are often in areas that were once industrialized, but have long since been abandoned. Abandoned or underused commercial and industrial parcels that usually contain some contaminated soil are referred to as *brownfields*. Given their location, such sites can become viable economic centers, which, when combined with transit services, could foster new development patterns for a metropolitan area (see section 3.6 on transit's link to sustainable development).

Approach: The major environmental issue concerning the use of brownfields for transit use is often the identification and disposal of hazardous materials. Formal procedures have been developed for identifying potentially hazardous materials sites (see, for example, Ref. 14). These procedures assign numerical values to risk factors based on conditions at the site. The factors are usually grouped into three categories:

1. Likelihood that a site has released or has the potential to release hazardous substances into the environment
2. Characteristics of the waste (e.g., toxicity and waste quantity)
3. People or sensitive environments (targets) affected by the release

Four major environmental hazards are often identified at these sites:

1. Ground water migration (drinking water)
2. Surface water migration (drinking water, human food chain, sensitive environments)
3. Soil exposure (resident population, nearby population, sensitive environments)
4. Air migration (population, sensitive environments)

Scores are calculated for one or more of these hazards, and a total score is assigned to each potential site. Those sites having the highest scores receive priority for mitigation.

Mitigation: The process for disposing of hazardous materials is most often defined by government regulation. Special handling and disposal of many of the hazardous materials are required. Readers are referred to Ref. 14 for more detailed discussion concerning this process.

Social and Economic

Issue: Transit projects can have a significant impact on the character and nature of the communities they serve due to both the large numbers of people they attract to stations and the enhanced value of land that often surrounds these stations. Social and economic impacts relate to land acquisition, community impacts, land use and development, economic impacts, and safety/security.

Approach: With respect to land use, the analysis should identify the characteristics and needs of persons and businesses to be displaced, inventory the availability of comparable replacement dwellings and sites, discuss potential relocation problems, and describe methods to mitigate adverse impacts. For community impacts, the analysis should include outreach to local planning agencies and the public. Specific impacts to be examined include physical and psychological barriers; changes in land use patterns, circulation patterns, and access to services; changes in population densities; and effects on neighborhood cohesiveness. If a proposed transit project is not compatible with surrounding land uses or might encourage land use and development inconsistent with local plans, goals, and objectives, the expected impacts on the area and alternative alignments need to be investigated. Transit projects could also have direct and indirect taxation impacts (due to increases in the value of land), cause substantial displacement of businesses and households, disrupt business activities, and influence regional construction costs. Transit projects can also create potential pedestrian and traffic safety hazards, as well as create rider and employee security hazards. Each of these potential impacts needs to be examined as part of a study, with much of the effort based on economic analysis and public outreach.

Mitigation: The mitigation strategies that can be considered for social and economic impacts range widely for the different types of impacts that might be present. One of the most useful approaches for analyzing social and economic impacts and determining appropriate mitigation strategies is called *community impact analysis* (CIA). CIA manuals provide a systematic approach for helping planners and engineers develop a mitigation strategy that best meets the needs of the community (see, for example, Refs. 15 to 17).

Visual Impacts/Aesthetics

Issue: The visual impact of a transit facility will often vary by mode. For example, buses will normally be considered as part of the general traffic stream and not be considered out of the ordinary. However, elevated guideways, such as those for peoplemovers, rail transit, or even new bus rapid transit services, could be viewed by the local community as an intrusion into the visual aesthetics of the surrounding environment. Also, light rail and tram facilities that run on urban streets often use overhead wires for their power source, which has become an issue in many urban areas because of how they change the aesthetics of the street environment.

Approach: The advent of computer simulation and visualization software has greatly contributed to visual assessment approaches. Most transit studies now provide computer-generated visuals showing how a particular design will affect local sites from different visual perspectives. These visual representations are presented to community groups and/or focus groups to determine the value placed on aesthetics by those most affected, and to identify the alternative having the least visual impact.

Mitigation: Potential mitigation strategies might include reduced or reoriented lighting, use of textured materials and colors that will be more conducive to neighborhood character, landscaping, reorienting guideway or station footprints to avoid visual effects, and use of building codes and guidelines that keep structures consistent with the local nature of the built environment.

Water Resources

Issue: Public transit projects can affect water resources in a number of ways. The location of guideways and stations near wetlands and other water bodies could potentially have a disruptive effect on these resources. Increasing runoff or altering surface and subsurface drainage patterns can also be of great concern. This is particularly true for large parking structures or lots where stormwater runoff can include vehicle-related pollutants such as lead, zinc, and cadmium.

Approach: Major transit projects undergo a very detailed hydrologic and runoff analysis, which examines the likely changes in drainage that will occur due to impermeable surfaces being added to the environment. A variety of software packages have been developed that allow the analyst to predict with some certainty what water resource impacts are likely to occur given the type of transit structure that is being built.

Mitigation: The primary means of mitigating water resource impact is to control for the magnitude and constituent components of the runoff generated from a transit site. Detention ponds can be used to store runoff so as to remove heavy metals from the water that reaches nearby water resources. Enclosed drainage systems provide the same protection. The most effective strategy is to avoid sensitive water resource areas to begin with.

3.2 Environmental Impacts of Transit Operations: Overview

The environmental impacts related to transit operations fall into two major categories—the physical phenomena associated with the operation of the vehicles themselves (such as fuel consumption), and the positive or negative environmental consequences related to travelers shifting to the transit mode (such as reduced automobile emissions because of drivers now taking transit). The first set of impacts is much easier to predict because they most often relate to known relationships between vehicle operations and resulting effects. For example, the engineer knows what the fuel consumption rate is for buses or what level of air pollutant emissions will likely occur due to vehicle operations. The second set of impacts necessarily require the engineer to estimate what new transit ridership is likely to occur given changes in the transit system, and how many of these new riders will be drawn from automobiles or possibly from other transit modes or even be newly generated trips.

To a first approximation, the magnitude of these second environmental impacts will be similar in proportion to the level to which new riders are attracted to the new transit service. Unlike the first set of impacts, this second set requires a network modeling approach that can estimate the new ridership based on the relative attractiveness of the new transit service compared to other travel options.

Sections 3.3 and 3.4 examine the most important environmental impacts associated with each type of transit system characteristic. The first describes those environmental impacts that are most related to transit vehicle technology, and the second examines those impacts that are more regional in nature and directly relate to the amount of transit ridership attracted to a new transit service.

3.3 Technology-based Environmental Impacts

The most important impacts in this category include air pollutant emissions, energy consumption, noise, and vibration.

Vehicle Air Pollutant Emissions

Issue: Diesel fuel has been the predominant fuel source for transit bus fleets for decades, with approximately 90 percent of the top 300 bus fleets in the United States relying on such fuel. Heavy-duty diesel engines are important sources of nitrogen oxides, a component of smog. In addition, the U.S. Environmental Protection Agency has linked diesel fuel to the production of fine particulate matter ($PM_{2.5}$) that has been linked to lung cancer and asthma.[18] In fact, the California Air Resources Board (CARB) has identified diesel particulate matter as a potential carcinogen. Diesel buses also emit significant levels of volatile organic compounds, as well as carbon dioxide and methane, both of which are greenhouse gases (methane has an estimated 21 times the CO_2 equivalency in terms of climate change potential). They also emit carbon monoxide and sulfur dioxide.[19] One can see why with many metropolitan areas developing strategies to meet air quality standards, the transit fleet is one of the sources of pollutants that has been targeted for conversion.

Analysis: The procedure for estimating vehicle emissions for transit operations is straight forward. As shown in Table 2, analysts have available to them emission rates for different types of pollutants, in this case, the equivalent CO_2 emissions for greenhouse gases emitted per mile by different fueled vehicles. From existing service plans or computer-based model estimates of future service, the analyst can estimate the number of vehicle-miles that the transit fleet will travel in a given year or, for that matter, any time period. By multiplying the activity variable, that is, vehicle-miles traveled, by the emissions rate, the analyst can predict the likely amount of emissions that will be associated with transit operations.

Table 2 Comparative Carbon Dioxide Emissions from Bus Fuels

Fuel	Bus Emissions (lb. CO_2/mile)
Gasoline	16.1
Petroleum diesel	13.3
CNG	11.7
B20 (20% biodiesel /80% diesel)	11.5
Ethanol from corn	11.0
Hydrogen from Natural Gas	7.3
B100 (100% biodiesel from soybeans)	3.7
Hydrogen from electrolysis	1.3

Note: From Ref. 19.

Mitigation: In response to public concerns about air pollutant emissions from transit vehicles and to laws concerning the conversion of transit fleets to alternative fuels, many transit properties have begun to convert their fleet to different fuels. The four primary technologies that seem to be most prevalent today include compressed natural gas, liquefied natural gas, hybrid diesel-electric, and fuel cells. Many of the newer buses being purchased to replace older vehicles are alternative-fueled vehicles. In some locations, such as in California, bus emission standards have been established that require meeting higher thresholds for low-emission engines (see Table 3). It seems likely that as more transit fleets are converted over time, the air pollution contribution of these fleets will decline.

Table 3 New California Bus Engine Standards (g/bhp-hr)

Model Year	Diesel Path		Alternative Fuel Path	
	NO_x	PM	NO_x	PM
2002	2.5 NO_x + NMHC	0.05	1.8 NO_x + NMHC	0.03
2004	0.5			
2007	0.2			
2008	15% of new purchases are zero emission buses (ZEBs) for large fleets (>200)	Same as for NO_x		
2012			15% of new purchases are zero emission buses (ZEBs) for large fleets (>200)	

*Although transit agencies on the alternative fuel path are not required to purchase engines certified to these optimal standards. CARB staff expects that they will do so in order to qualify for incentive funding. Source: CARB—adopted 2/24/2000. From Ref. 19.

Energy Consumption

Issue: Because public transportation systems carry more people per vehicle than the automobile, the relative energy efficiency of the different modes is often of interest to those concerned with reducing energy consumption. The calculation of such efficiency is based on assumptions such as the number of people carried per vehicle (known as the load factor), the average miles traveled per vehicle, and the fuel-consumption characteristics of the typical fleet. In particular, there are often widely varying claims about the relative energy efficiency of public transit compared to automobile travel. When comparing the energy efficiency of public transit on energy per passenger carried, public transit shows significantly greater energy efficiency. When examining this relationship on the basis of energy efficiency per passenger-mile traveled, the differences are not so great (at least in the United States). Table 4 shows estimates of relative energy efficiency for selected transportation modes in the United States on a per passenger-mile basis.

Analysis: The analysis process for estimating energy consumption is very similar in approach as that for vehicle emissions. One simply multiplies the vehicle fuel economy rate by the number of miles traveled to get the total amount of fuel consumed—or the figure can remain as gallons consumed per mile. Comparative energy consumption figures are usually converted to British thermal units (Btus), and often divided by the average number of passengers carried for each mode resulting in Btus per passenger estimate (or divide by total passenger-miles to get Btus per passenger-mile). Models have been developed that examine in greater detail the energy consumption of heavy vehicles such as diesel buses. These models include variables relating to such things as engine performance, load characteristics, and road and wind friction. In most cases, such models are used in research and policy studies, not in analyses conducted by individual transit properties.

Table 4 U.S. Passenger Transportation Energy Consumption

Transportation Mode	Load Factor (passengers per vehicle)	Fuel Consumption (Btus per passenger-mile)
Cars	1.57	3,549
Personal trucks	1.72	4,008
Motorcycles	1.22	2,049
Buses (transit)	8.7	4,160
Air	N/A	3,587
Rail (intercity Amtrak)	17.2	2,935
Rail (transit light & heavy)	21.7	3,228
Rail (commuter)	33.4	2,571

Note: From Ref. 20.

Mitigation: Although mitigation strategies for energy consumption can include changes in transit operations strategies and in how vehicles are maintained, the primary strategy for the industry is to consider the use of alternative fuels that result in reduced demand for petroleum-based fuels. As was noted for vehicle pollutant emissions, a similar strategy is often considered for reducing air quality impact. Thus, the challenge for the transit industry is finding the fuel that achieves both objectives in a cost-effective way.

Noise

Issue: Transit vehicles and facilities generate noise in a variety of ways. For example, the interaction between a vehicle's wheels and the guideway (such as rails) can create squeal that is irritating to those nearby. Other sources of noise might include vehicle breaking, horns and crossing gate bells, ventilation systems, tire/roadway interactions, vehicle start-up and idling, and the general background noise that emanates from facilities such as storage or maintenance yards. Similar to noise analyses for other surface transportation modes, transit noise is measured with various indicators, including the following:

- The *A-weighted sound level* describes a receiver's noise at any moment in time.
- The *maximum sound level* (L_{max}) describes the loudest point during a single noise event.
- The *sound exposure level (SEL)* describes a receiver's cumulative noise exposure from a single noise event.
- The *hourly equivalent sound level* ($L_{eq}(h)$) describes a receiver's cumulative noise exposure from all events over a one-hour period.
- The *day-night average sound level* (L_{dn}) describes a receiver's cumulative noise exposure from all events over a full 24 hours, with events between 10 p.m. and 7 a.m. increased by 10 dB to account for greater nighttime sensitivity to noise.

Figures 3 and 4 show the relative levels of human annoyance due to noise.[21] The impact of these values will vary by specific context and the degree to which noise is part of the ambient conditions. It is common when building new transit facilities to conduct detailed noise assessments of sensitive areas along the proposed alignment.

Approach: Figure 5 shows the general approach toward noise assessment for transit project analysis. Note that the procedure requires the analyst to define project characteristics, typical operations, the factors that might influence sound propagation, and the characteristics of the study area. More detailed noise assessments might be necessary for sites that are considered at high risk.[21]

30 Public Transportation and the Environment

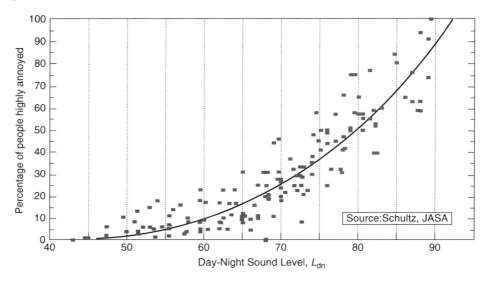

Figure 3 Community annoyance with sound levels L_{dn}. (From Ref. 21.)

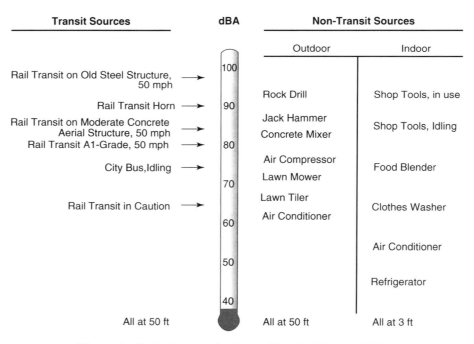

Figure 4 Typical A-weighted sound levels. (From Ref. 21.)

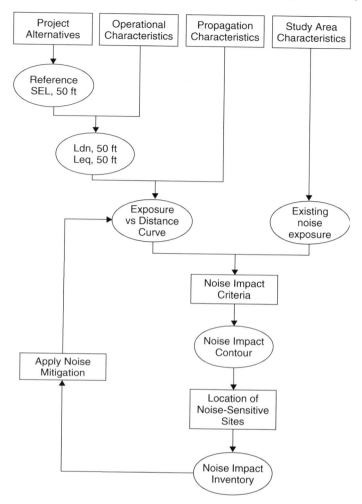

Figure 5 General procedure for noise assessment. (From Ref. 21.)

Mitigation: Mitigation at noise-sensitive sites can be focused on three areas—at the noise source (i.e., the vehicle or guideway), along the pathway between the source and receiver, and at the receiver. Table 5 shows the different types of mitigation strategies that could be considered for each of these categories, along with a range of expected noise reductions.

Vibration

Issue: Vibration is a greater concern for transit than it is for other surface transportation projects. The movement of vehicles along a guideway and the variable motion of this movement creates dynamic stresses on the guideway itself, resulting in vibrations that can reach nearby buildings. Such vibration can

Table 5 Potential Transit Noise Mitigation Measures

Application	Mitigation Measure		Effectiveness
SOURCE	Stringent vehicle & equipment noise specifications		Varied
	Operation restrictions		Varied
	Resilient or damped wheels*	For rolling noise on tangent track	2 dB
		For wheel squeal on curved track	10–20 dB
	Vehicle skirts*		6–10 dB
	Undercar absorption		5 dB
	Spin-slide control (prevents flats)*		**
	Wheel truing (eliminates wheel flats)*		**
	Rail grinding (eliminates corrugations)*		**
	Turn radii greater than 1,000 ft*		(Avoids squeal)
	Rail lubrication on sharp curves*		(Reduces squeal)
	Movable-point frogs (reduced rail gaps at crossovers)*		(Reduces impact noise)
	Engine compartment treatments (buses)		6–10 dB
PATH	Sound barriers close to vehicles		6–15 dB
	Sound barriers at ROW line		3–10 dB
	Alteration of horizontal & vertical alignments		Varied
	Acquisition of buffer zones		Varied
	Ballast on at-grade guideway*		3 dB
	Ballast on aerial guideway*		5 dB
	Resilient track support on aerial guideway		Varied
RECEIVER	Acquisition of property rights for construction of sound barriers		5–10 dB
	Building noise insulation		5–20 dB

*Applies to rail projects only.
**These mitigation measures work to maintain a rail system in its as-new condition. Without incorporating them into the system, noise levels could increase up to 10 dB.
Note: From Ref. 21.

be a particularly important concern to building uses such as hospitals, research laboratories, schools, and performing arts centers.

Approach: The general approach to analyzing vibration impacts includes the following:[21]

1. Screen the proposed transit alignment to identify areas where there is the potential of impact from ground-borne vibration.
2. Define the curves of ground-surface vibration level as a function of distance (found in the technical manual).
3. Estimate vibration levels for specific buildings or groups of buildings (in some cases, it might be necessary to conduct ambient vibration studies at these buildings).
4. Where impact criteria are exceeded, review potential mitigation measures and assemble a list of feasible approaches to vibration control.
5. Refine the impact assessment and develop detailed vibration mitigation measures where needed.

Mitigation: Possible mitigation strategies for reducing the level of exposure to transit-induced vibration include such things as developing maintenance programs that address degradation of vehicle components (e.g., degradation of wheel and rail surface interaction can increase vibration by up to 20 dB compared to new systems), planning and designing special track work in areas where vibration is likely to occur, requiring through vehicle specifications that certain vibration thresholds must be maintained (such as the vertical resonance frequency), modifying nearby buildings to reduce vibration, reducing vehicle speed in sensitive areas, and establishing buffer zones that will protect future development from transit-induced vibration.

3.4 Ridership-based Environmental Impacts

The environmental impacts in this category primarily relate to the larger-scale environmental consequences of attracting new riders to a transit system or facility, with many of these new riders coming from other modes of transportation. The most important analysis questions for these types of impacts ask the following. How many riders will be attracted to new transit services? Where will these new riders come from? What are the resulting regional changes in travel and thus environmental impacts associated with this shift? The two major environmental impacts that are most relevant to this type of analysis are regional air quality and energy consumption. In both cases, the change in the impact is closely related to how many new transit riders shift from using the automobile for the same trip purpose. Once this change has been estimated, one can use approaches similar to those described in the previous section to determine the reduction in air pollutants or in energy consumption. That is, multiply the emission or fuel consumption rate per mile by the number of automobile vehicle-miles now not driven, thus providing an aggregate measure of the regional reduction in the impact being considered.

Many books and articles have been written on the approach used to model travel behavior in an urban transportation network (see, for example, Refs. 1 and 22 to 24). The important point of departure for this approach is that travelers choose which mode to use based on a number of factors relating to the relative utility of each mode compared to all other modes available for the trip. This utility can include such things as the financial cost of travel, travel time, ease and comfort associated with each mode (e.g., is there a transfer?), and the availability of a mode for reaching a particular destination.

From an environmental perspective, the key policy question is, how can public transportation be made more appealing to potential riders, either through improvements to transit itself, or possibly by making other modes more expensive or not as convenient? In other words, the question is, what can be done to increase the utility of transit modes relative to using the automobile? It is important to note that from a regional modeling perspective, other factors relating to the individual traveler (such as income, family status, car ownership, etc.) are important determinants in predicting aggregate travel decisions as well.

Table 6 lists some of the public policies that can affect the attractiveness of transit as a travel mode. From an environmental perspective, the degree to which new riders can be attracted from single-occupant automobiles is the level of environmental benefit that is likely to occur from the investment. As is suggested in Table 6, the ability to attract such riders is part of a much larger policy context than simply providing new transit service.[25] Supporting strategies are necessary to reinforce the desirability of choosing transit over the automobile. In the United States and other developed nations, this means primarily making automobile use more expensive through such things as parking prices, areawide pricing schemes (such as the London congestion pricing program), and highway tolls.

3.5 Mitigating the Environmental Impacts

As described in the previous sections, public transit facilities and services can have a potentially very wide range of environmental impacts. Environmental studies and assessments determine which of these impacts are likely to have the most important consequences to the metropolitan area and local communities. When making a decision to invest in a transit project, decision makers must also identify the mitigation program that will be put in place to reduce negative impacts.

An example from the Bay Area Rapid Transit (BART) system in San Francisco illustrates this point. As part of the final supplemental environmental impact report for a rapid rail line extension, the BART board of directors identified the following "significant and unavoidable" impacts, which became part of its mitigation plan for the project.[26]

Table 6 Policies That Influence Mode Choice

Transportation Investment Policy

- Infrastructure spending directly affects the relative attractiveness of each mode.
- Transit operating assistance can help to maintain, improve, or expand services.
- Research and development funding provides innovations in the provision of transportation services.

Transportation Pricing Policy

- Taxes and tolls make automobile use more expensive.
- Fare policy determines cost of transit trips.
- Local policies dictate taxi fare and, indirectly, service levels.
- Local parking pricing and availability are very important components of the cost of driving.

Environmental Policy

- Federal/state emissions standards increase new car prices.
- Local air quality mandates require programs to reduce single-occupant vehicle use.
- Local policies influence development patterns and transportation planning.

Energy Policy

- Minimum average fuel economy standards increase new car prices, decrease operating costs.
- Alternative fuel vehicles are unlikely to affect modal choice.

Tax Policy

- Income taxes affect economic activity and disposable income, thereby influencing the affordability of various travel choices.
- Preferential parking cost deductions promote automobile commuting over transit.
- Sales taxes affect automobile costs, and may support public transit.
- Mortgage interest deductions influence housing location choice.
- Property taxes may support local roadway infrastructure.

Land Use Policy

- Provisions of zoning laws (lot size, use) affect the viability of public transit.
- Design reviews and other restrictions can require definitive plans for addressing transportation issues in new developments.

1. Biological impacts with the cumulative loss of grassland and the loss of habitat for the burrowing owl.
2. Aesthetic impacts to the visual aesthetics with construction of soundwalls to reduce noise impacts from the BART extension. Temporary visual impacts will also occur from the construction of the extension on the Fremont Central Park area.
3. Traffic impacts will occur with the addition of the optional Irvington station in that the capacity to traffic volume and level of service at Osgood Road/Durham Road/Auto Mall Parkway and at the intersection of Mission Boulevard/Warm Springs Boulèvard around the proposed station will be at unacceptable levels. Traffic level of service would also deteriorate on northbound I-880, south of the Irvington station at Mission Boulevard.
4. Vibration impact will occur with the operation of the project and may affect up to 134 residences.
5. Electricity demand of the project during operation will significantly impact peak- and base-period electrical transmission system capacity.
6. Geological fault could involve potential damage to the rail and stations and injury to patrons.
7. Increased populations densities could occur because of the project, which could expose people and structures to seismic hazards.

By committing to the mitigation of these impacts, BART is linking the results of a fairly lengthy environmental assessment process and the input received from concerned citizens with actions specifically aimed at reducing environmental harm.

3.6 Environmental Stewardship and Sustainable Development

As already noted, transit facilities and operations can have important long-lasting environmental consequences, especially as they relate to land-use decisions. This section presents two aspects of this longer-term perspective—environmental stewardship and transit's linkage to sustainable development.

Transit Environmental Stewardship

Transit agencies undertake numerous activities that cumulatively could have a significant effect on environmental quality. Internationally, the International Organization for Standardization (ISO) has developed a set of standards known as the ISO 14000 series that outlines the steps that organizations can take to adopt an environmentally conscious approach to managing the organization's assets and activities.[27] One of the most important elements of such an approach is the creation and use of an environmental management system (EMS). An *environmental management system* is a set of processes and practices that enable an organization to reduce its environmental impacts and increase its operating efficiency.[28]

> **Sound Transit Environmental Policy**
>
> Sound Transit is committed to the protection of the environment for present and future generations as we provide high capacity transit to the Puget Sound Region. Sound Transit has been a catalyst and model for engaging federal and state partners to resolve environmental issues that apply to our program. We will continue to be an environmental leader in the State of Washington through the integration of the following principles into our daily business practices:
>
> - We will be in full compliance with all environmental laws and regulations. We will strive to exceed compliance by the continual improvement of our environmental performance through cost-effective innovation and self-assessment.
>
> - We will restore the environment by providing mitigation and corrective action, and will monitor to ensure that environmental commitments are implemented. We will improve our ability to manage and account for environmental risk.
>
> - We will avoid environmental degradation by minimizing releases to air, water, and land. We will prevent pollution and conserve resources by reducing waste, reusing materials, recycling, and preferentially purchasing materials with recycled content.
>
> - We will increase the awareness of environmental issues among agency employees through education and training. We will continue to educate the public about the environmental benefits of our transit system. We will build relationships with our contractors, vendors, consultants, and transit partners during planning, design, construction, and operation to protect and enhance the environment.
>
> - In order to implement this Policy, Sound Transit will establish and maintain an Environmental Management System (EMS) with environmental objectives and targets that are measurable, meaningful and understandable. The goals and progress of this Policy and the EMS will be communicated to agency board members, officers, employees and the public.

Figure 6 Sound Transit's environmental policy. (From Ref. 29.)

Part of the EMS is to have a clear statement on the agency's environmental policies and a vision associated with how agency activities will be managed from the perspective of environmental consequences. Figure 6, for example, shows the environmental policy that has been adopted by Sound Transit, the regional transit agency for the Seattle metropolitan area.[29] As can be seen, the policy includes a wide range of efforts that are aimed at not only avoiding environmental impacts, but also restoring environmental characteristics through the management of environmental risk.

Part of Sound Transit's implementation of its environmental policy has been to identify potential areas where the agency can become more environmentally sensitive. Sound Transit has identified some possible actions:

- *Green building and design:* Incorporating green building practices into capital project design guidelines

- *Environmentally Preferable Purchasing:* Identifying green products and vendors that meet environmentally friendly criteria
- *Waste Prevention and Recycling:* Reducing, reusing, and recycling products
- *Climate Change:* Building high-capacity transit facilities and services to reduce reliance on single-occupant cars; purchasing hybrid vehicles and alternative fuels for revenue and nonrevenue fleet
- *Energy Efficiency:* Identifying ways to conserve energy and contribute to the viability of renewable energy sources

In addition, Sound Transit has adopted an approach toward the building of facilities that will lessen the development impact by using natural methods of stormwater management. Called low-impact development, or *LID*, the strategy is founded on using more trees, open space, and plantings for managing stormwater runoff. Sound Transit has incorporated several low-impact development strategies in its projects.

Utah Transit, the regional transit authority in Salt Lake City, has also aggressively adopted an environmental management system approach toward reducing the environmental impacts of its activities.[30] Its EMS efforts have identified several areas of agency activities where more environmentally conscious approaches could be adopted by agency staff:

Combustion Sources

- Air heating and supply
- Mobile transportation, such as forklift or carts
- Construction activities
- Excavation or grading
- Drilling or blasting
- Rock crushing
- Demolition
- Welding or soldering
- Painting
- Asphalt paving
- Use or storage of chemicals or fuels
- Transfer of bulk materials
- Disposal of chemical wastes
- Disposal of general wastes
- Storage tanks

Building Maintenance Activities

- Architectural paint removal
- Architectural painting

- Hydroblasting
- Sandblasting
- Surface preparation/treatments, such as floors and roof repair
- Purging or repair of distribution lines such as those for fuel, oil, or solvents
- Use of chemicals, solvents, caustics, acids, oils, etc.
- Use of herbicides, pesticides, or insecticides

Business or Work-related Activities

- Use or receipt of chemical materials (other than janitorial or cleaning materials)
- Generation and disposal of chemical wastes
- Application of sealers, adhesives, coatings, or paints
- Welding, soldering, brazing, or similar activities
- Use of caustics or acids
- Use of combustion gases
- Medical waste
- Discharge to storm drains

Importantly, Utah Transit conducted a cost assessment of all the actions identified by the EMS that would reduce environmental impacts, and found that for calendar year 2005, the savings to the agency were estimated to be $1,300,000, and the estimated annualized savings in the future were $814,000. The biggest savings came from changing the agency's bus-idling policy that reduced the amount of fuel consumed during the year. In addition, Utah Transit estimated that its EMS efforts reduced CO_2 emissions by 3.04 million lb./yr., NO_x emissions by 91,000 lb./yr., and particulate matter by 2,500 lb./yr.[30] The important message coming from this experience was that implementing actions to reduce environmental impacts also resulted in cost savings to the agency.

It seems likely that as the transportation industry becomes more attuned to considering environmental factors in organizational activities, the concept of an environmental management system will become an important element of agency management. This will be particularly true for transit agencies, where the very nature of the services provided and the infrastructure used relate so strongly to resulting environmental conditions. The use of an EMS will be a critical factor in developing a strong environmental stewardship approach toward transit management.

Transit's Linkage to Sustainable Development

Much of the urban development that occurs over the next several years will be in place for many decades to come. The land-use patterns and overall urban form that result from the cumulative development decisions of today will go a long way toward defining the environmental footprint of metropolitan areas in the future. In many cities of the world, the symbiotic relationship between development decisions and the availability of transit service is the basis for site design and development plan approvals. In other cities, such as in many of the fast-growing cities in the United States, this relationship is not so clear and is, in fact, often missing.

Development patterns in fast-growing cities tend to rely on the predominant mode of transportation, the automobile, which encourages development patterns that are often widely spread out, resulting in a very large level of daily vehicle miles traveled. The implications of such development patterns for environmental quality are significant. The more traveling that occurs, the greater the consumption of fuel and the more pollutants emitted into the air. The greater the dependence on the automobile for transportation, the more significant become the issues associated with social equity for those not having automobiles or unable to drive the vehicles that are available (such as the elderly or handicapped).

In response, many communities are encouraging a closer linkage between development decisions and transit services. This linkage occurs not only during the general planning stage where transit routes and future developments are contemplated, but also during urban design phases of development approvals, where more specific decisions must be made relating to individual development sites, such as the position or footprint of the buildings on-site, the circulation network provided internal to the site, and the type and level of access provided to transit services.

As noted earlier in the chapter, one of the important concepts at the broader planning level for linking development and transit services is called *transit-oriented development (TOD)*. The basic thrust of TOD policies is to provide good transit access to development sites, thus making the sites more attractive and accessible to the rest of the region, while at the same time developing a transit ridership market by encouraging people to live near a transit station. A recent TOD policy adopted by the Regional Transit District (RTD) in Denver, Colorado stated the following:

> By focusing compact development around transit stations, TOD capitalizes on the value of public infrastructure investments and promotes sustainability. These development synergies promote increased transit ridership for transit agencies. In addition to increased ridership, TOD also is a successful tool for promoting local economic development, helping communities plan for sustainable growth, and increasing the overall quality of life in a region.[31]

Although the characteristics of TOD can vary from one community to another, in general, the following criteria are used to define a TOD area.[32]

- Walkable design with pedestrian as the highest priority
- Transit station as prominent feature of town center
- A regional node containing a mixture of uses in close proximity including office, residential, retail, and civic uses

- High-density, high-quality development within 10-minute walk circle surrounding transit station
- Collector support transit systems including trolleys, streetcars, light rail, and buses
- Designed to include the easy use of bicycles, scooters, and rollerblades as daily support transportation systems
- Reduced and managed parking inside 10-minute walk circle around town center/transit station

One of the challenges in fostering a coordinated land-use and transit investment strategy is the institutional process used to establish the respective responsibilities for the different agencies and groups that will be involved in the TOD process. In addition, the TOD process needs to be closely linked with the transit project development process as well to make sure that one does not interfere with the progress being made by the other. Figure 7 presents a model from the Denver RTD of how the TOD process relates to stages in the project development process for transit facilities.[31] Even though the implementing mechanisms for TOD, that is, zoning changes, public/private partnerships, and design/construction approaches, are shown as occurring in Phase 3 of project development, in reality one needs to begin structuring many of these strategies much earlier given that it often takes considerable time to get approval from decision-making authority to use these mechanisms for such a purpose.

In the United States, the importance of TOD as part of a community strategy for transit investment has become an important criterion for receiving government support for transit projects. For example, federal government support for transit capital projects requires an assessment of the concentration of development proposed around transit centers, the plans that will be implemented to increase corridor and station area development including parking policies, proposed improvements to pedestrian access facilities, the use of zoning ordinances to reinforce the desired land use patterns, and the efforts that have been made

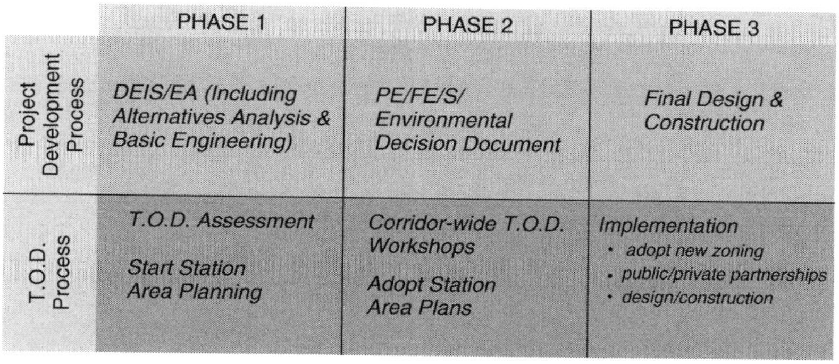

	PHASE 1	PHASE 2	PHASE 3
Project Development Process	DEIS/EA (Including Alternatives Analysis & Basic Engineering)	PE/FE/S/ Environmental Decision Document	Final Design & Construction
T.O.D. Process	T.O.D. Assessment Start Station Area Planning	Corridor-wide T.O.D. Workshops Adopt Station Area Plans	Implementation • adopt new zoning • public/private partnerships • design/construction

Figure 7 TOD relationship to project development process. (Adapted from Ref. 31.)

to engage the development community in the implementation of the proposed development plans.[33]

The environmental stewardship responsibilities of transit systems will increasingly lead them into efforts at more closely linking urban development with transit facility designs. Most of the major transit agencies in the United States and Canada have adopted formal policies and guidelines on how this linkage should occur. Most of the major transit agencies in other parts of the world incorporate this linkage as a matter of course in the development of transit projects. Good examples of TOD guidance and organizing principles can be found in Refs. 34 to 39. Figure 8 presents a checklist that can be used to design transit-oriented development around a proposed transit station.

Within an easy walk of a major transit stop (e.g., 1/4 to 1/2 mile walk), consider the following:

Land Use

- ❏ Are key sites designated for "transit-friendly" land uses and densities (i.e., walkable, mixed-use, not dominated by activities associated with significant automobile use)?
- ❏ Are "transit-friendly" land uses permitted outright, not requiring special approval?
- ❏ Are higher densities allowed near transit?
- ❏ Are multiple compatible uses permitted within buildings near transit?
- ❏ Is the mix of uses generating pedestrian traffic concentrated within walking distance of transit?
- ❏ Are auto-oriented uses discouraged or prohibited near transit?

Site Design

- ❏ Are buildings and primary entrances sited to be easily accessible from the street?
- ❏ Do the designs of areas and buildings allow direct pedestrian movements between transit, mixed land uses, and surrounding areas?
- ❏ Does the site's design allow for the intensification of densities over time?
  - Are the first floor uses "active" and pedestrian-oriented?
  - Are amenities provided to create an interesting and enjoyable pedestrian environment along and between buildings?
  - Are there sidewalks along the site frontage? Do they connect to sidewalks and streets on adjacent and nearby properties?
  - Are there trees sheltering streets and sidewalks? Is there pedestrian-scale lighting?

Street Patterns and parking

- ❏ Are parking requirements reduced in close proximity to transit, compared to the norm?
  - Is structured parking encouraged rather than surface lots in higher density areas?
  - Is most of the parking located to the side or to the rear of the buildings?
  - Are street patterns based on a grid/interconnected system that simplifies access?
  - Are pedestrian routes buffered from fast-moving traffic and expanses of parking?
  - Are there convenient crosswalks to other uses on-and off-site?
  - Can residents and employees safely walk or bicycle to a store, post office, etc.?
- ❏ Does the site's street pattern connect with streets in adjacent developments?

Figure 8 TOD guidelines. (Adapted from Ref. 34.)

4 CONCLUSION

This chapter has presented an overview of many of the dimensions of environmentally conscious public transportation. As described, environmentally conscious public transportation is manifested in many different ways, ranging from how a transit facility is designed to the degree to which transit investment can influence sustainable development patterns. In some ways, public transportation is not very different from other transportation modes in terms of the types of environmental impacts that need to be considered and the analysis approaches that are used. However, four key observations with respect to environmentally conscious transit merit special attention:

1. Although transit facilities and vehicles have their own near-term and direct impact on the environment, the most important contribution to environmental quality from public transportation is its ability to replace other, more polluting modes of travel as a primary travel choice for travelers. This suggests that attention should be focused not only on how to make transit service itself more appealing, but also how to lower the appeal of the automobile. The most successful metropolitan areas in terms of enhancing environmental quality through transit actions have used a combination of pricing, land-use, urban design and financial incentives to make transit a viable alternative for travelers.

2. Perhaps the most important role for public transportation in the context of long-term environmental sustainability is its enabling influence on development patterns that themselves reflect environmentally sound practices. The concept of transit-oriented development is an example of how transit investment and development can be considered jointly to provide a more environmentally sound approach to community growth.

3. Similar to other modes of transportation, technology innovations (in particular as they relate to fuel types) are an important consideration for future transit operations. Many transit fleets are converting to alternative fueled vehicles, and more are likely to follow in the coming decade. This transformation of the fuel of choice becomes an important step to a more environmentally conscious approach to transit service provision.

4. Transit agencies by their very nature are usually major owners of property and buildings, and conduct many maintenance procedures that could have potentially important impacts on the surrounding environment. Many transit agencies have adopted an environmental stewardship perspective on their day-to-day activities, supported with environmental management systems that permit them to reduce their impact on the environment. This concept is one that will likely become more prevalent in future years throughout the transit industry.

Another way of saying these four observations is that an environmentally conscious transit strategy will only be effective if it is integrated with sustainable development policies and integrated throughout every agency function. Mitigating the impacts of guideway or station construction is necessary, but not the sole component of a true environmentally conscious transit program. As described in this chapter, environmentally conscious transit needs to be linked strongly to urban design, community development, and management of the daily operations of the transit agency. By so doing, public transit can become a major contributor to the much broader goal of developing sustainable communities.

REFERENCES

1. M. D. Meyer and E. Miller, *Urban Transportation Planning: A Decision-Oriented Approach*, 2nd edition, McGraw-Hill, New York, 2001.
2. R. Cervero, *The Transit Metropolis, A Global Inquiry*, Island Press, Washington D.C, 1998.
3. M. Safdie, *The City after the Automobile*, Westview, Boulder, CO, 1997.
4. J. H. Crawford, *Car Free Cities*, International Books, Utrecht, The Netherlands, 2000.
5. Transportation Research Board, *Transportation, Urban Form, and the Environment*, Special Report 231. National Research Council, Washington D.C., 1991.
6. Transportation Research Board, *Making Transit Work, Insight from Western Europe, Canada, and the United States*, Special Report 257. National Research Council, Washington D.C., 2001.
7. Parsons Brinckerhoff Quade & Douglas, *Transit and Urban Form, Transit Cooperative Research Program Report 16*, vols. 1 and 2, National Research Council, Washington D.C., 1996.
8. Parsons Brinckerhoff Quade & Douglas, *Consequences of the Interstate Highway System for Transit: Summary of Findings, Transit Cooperative Research Program Report 42*, National Research Council, Washington D.C., 1998.
9. M. Bernick and R. Cervero, *Transit Villages in the 21st Century*, McGraw-Hill, New York, 1997.
10. V. R. Vuchic, *Transportation for Livable Cities*, Center for Urban Policy Research, Rutgers University, New Brunswick, NJ, 1999.
11. R. Dunphy, D. Myerson, and M. Pawlukiewicz, *Ten Principles for Successful Development Around Transit*, ULI—The Urban Land Institute, Washington, D.C., 2003.
12. Oregon Department of Transportation, "Oregon Transportation Plan," www.oregon.gov/ODOT/TD/TP/docs/ortransplanupdate/06otp/06otpVol1sep.pdf, September 20, 2006. p. II–14. Accessed on January 26, 2007.
13. Council of Environmental Quality, "Incorporating environmental justice into NEPA," www.whitehouse.gov/CEQ/, Washington, D.C.: Dec. 1997.
14. Environmental Protection Agency, *The Hazard Ranking System Guidance Manual; Interim Final*. Report NTIS PB92-963377, EPA 9345.1-07. Washington, D.C.: November 1992.

References

15. Florida Department of Transportation, "Community Impact Assessment, Chapter 9," www.dot.state.fl.us/emo/pubs/pdeman/updated/Chapter%209%20PD&E%20Manual%20111003.pdf. Accessed January 11, 2007.
16. Florida Department of Transportation, "Community Impact Assessment and Environmental Justice for Transit Agencies: A Reference," Center for Urban Transportation Research. January 2002.
17. Transportation Research Board, "Third National Community Impact Assessment Conference," Proceedings of a conference held in Madison, Wisconsin, August 2002. Transportation Research Board Circular E-C054. July 2003.
18. Bradley & Associates, *Hybrid-Electric Transit Buses: Status, Issues and Benefits, Transit Cooperative Research Program Report 59*, Transportation Research Board, National Academy Press, Washington, D.C., 2000.
19. S. Feigon, D. Hoyt, L. McNally, and R. Mooney-Bullock, *Travel Matters: Mitigating Climate Change with Sustainable Surface Transportation, Transit Cooperative Research Program Report 93*, Transportation Research Board, National Academy Press, Washington, D.C., 2002.
20. S. Davis and S. Diegel, *Transportation Energy Data Book*, 25 ed., Oak Ridge National Laboratory, U.S. Department of Energy, Center for Transportation Analysis, December 2005.
21. C. Hanson, D. Towers, and L. Meister, *Transit Noise and Vibration Impact Assessment*, Report FTA-VA-90-1003-06. Washington, D.C.: Federal Transit Administration, May 2006.
22. N. Oppenheim, Urban Travel Demand Modeling, From Individual Choices to General Equilibrium. New York: John Wiley & Sons, 1995.
23. K. Goulias, ed., Transportation Systems Planning, Methods and Applications, CRC Press, Boca Raton, FL, 2003.
24. J. Ortuzar, and L. Willumsen, *Modelling Transport*, 2nd ed, John Wiley & Sons, New York, 1994.
25. Charles River Associates, *Building Transit Ridership An Exploration of Transit's Market Share and the Public Policies that Influence It, Transit Cooperative Research Report 27*, Transportation Research Board. National Academy Press, Washington, D.C., 1997.
26. Memorandum from D. Eidam to Chair and Commissioners, "Approval of Project for Future Consideration of Funding—BART Fremont to Warm Springs Extension (TCRP #1) (Final Supplemental EIR)," Resolution E-03-33. September 15, 2003.
27. International Organization for Standardization, "Environmental Management, The ISO 14000 Family of International Standards," www.iso.org/iso/en/prods-services/otherpubs/iso14000/index.html.
28. Federal Transit Administration, "Environmental Management Systems Training & Assistance, Final Report," www.fta.dot.gov/library/FTA_EMS/EMS_Final_Report.pdf, Federal Transit Administration, Washington, D.C., 2006.
29. Sound Transit, "Environmental Initiatives," www.soundtransit.org/x3685.xml. Accessed on March 1, 2007.
30. Utah Transit, "Environmental Management," www.utabus.com/utaInfo/environmentalManagement/factSheet.aspx, Accessed on Febuary 25, 2007.

31. Regional Transit District, "Strategic Plan for Transit Oriented Development," www.rtd-denver.com/Projects/TOD/TODStrategicPlan.pdf Accessed on Jan. 18, 2007,
32. Transit Oriented Development Organization, "Transit Oriented Development," www.transitorienteddevelopment.org/. Accessed on February 3, 2007.
33. Federal Transit Administration, "Guidelines and Standards for Assessing Transit-Supportive Land Use," www.fta.dot.gov/documents/FTA_LU_Contractor_Guidelines_FY04_complete1.pdf. Accessed on January 15, 2007.
34. Calgary Land Use Planning and Policy, "Transit Oriented Development Policy Guidelines," www.calgarytransit.com/Approved%20TODPG%20041206.pdf. Accessed on January 2007,
35. "Transit-Oriented Development Design," www.dot.ca.gov/hq/MassTrans/doc_pdf/TOD2/TODC_Appendix.pdf. Accessed on January 15, 2007,
36. Bay Area Rapid Transit District, "BART Transit-Oriented Development Guidelines," www.bart.gov/docs/planning/TOD_Guidlines.pdf. Accessed on January 28, 2007.
37. Puget Sound Regional Council, "Implementing Transit-Oriented Development in Station Communities," www.psrc.org/projects/tod/docs/part3.pdf. Accessed on January 7, 2007.
38. Georgia Department of Community Affairs, "TOD Small Area Plans," www.dca.state.ga.us/intra_nonpub/Toolkit/Guides/TODSmAreaPlns.pdf. Accessed on Jan. 6 2007,
39. R. Cervero, "Transit-Oriented Development and Joint Development in the United States: A Literature Review," *Transit Cooperative Research Program Research Results Digest*. **52** (October 2002).

CHAPTER 3

TRANSPORTATION AND AIR QUALITY

Mohan M. Venigalla, Ph.D., P.E.
Department of Civil Environmental and Infrastructure Engineering
George Mason University
Fairfax, Virginia

1	IMPACT OF TRANSPORTATION ACTIVITY ON AIR QUALITY	47	5 AIR QUALITY MONITORING	52
2	CRITERIA POLLUTANTS, AIR TOXICS, AND OTHER HARMFUL SUBSTANCES	48	6 AIR QUALITY DESIGNATIONS	53
3	EMISSION CONTROL STRATEGIES	50	7 STATE IMPLEMENTATION PLANS (SIPS)	53
4	NATIONAL AMBIENT AIR QUALITY STANDARDS (NAAQS)	52	8 THE CONFORMITY RULE	53
			9 TRANSPORTATION AND AIR QUALITY PLANNING	54

Transportation activities have a direct impact on the quality of ambient air. The proper phrase to describe the science that studies this impact is "transportation and air quality" rather than 'transportation and air pollution." For, a majority of the substances that may be considered as *air pollutants* are, in fact, naturally present in the atmosphere. The predominant effect transportation sources have on the atmospheric air is to affect the quality of the ambient air. Presented in this chapter is a broad outline of the underlying science and practices related to transportation air quality.

1 IMPACT OF TRANSPORTATION ACTIVITY ON AIR QUALITY

The primary goal of transportation means and services is to mobilize people and goods in a safe and efficient manner from their origin to destination. At the same time, the adverse effects of transportation activities have caused growing levels of motorization and congestion. The total vehicle-miles that people travel in the United States increased 178 percent between 1970 and 2005 and continues to increase at a rate of 2 to 3 percent each year.[1]

Excessive reliance on internal combustion engines and fossil fuels as a source of energy in the transportation sector has also adversely affected the quality of ambient air. Direct or primary impacts of transportation activities include economic activity and congestion on roads, which are relatively easy to understand and measure. The effect on air quality may be characterized as an indirect or secondary or tertiary impact of transportation activities. These indirect impacts are difficult to measure. The additive or even multiplicative long-term cumulative effects of these activities are even more complicated to comprehend and certainly very difficult to measure. The relationship between the air quality and transportation activities is complex and multidimensional. Controversy surrounding the environmental policy related to transportation can be directly attributable to such complexities.

The U.S. EPA estimates that motor vehicles are responsible for nearly one half of smog-forming volatile organic compounds (VOCs), more than half of the nitrogen oxide (NO_x) emissions, and about half of the toxic air pollutant emissions in the United States. Motor vehicles, including nonroad vehicles, now account for 75 percent of carbon monoxide emissions nationwide.

In the internal combustion engine, energy is produced due to rapid oxidation of air–fuel mix in a confined space. Complexities associated with the oxidation process and the exhaust mechanism result in the production of a variety of gaseous substances and particulate matter. Included among them are unburnt fuel such as volatile organic compounds (VOC) (also referred as hydrocarbons, or HC), carbon monoxide (CO), and oxides of nitrogen (NO_x).

2 CRITERIA POLLUTANTS, AIR TOXICS, AND OTHER HARMFUL SUBSTANCES

EPA designated the following six elements as *criteria pollutants*:

1. Carbon monoxide (CO)
2. Oxides of nitrogen (NO_x)
3. Ground-level ozone (O_3)
4. Fine particulate matter (PM_{10} and $PM_{2.5}$)
5. Sulfur dioxide (SO_2)
6. Airborne lead (Pb)

Carbon monoxide, CO, is formed mainly due to incomplete fuel burning. CO is a stable, colorless, odorless, and poisonous gas. It reacts with blood oxygen to form CO_2 and hemoglobin (Hb) to form carboxy-hemoglobin. These reactions reduce the oxygen-carrying capacity of blood. At lower doses CO causes visual impairment and reduced work capacity. Exposure to higher levels of CO is fatal. Main sources of CO in the atmosphere include motor vehicles, industrial processes, and wildfires.

Oxides of nitrogen (NO_x) are formed due to fuel burning at high temperatures. Physical properties of these highly reactive poisonous gases include a brownish tint. Exposure to NO_x causes irritation to lungs. Further, it lowers resistance to respiratory illnesses. Excessive presence of NO_x causes brownish tint to the atmosphere. Sources of NO_x include motor vehicles, electric utilities, and industrial boilers.

The term *particulate matter* (PM) refers to solid or liquid particles in atmosphere with diameters less than a specified number of microns (e.g., PM10 refers to particulate matter smaller than 10 microns). Particles in the atmosphere are of variety of shapes, sizes and material (e.g., fugitive dust). The sources of atmospheric PM include a variety of natural and human activities such as of fuel burning and natural environment. Although the PM may or may not be able to be seen by the naked eye, it can act as an allergen and cause damage to lung tissues and visibility problems.

Sulfur dioxide (SO_2) is formed due to the oxidation of sulfur (S), which is present in the fuel. SO_2 is a colorless gas with pungent smell. When rainwater mixes with SO_2 in the atmosphere, sulfuric acid (H_2SO_4) is formed, which results in *acid precipitation*. Point sources such as coal-powered boilers are main source of SO_2. Sulfur content is also present in diesel fuel. Motor vehicles using diesel fuels (e.g., trucks and trains) emit SO_2. Emissions of SO_2 from the gasoline-powered motor vehicles is negligible.

Lead (Pb) is a highly toxic metal that produces a range of adverse health effects, particularly in young children. It accumulates in blood, bones, and soft tissues. It affects kidneys, liver, and nervous system functions. Excessive exposure to lead causes neurological impairment. Children are affected at even very low doses. Lead can also harm wildlife. Sources of lead include motor vehicles, smelters, battery plants, and volcanic eruptions. Even though lead has been removed in U.S. gasoline supplies, in several parts of the world leaded gasoline is still in use.

Ground-level ozone (O_3) is not emitted directly and is the most complex of the six criteria pollutants. Rather, it is formed due to photochemical reaction of atmospheric gases, mainly NO_x and volatile organic compounds (VOC) in the presence of high-intensity sunlight. NO_x and VOC are called precursors to O_3. Ozone is a relatively stable and odorless gas. Exposure to O_3 complicates respiratory problems for adults and children. Prolonged exposure results in reduced lung function. Studies show that exposure to O_3 also affects healthy people. Furthermore, atmospheric ozone causes smog formation. The sources of O_3 precursors include motor vehicles, thousands of VOC gases, bakeries, industries, gas stations, and lawn mowers.

Like all organic compounds, VOCs contain carbon. Organic chemicals are the basic chemicals found in all living things and all products derived from living things. Many organic compounds that we use do not occur in nature, but are synthesized by chemists in laboratories. Volatile chemicals produce vapors easily at room temperature. VOCs include gasoline; industrial chemicals, such as benzene; solvents, such as toluene and xylene; and perchloroethylene (the principal dry-cleaning solvent). VOCs are released from burning fuel, such as gasoline, wood, coal, or natural gas, and from solvents, paints, glues, and other products used at home or work. Vehicle emissions are an important source of VOCs. Many VOCs are also HAPs.

Although the formation of carbon dioxide (CO_2), a principal greenhouse gas, occurs naturally, it is also emitted as a result of human activity (e.g., burning of coal, oil, and natural gas). If inhaled in high concentrations, CO_2 can be toxic and can cause increased breathing rate, unconsciousness, and other serious health problems.

Chlorofluorocarbons (CFCs) are chemicals used in industry, refrigeration and air conditioning systems, and consumer products. When released into the air, CFCs rise into the stratosphere. In the stratosphere, CFCs react with other chemicals and reduce the stratospheric ozone layer, which protects Earth's surface from the sun. Reducing CFC emissions and eliminating the production and use of ozone-destroying chemicals is very important to protecting Earth's stratosphere.

Air toxics are chemicals that cause serious health and environmental effects. Health effects include cancer, birth defects, nervous system problems, and death due to massive accidental releases, such as occurred at a pesticide plant in Bhopal, India, in the mid-1980s. HAPs are released by sources such as chemical plants, dry cleaners, printing plants, and motor vehicles (e.g., cars, trucks, buses, and planes). The EPA designated 33 pollutants as air toxics. Included in this list are chloroform, hexachlorobenzene and compounds of arsenic beryllium, lead, cadmium and nickel. EPA sets standards for air toxics emissions, which state and local programs are responsible for carrying out. In addition, some state and local programs have their own air toxics rules.

3 EMISSION CONTROL STRATEGIES

Today, motor vehicles are responsible for nearly half of smog-forming volatile organic compounds (VOCs), more than half of the nitrogen oxide (NO_x) emissions, and about half of the toxic air pollutant emissions in the United States. Motor vehicles, including nonroad vehicles, now account for 75 percent of carbon monoxide emissions nationwide.

The most effective way to reduce air pollutants is to establish strict emission controls at the source of pollution. The Clean Air Act of 1990 takes a comprehensive approach to reducing pollution from these sources by requiring manufacturers to build cleaner engines; refiners to produce cleaner fuels; and certain areas with air pollution problems to adopt and run passenger vehicle inspection and maintenance programs. EPA has issued a series of regulations affecting passenger cars, diesel trucks and buses, and nonroad equipment (recreational vehicles, lawn and garden equipment, etc.) that will dramatically reduce emissions as people buy new vehicles and equipment. Here are some of the control strategies for reducing vehicle emissions:

- Cleaner cars
- Reformulated gasoline
- Low-sulfur fuels
- Alternative fuels
- Cleaner trucks, buses and nonroad equipment
- Transportation policies
- Inspection and maintenance programs

EPA is authorized by the Clean Air Act to issue a series of rules to reduce pollution from vehicle exhaust, refueling emissions, and evaporating gasoline. As a result, emissions from a new car purchased today are well over 90 percent cleaner than a new vehicle purchased in 1970. This applies to SUVs and pickup trucks, as well. Since 2004, all new passenger vehicles—including SUVs, minivans, vans and pick-up trucks—have had to meet more stringent tailpipe emission standards.

Reformulated gasoline reduces emissions of toxic air pollutants, such as benzene, as well as pollutants that contribute to smog. The Clean Air Act requires certain metropolitan areas with the worst ground-level ozone pollution to use gasoline that has been reformulated to reduce air pollution. Other areas, including the District of Columbia and 17 states, with ground-level ozone levels exceeding the public health standards, have voluntarily chosen to use reformulated gasoline.

In order to reduce pollutant emissions, the Clean Air Act encourages the increased use of alternative transportation fuels such as natural gas, propane, methanol, ethanol, electricity, and biodiesel. These fuels can be cleaner than gasoline or diesel and can reduce emissions of harmful pollutants. Renewable alternative fuels are made from biomass materials like wood, waste paper, grasses, vegetable oils, and corn. They are biodegradable and reduce carbon dioxide emissions. In addition, most alternative fuels are produced domestically, which is better for our economy and energy security, and helps offset the cost of imported oil.

Diesel engines are more durable and are more fuel efficient than gasoline engines, but can pollute significantly more. Heavy-duty trucks and buses account for about one-third of nitrogen oxides emissions and one-quarter of particle pollution emissions from transportation sources. In some large cities, the contribution is even greater. Similarly, nonroad diesel engines such as construction and agricultural equipment emit large quantities of harmful particle pollution and nitrogen oxides, which contribute to ground-level ozone and other pervasive air quality problems.

Other transportation policies aimed at reducing emissions include such strategies as the conformity process and inspection and maintenance program. To understand the conformity process, it is important to understand the purpose of national ambient air quality standards (NAAQS) and the air quality designations.

4 NATIONAL AMBIENT AIR QUALITY STANDARDS (NAAQS)

The Clean Air Act of 1990 authorizes the U.S. Environmental Protection Agency to establish standards (NAAQS) for pollutants considered harmful to public health and the environment. *Primary standards* set limits to protect public health, including the health of sensitive populations such as asthmatics, children, and the elderly. *Secondary standards* set limits to protect public welfare, including protection against decreased visibility, damage to animals, crops, vegetation, and buildings.

The EPA Office of Air Quality Planning and Standards (OAQPS) has set National Ambient Air Quality Standards for six criteria principal pollutants (Table 1).[2]

5 AIR QUALITY MONITORING

The quality of the geographic region's air is monitored for each of the criteria pollutants with equipment and methods approved by EPA and the state environmental agency. Air quality data are recorded at various monitoring sites:

- State and Local Air Monitoring Stations (SLAMS—over 4,000 stations)
- National Air Monitoring Stations (NAMS—over 1,080 stations)

Table 1 National Ambient Air Quality Standards

Pollutant	Primary Stds.	Averaging Times	Secondary Stds.
Carbon monoxide	9 ppm (10 mg/m^3)	8 hours	None
	35 ppm (40 mg/m^3)	1 hours	None
Lead	1.5 µg/m^3	Quarterly average	Same as primary
Nitrogen dioxide	0.053 ppm (100 µg/m^3)	Annual (arithmetic mean)	Same as primary
Particulate matter (PM$_{10}$)	Revoked 150 µg/m^3	Annual (arithmetic mean) 24 hours	
Particulate matter (PM$_{2.5}$)	15.0 µg/m^3 35 µg/m^3	Annual (arithmetic mean) 24 hours	Same as primary
Ozone	0.08 ppm	8 hours	Same as primary
	0.12 ppm	1 hour (Applies only in limited areas)	Same as primary
Sulfur oxides	0.03 ppm	Annual (arithmetic mean)	———
	0.14 ppm	24 hours	———
	———	3 hours	0.5 ppm (1,300 µg/m^3)

Note: From Ref. 2.

- Special Purpose Monitoring Stations (SPAMS)
- Photochemical Assessment Monitoring Stations (PAMS)

6 AIR QUALITY DESIGNATIONS

The monitored concentrations of the pollutants are analyzed vis-à-vis the standards set forth as the basis for air quality designations and subclassifications. EPA then designates a geographic area as an *attainment* area or *nonattainment* area for a given criteria pollutant. This designation is based on specified number of *exceedences* and/or *violations* of NAAQS. A nonattainment area is further classified as follows:

- Extreme and severe
- Serious
- Moderate
- Marginal

7 STATE IMPLEMENTATION PLANS (SIPS)

States with one or more nonattainment areas, are required by law to submit state implementation plans (SIP) for phased achievement of into the attainment status. SIPs are collections of regulations and measures used by a state to reduce emissions from stationary, area, and mobile sources, and demonstrate attainment and maintenance of air quality standards. Included in a typical SIP are emissions budgets, which is that portion of allowable emissions defined in a SIP allocated to on-road (highway and transit) vehicle emissions.[3]

8 THE CONFORMITY RULE

The Clean Air Act Amendments of 1990 requires that the transportation projects should be in conformity with the SIPs. Transportation projects such as construction of highways and transit rail lines cannot be federally funded or approved unless they are consistent with state air quality goals set forth in SIPs. Furthermore, transportation projects must not cause or contribute to new violations of the air quality standards, worsen existing violations, or delay attainment of air quality standards. Transportation conformity ensures that federal funding and approval goes to those transportation activities that are consistent with air quality goals, and can have a significant impact on the transportation planning process. Transportation officials must be involved in the air quality planning process to ensure that emissions inventories, emissions budgets, and transportation control measures (TCMs) are appropriate and consistent with the transportation vision of a region. If transportation conformity cannot be determined, projects and programs may be delayed.

9 TRANSPORTATION AND AIR QUALITY PLANNING

In recent years, the planning process has given considerable emphasis to the assessment of the effect of transportation alternatives on the environmental consequences, especially air quality impacts. The Environmental Protection Agency (EPA) developed several versions of a model over time for estimating emission factors of air pollutants from mobile sources in terms of grams per mile. One of the recent versions of this model—MOBILE5—was used widely during 1990s. Currently, the latest version, MOBILE6, is being used. In California, a different emission factor model called EMFAC is used. These emission factor models need a variety of travel-related measures for the estimation of emissions from vehicular travel, and this need uncovered several deficiencies of the traditional travel forecasting models and led to various refinements and advancement of the modeling procedure. The integration of travel models with emission factors models is illustrated in Figure 1.[4]

More and more, the regulatory burdens of air quality modeling related to transportation projects are shouldered by transportation planners. Air quality models are constantly being updated with new knowledge gained on transportation related emissions. The EPA is currently developing the next generation emission factor model called *MOVES* (for Motor Vehicles Emissions Simulator). The current state of the art in emissions modeling requires more from the transportation planning community than ever before. For example, the concept of trip ends and trip chaining are easily extended to deriving travel-related inputs to the emission factor models.[5] Although model improvement efforts undoubtedly improve the state of the practice, additional burdens are placed on transportation modeling community to develop innovative methods to derive travel-related inputs to emissions model. In order to accommodate the needs of the transportation

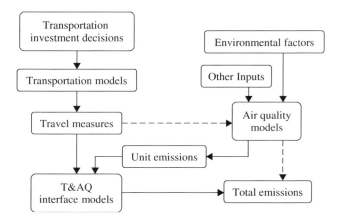

Figure 1 Schematic representation shows TDM integrated with air quality modeling.

related air quality modeling in the foreseeable future, the transportation planners are expected to develop new methods or adapt existing methods.

REFERENCES

1. "Cars, Trucks, Buses, and "Nonroad" Equipment," http://www.epa.gov/air/caa/peg/cars-trucks.html.
2. http://www.epa.gov/art/criteria.html. Accessed on April 10, 2007.
3. "Air Quality Planning for Transportation Officials: An Introduction," http://www.fhwa.dot.gov/environment/aqplan/aqintro.htm.
4. A. Chatterjee and M. M. Venigalla, "Travel Demand Forecasting for Urban Transportation Planning," Chapter 7 in *Handbook of Transportation Engineering*. Myer Kutz, ed., McGraw-Hill Publications, New York, 2003.
5. S. Chalumuri, *Emission Impacts of Personal Travel Variables*. A thesis submitted in partial fulfillment of master of science degree in civil and infrastructure engineering department, George Mason University, Fairfax, VA 2003.
6. U.S. EPA, "Understanding the Clean Air Act," http://www.epa.gov/air/caa/peg/understand.html. Accessed on April 10, 2007.
7. U.S. EPA, "The Bridge to Clean Air—Transportation and Conformity," An educational brochure, http://www.epa.gov/otaq/stateresources/transconf/conf-brochure.pdf. Accessed on April 10, 2007.
8. Arnold M. Howitt, and E. M. Moore, Linking Transportation and Air Quality Planning, EPA420-R-99-011. A Report to the U.S. EPA and the Federal Highway Administration, March 1999.
9. A. Chatterjee, J. W. Philpot, T. F. Wholley, R. Guensler, D. Hartgen, R. A. Margiotta, and P. R. Stopher, Improving Transportation Data for Mobile Source Emission Estimates, National Cooperative Highway Research Program Report 394, 1997.

CHAPTER 4

THE SOCIAL COST OF MOTOR VEHICLE USE IN THE UNITED STATES

Mark A. Delucchi
Institute of Transportation Studies
University of California, Davis
Davis, California

1	**BACKGROUND**	58
2	**THE PURPOSE OF A SOCIAL-COST ANALYSIS OF MOTOR VEHICLE USE**	58
	2.1 Evaluate Costs of Transportation Projects, Policies, and Long-Range Scenarios	62
	2.2 Establish Efficient Prices for and Ensure Efficient Use of Transportation Services	63
	2.3 Prioritize Research and Funding	63
3	**CONTEXT OF SOCIAL-COST ANALYSIS**	64
4	**THE CONCEPTUAL FRAMEWORK**	65
	4.1 Annualized Cost of Motor Vehicle Use in the United States	65
	4.2 What Counts as a Cost of Motor Vehicle Use or Infrastructure?	66
	4.3 How to Interpret the Cost of All Motor Vehicle Use in the United States	66
	4.4 Benefits versus Costs	67
	4.5 Minor Conceptual Issues on Social-Cost Analysis	69
	4.6 Classification of Components of the Total Social Cost	69
5	**COMPONENTS OF THE SOCIAL COST OF MOTOR VEHICLE USE**	72
	5.1 Column 1 of Table 1: Personal Nonmonetary Costs of Motor Vehicle Use	72
	5.2 Column 2 of Table 1: Motor Vehicle Goods and Services Produced and Priced in the Private Sector (Estimated Net of Producer Surplus and Taxes and Fees)	73
	5.3 Column 3 of Table 1: Motor Vehicle Goods and Services Bundled in the Private Sector	76
	5.4 Column 4 of Table 1: Motor Vehicle Infrastructure and Services Provided by the Public Sector	78
	5.5 Column 5 of Table 1: Monetary Externalities of Motor Vehicle Use	80
	5.6 Column 6a of Table 1: Nonmonetary Externalities of Motor Vehicle Use	80
	5.7 Column 6b of Table 1: Nonmonetary Costs of Motor Vehicle Infrastructure	82
	5.8 Summary Observations Regarding Table 1	83
	5.9 Quality of the Estimates	86
6	**RESULTS OF THE ANALYSIS**	88
	6.1 Allocation of Costs to Individual Vehicle Categories	88
	6.2 How the Results of This Analysis Should Not Be Used	90
7	**SUMMARY**	93

1 BACKGROUND

Every year, U.S. drivers spend hundreds of billions of dollars on highway transportation. They pay for vehicles, maintenance, repairs, fuel, lubricants, tires, parts, insurance, parking, tolls, registration, fees, and other items. These expenditures buy Americans considerable personal mobility and economic productivity.

But the use of motor vehicles costs society more than the hundreds of billions of dollars spent on explicitly priced motor vehicle goods and services in the private sector. Some of the motor vehicle goods and services provided in the private sector are not priced explicitly, but rather are *bundled* in the prices of nontransportation goods and services. For example, "free" parking at a shopping mall is unpriced, but it is not costless; the cost is included—bundled—in the price of goods and services sold at the mall.

In addition to these priced or bundled private-sector costs, there are public-sector costs: the tens of billions of dollars spent every year to build and maintain roads and to provide a wide range of services that support the use of motor vehicles. These services include police protection, the judicial and legal system, the prison system, fire protection, environmental regulation, energy research and regulation, military protection of oil supplies, and more.

And finally, beyond these *monetary* public and private-sector cost are the *nonmonetary* costs of motor vehicle use—those costs that are not valued in dollars in normal market transactions. There are a wide variety of nonmonetary costs, including the health effects of air pollution, pain and suffering due to accidents, and travel time. Some of these nonmonetary costs, such as air pollution, are externalities; others, such as travel time in uncongested conditions, are what we will call personal nonmonetary costs.

The total national social cost of motor vehicle use is the sum of all of the costs mentioned previously: explicitly priced private-sector costs, bundled private-sector costs, public-sector costs, external costs, and personal nonmonetary costs. These costs are listed and classified more rigorously in Table 1.

My colleagues at the University of California and I have done a detailed analysis of some of the costs of motor vehicle use in the United States.[1] In this chapter, we explain the purpose of estimating the total social cost of motor vehicle use, briefly review some of the pertinent research, explain the conceptual framework and cost classification, and present and discuss our cost estimates.

2 THE PURPOSE OF A SOCIAL-COST ANALYSIS OF MOTOR VEHICLE USE

Researchers have performed social-cost analyses for a variety reasons, and have used them in a variety of ways to support a wide range of policy positions. Some researchers have used social-cost analyses to argue that motor vehicles and gasoline are terrifically underpriced, while others have used them to downplay the need for drastic policy intervention in the transportation sector. In any case,

Table 1 Classification of the Costs of Motor Vehicle Use

Personal Costs	Private-sector Costs	Public-sector Costs	External Costs (except 6b)
MPC or MPV might be misestimated, because of poor information or irrational behavior.	Prices are not optimal because of imperfect standards (MCC ≠ MDC), distortionary taxes, subsidies, price controls, quotas, imperfect competition (P ≠ MPC), or poor information. Bundling decision can be distorted or determined by regulations, taxes, poor information.	User taxes and fees ≠ MPC, and B/C not maximized, because of nonefficiency objectives.	These are unpriced (MPC ≠ MSC in markets with externalities) because of the absence of fully enforced individual or collective property rights, or the absence of optimal Pigovian taxes.

Nonmonetary	Monetary Costs			Nonmonetary costs
(1) Personal NonMonetary Costs of MV Use	(2) MV Goods and Services Produced and Priced in the Private Sector (Estimated Net of Producer Surplus and Taxes and Fees)	(3) MV Goods and Services Bundled in the Private Sector	(4) MV Infrastructure and Services Provided by the Public Sector	(5) Monetary Externalities of MV Use	(6a) Nonmonetary Externalities of MV Use
• Travel time, excluding travel delay imposed by others, that displaces unpaid activities • Accidental pain, suffering, death, and lost nonmarket productivity, inflicted on oneself	*These kinds of costs usually are included in GNP-type accounts:* • Annualized cost of the fleet (excluding vehicles replaced as a result of motor-vehicle accidents) • Cost of transactions for used cars • Parts, supplies, maintenance, repair, cleaning, storage, renting, towing, etc. (excluding parts and services in repair of vehicles damaged in accidents)	• Annualized cost of nonresidential offstreet parking included in the price of goods or services or offered as an employee benefit • Annualized cost of off-street residential parking included in the price of housing	• Annualized cost of public highways, including on-street parking • Annualized cost of municipal and institutional off-street parking • Highway law enforcement and safety • Regulation and control of MV air, water, and solid waste pollution	• Monetary costs of travel delay imposed by others: extra consumption of fuel, and foregone paid work • Accident costs *not* accounted for by economically responsible party: property damage, medical, productivity, and legal and administrative costs	• Accidental pain, suffering, death, and lost nonmarket productivity, not accounted for by the economically responsible party • Travel delay, imposed by other drivers, that displaces unpaid activities • Air pollution: effects on human health, crops, materials, and visibility** • Global warming due to fuel-cycle emissions of greenhouse gases (U.S. damages only)

(*continues*)

Table 1 (*continued*)

Personal Costs	Private-sector Costs	Public-sector Costs	External Costs (except 6b)	
• Personal time spent working on MVs and garages, refueling MVs, and buying and disposing of MVs and parts	• Motor fuel and lubricating oil, excluding cost of fuel use attributable to delay • Motor-vehicle insurance: administrative and management costs • Priced private commercial and residential parking, excluding parking taxes	• Annualized cost of roads provided or paid for by the private sector and recovered in the price of structures, goods, or services	• MV and energy technology R&D	• Macroeconomic adjustment losses of GDP due to oil-price shocks • Pecuniary externality: increased payments to foreign countries for non-transport oil, due to ordinary price effect of using petroleum for motor vehicles • Monetary, non-public-sector costs of fires and net crimes* related to using or having MV goods, services, or infrastructure	• Noise from motor vehicles • Water pollution: effects of leaking storage tanks, oil spills, urban runoff, road deicing • Nonmonetary costs of fires and net crimes* related to using or having MV goods, services, or infrastructure • Air pollution damages to ecosystems other than forests, costs of motor vehicle waste, vibration damages, fear of MVs and MV-related crime (not quantified here)

	Usually not included in GNP-type account:		(6b) **Nonmonetary Impacts of MV Infrastructure**[#] (not quantified here)
• MV noise and air pollution inflicted on oneself	• Travel time, excluding travel delay imposed by others, that displaces paid work • Overhead expenses of business and government fleets • Private monetary costs of motor vehicle accidents, including user payments for cost of motor-vehicle accidents inflicted on others, but excluding insurance administration costs	• Police protection (excl. highway patrol), court and corrections system (net of cost of substitute crimes) • Fire protection • Motor vehicle-related costs of other agencies • Military expenditures related to the use of Persian-Gulf oil by motor vehicles • Annualized cost of the Strategic Petroleum Reserve	• Land-use damage: habitat, species loss due to highways, MV infrastructure • The socially divisive effect of roads as physical barriers in communities • Esthetics of highways and vehicle and service establishments

MPC = marginal private cost; MPV = marginal private value; P = price, MCC = marginal control cost; MDC = marginal damage cost; B/C = dollar benefit/cost ratio of investment; MSC = marginal social cost; MV = motor vehicle; GNP = gross national product; R&D = research and development.
*These really should be classified not as external costs, within an economic framework, but rather as costs of illegal or immoral behavior, within a framework that encompasses more than just economic criteria. However, regardless of how these are *classified*, they in fact are related to using or having motor-vehicle goods, services, or infrastructure.
**The cost of crop loss, and some of the components of other air-pollution costs (e.g., the cost of medical treatment of sickness caused by air pollution), technically should be classified as monetary externalities.
[#] Although these are nonmonetary environmental and social costs of total motor vehicle use, they are not costs of marginal motor vehicle use, and hence technically are not externalities.

social-cost analyses usually excite considerable interest, if only because nearly all of us use motor vehicles.

By itself, however, a social-cost analysis does not determine whether using motor vehicles is good or bad, or better or worse than some alternative, or whether it is wise to tax gasoline or restrict automobile use or encourage travel in trains. Rather, a social-cost analysis is but one of many pieces of information that might be useful to transportation analysts and policy makers.

A social-cost analysis can provide several kinds of information that can be used for several purposes. A social-cost analysis can provide the following:

1. General cost data, references, methods, and cost models, Cost models relate total dollar cost to transportation quantities, such as vehicle-miles of travel, trips, vehicles, fuel consumption, highway-miles, or parking spaces, and to nontransportation parameters such as weather or geography.
2. Marginal unit-cost estimates derived from detailed cost models (e.g., $/kg of pollutant emitted).
3. Simple estimates of total cost and average cost. (*Average cost* is total cost divided by total quantity).

These data, models, unit costs, and results can help analysts: (1) evaluate the costs of transportation projects, policies, and long-range scenarios; (2) establish efficient prices for and ensure efficient use of transportation services and commodities; and (3) prioritize research and funding. The following sections examine the three uses for a social-cost analysis.

2.1 Evaluate Costs of Transportation Projects, Policies, and Long-Range Scenarios

In cost–benefit analyses, policy evaluations, and scenario analyses, analysts must quantify changes to and impacts of transportation systems. The extent to which a generic national social-cost analysis (such as is presented here) can be of use in the evaluation of specific projects or policies depends, of course, on the detail and quality of the social-cost analysis. At a minimum, a detailed, original social-cost analysis can be mined as a source of data, methods, and models for cost evaluations of specific projects. Beyond this, if costs are a linear function of quantity, and invariant with respect to location, then estimates of national total or average cost, which any social-cost analysis will produce, may be used to estimate the incremental costs for specific projects, policies, or scenarios*.

---

* The average unit cost is equal to the total cost of the entire system divided by some measure of total use (quantity, or output), and so is expressed in terms of $/vehicle-mile of travel (VMT), $/trip, $/vehicle, etc. The marginal or incremental unit cost is the cost of an increment to the total system divided by the incremental quantity. Given this, we may scale an estimate of the total social cost of the entire system to an estimate of the cost of an increment to the system only if average unit costs are close to marginal unit costs.

(Average-cost estimates are more likely to be useful for long-range, broad-brush scenario analysis than for specific project evaluations.) Otherwise, analysts must estimate the actual nonlinear cost functions for the project, policy, or scenario at hand. The social-cost analysis presented here is based on total-cost models for noise, air pollution, government expenditures, accidents, travel time, and a few other components of the social cost.

2.2 Establish Efficient Prices for and Ensure Efficient Use of Transportation Services

Social-cost analysis can be used to establish true costs for transportation resources or impacts that at present either are not priced but in principle should be (e.g., emissions from motor vehicles) or else are priced but not efficiently (e.g., roads).

An efficient price is equal to marginal cost, which is the slope of the total-cost function. Hence, any cost *models* in a social-cost analysis in principle may be employed to estimate marginal-cost prices. (As already mentioned, we have estimated total-cost functions for some of the many cost items in our own social-cost analysis[2].) Beyond this, the average-cost results of a social-cost analysis might give analysts some idea of the magnitude of the gap between current prices (which might be zero, as in the case of pollution) and theoretically optimal prices, and inform discussions of the types of policies that might narrow the gap and induce people to use transportation resources more efficiently. To the extent that total-cost functions for the pricing problem at hand are thought to be similar to any simple linear national cost functions of a social-cost analysis, the average-cost results of the national social-cost analysis may be used to approximate prices for the problem at hand.

2.3 Prioritize Research and Funding

Prioritize efforts to reduce the costs or increase the benefits of transportation. The total-cost or average-cost results of a social-cost analysis can help analysts and policy makers rank costs (Is road dust more damaging than ozone?), track costs over time (Is the cost of air pollution going down?), and compare the costs of pollution control with the benefits of control (Are expenditures on pollution control devices greater or less than the value of the pollution eliminated?). This information can help people decide how to fund research and development to improve the performance and reduce the costs of transportation. For example, if one is considering funding research into the sources, effects, and mitigation of pollution, it might be useful to know that road-dust particulate matter might be an order of magnitude more costly than is ozone attributable to motor vehicles.

We present our analysis and estimates with these relatively modest purposes in mind, not to promote a particular policy agenda regarding the use of motor vehicles, and certainly not to forward any particular position about what, for example, gasoline taxes "should be," or whether the nation should invest more or less in motor vehicle use than it is now. To this list one perhaps might add a fourth: simply to know what the costs are now and were in the past. However, this is an additional purpose only if the knowledge is valued intrinsically, and not instrumentally. If the knowledge is valued instrumentally, then its *use* must be one of the three described above.

3 CONTEXT OF SOCIAL-COST ANALYSIS

Interest in full social-cost accounting and socially efficient pricing has developed relatively recently. From the 1920s to the 1960s, major decisions about building and financing highways were left to technical experts—chiefly, engineers who rarely, if ever, performed social cost–benefit analyses. Starting in the late 1960s, however, "a growing awareness of the human and environmental costs of roads, dams, and other infrastructure projects brought the public's faith in experts to an end".[3] It was a short step from awareness to quantification of the costs not normally included in the narrow financial calculations of the technical experts of the past.

Today, the call for full-social-cost accounting and efficient pricing is being sounded in many sectors of the economy, from transportation to the chemical industry.[4] In transportation, discussions of efficient pricing and full-social cost accounting now are routine. For example, in a summary of views on high-speed ground transportation in the U. S., two of the four authors suggest that the cost of high-speed rail (HSR) should be compared with the full, unsubsidized costs of the alternatives, including auto and air travel.[5,6]

Not surprisingly, however, there is little agreement about the proper items in a social-cost analysis, the magnitude of the major components of the social cost, or the extent to which present prices are not optimal. On the one hand, many analyses argue that the "unpaid" or external costs of motor-vehicle use are quite large—perhaps hundreds of billions of dollars per year—and hence that automobile use is heavily subsidized and underpriced.[7–15] On the other hand, not unexpectedly, others have argued that this is not true. For example, the National Research Council (NRC), in its review and analysis of automotive fuel economy, claims that "some economists argue that the societal costs of the externalities associated with the use of gasoline (e.g., national security and environmental impacts) are reflected in the price and that no additional efforts to reduce automotive fuel consumption are warranted."[16] Green makes essentially the same argument.[17]

Beshers and Dougher make the narrower claim that road-user tax and fee payments at least equal government expenditures related to motor-vehicle use.[18,19] However, Morris and DeCicco revise Dougher's accounting, deducting general taxes from the revenue side and adding some motor-vehicle-related services to the expenditure side, and find that revenues from users fall short of government expenditures by 22 percent.[19,20] Similarly, the most recent highway-cost allocation study by the Federal Highway Administration (FHWA) and others indicates that highway user fees are about 20 percent below highway-related expenditures, for all levels of government and all vehicle classes in the United States in 2000.[21]

We could cite other examples. This extraordinary disagreement exists because of the wide range of conceptual frameworks, methods, data, and assumptions. There are detailed, original, and conceptually correct analyses of some of the components of the total social cost (e.g., air pollution,[22,23] accidents,[24] oil use,[25] and freight transport,[26,27]), analyses of social or external costs in particular localities in the United States,[2,11] original and conceptually correct analyses of the external costs of transport in Europe,[28,29] and detailed but old analyses of the social costs of transportation in the United States[30]. However, prior to our work nobody had done a more up-to-date, detailed, and conceptually sound analysis of the major social costs of motor-vehicle use in the United States.

4 THE CONCEPTUAL FRAMEWORK

4.1 Annualized Cost of Motor Vehicle Use in the United States

When we speak of the social cost of motor vehicle use, we mean *the annualized social cost of motor vehicle use in the United States based on 1990 to 1991 cost levels*. The annualized cost of motor vehicle use, based on 1990 to 1991 data, is equal to the sum of the following:

- 1990 to 1991 periodic or *operating costs*, such as fuel, vehicle maintenance, highway maintenance, salaries of police officers, travel-time, noise, injuries from accidents, and disease from air pollution.
- The 1990 to 91 replacement value of all capital, such as highways, parking lots, and residential garages (i.e., items that provide a stream of services), converted into an equivalent stream of annual costs (annualized) over the life of the capital, on the basis of real discount rates.

This annualization method—whereby the total yearly cost is equal to periodic "operations and maintenance costs" plus annualized capital replacement costs—is just the obverse of evaluating the net present value of alternative investment options (in transportation or any other arena). In essence, the yearly social cost of motor vehicle use, as we estimate it, is the yearly cost stream of the whole motor vehicle system, analyzed as if it were one large transportation alternative among several. Of course, the *scale* that we have chosen—all motor vehicle use—is just a convenient point of reference. (That is, one just as well could view the analysis presented here as an analysis of a generic motor-vehicle-use project, or alternative, scaled up to the level of all motor vehicle use in the United States)

4.2 What Counts as a Cost of Motor Vehicle Use or Infrastructure?

In economic analysis, *cost* means *opportunity cost*. The opportunity cost of action *A* is the opportunity you forgo—what you give up, or use, or consume as a result of doing *A*. For some resource *R* to count as a cost of motor vehicle use, it must be true that a change in motor vehicle use *will result* in a change in use of *R*. Thus, gasoline is cost of motor-vehicle use because a change in motor-vehicle use will result in a change in gasoline use, all else being equal. But general spending on health and education is not a cost of motor-vehicle use because a change in motor-vehicle use will not result in a change in resources devoted to health or education.

However, for the purposes of planning, evaluating, or pricing, we care not only whether something is a cost of motor vehicle use, but, if it is a cost, exactly how it is related to motor vehicle use. For example, pollution is a direct, immediate cost of motor vehicle use: you change motor vehicle use a little, and you immediately change pollution a little. But defense expenditures in the Persian Gulf, if they are a cost of motor vehicle use at all, are an indirect, long-term, and tenuous one. This is discussed more below.

4.3 How to Interpret the Cost of All Motor Vehicle Use in the United States

If one wishes to apply the estimates of the total cost of all motor vehicle use, or to understand the basis for deciding what is included in our list of costs in Table1, then one might ask what is meant by the cost of *all* motor-vehicle use: All motor-vehicle use compared to what?

In normal cost-benefit analysis of transportation projects, one estimates costs and benefits relative to a well defined *no-project* alternative, or base case. For example, one might compare a highway-expansion project with a light-rail project relative to a base case of "business as usual" improvement in the management of the existing infrastructure. But if the "project" is all motor vehicle use, what is the base case—the world without motor vehicle use? Of course, one must be more specific about the base case than this, because the estimated costs and benefits will depend greatly on the details. A day-time parking-management plant that reduces VMT by 10 percent will result in costs and benefits quite different from those of, say, a congestion pricing scheme on a toll bridge that also reduces VMT by 10 percent.

In this analysis, the world without motor vehicle use is presumed to be the same as the world with motor vehicle use, *except* that in the former, people don't use motor vehicles. This means that the benefits of motor vehicle use—the access provided—are presumed to be the same in both worlds. Put another way, the total social cost of motor vehicle use is the welfare difference between the present (in our case, ca. 1991) motor-vehicle system, and a system that provides exactly the same services (i.e., moves people and goods to and from the same places as do motor vehicles) but without time, manpower, materials, or energy—in short, without cost. Of course, if there were a costless transportation system, people would make more and longer trips, and settlement would be more dispersed. Conceptually, we ignore this effect in the baseline "no-motor-vehicle" case.

This costless transportation baseline is just a frame of reference, a conceptual baseline with respect to which total costs trends can be estimated, or the total costs of one system (say, passenger vehicles) compared with the costs of another (say, passenger trains). Moreover, it is relevant only to understanding the meaning of the total cost estimates themselves; it is not relevant if one is interested specifically in the data, methods, and marginal-cost models of the social-cost analysis, for the purpose of estimating efficient prices (say, for motor vehicle emissions), or doing cost-benefit analysis of specific projects.

This last point, obvious though it may be, probably cannot be overemphasized. If one is interested in, say, establishing so-called *Pigovian* taxes (named after the English economist A. C. Pigou, who made significant contributions to the economic analysis of social welfare) to internalize the damages from motor-vehicle emissions, then one probably will wish to examine the details of the damage-function model that produces estimates of the $/kg cost of emissions, as a function of the change in emissions. One will not care about our estimate of the total dollar damages due to air pollution from motor vehicles in 1990. Thus, insofar as one is interested in the details of a social-cost analysis, and not in the total-cost estimates themselves, the question "Total cost compared to what?" never arises.

4.4 Benefits versus Costs

In this project, we estimate the dollar social cost but not the dollar social benefit of motor vehicle use. More precisely, we identify, classify, and quantify many—we hope virtually all—of the impacts and resources of motor vehicle use. The social cost of motor vehicle use is the value of the resources devoted to motor vehicle use. (In this context, *resources* should be broadly construed to include health, esthetic, environmental, and similar impacts of motor vehicle use.) In Figure 1, the total social cost is the area under the social supply curve, $S*$ (region $Ox*Q*$ if we are at the social optimum, with all externalities internalized; region $Ox'Q$ if we are at the private market optimum, with external costs extant).

The social benefit of motor vehicle use is the value that beneficiaries ascribe to motor vehicle use—in economic parlance, the total willingness to pay for motor vehicle use. Total willingness to pay is the area under the demand curve, D, of Figure 1 (region $OAx*Q*$). The difference between the total benefit and the total cost, region $OAx*$ of Figure 1, is the net benefit of motor vehicle use. (The net benefit can be negative, of course.) *Net social benefit*, or the ratio of social benefit to social cost, is the ultimate measure of economic worth. In cost-benefit analysis, the preferred package of policies or investments is the one

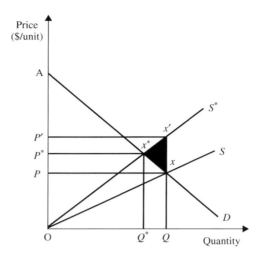

Figure 1 Social cost-benefit analysis of motor vehicle use.

that generates the highest net benefits for the available budget. For a general review of cost–benefit analysis, see Ref. 31. For a discussion of some of the more problematic aspects of cost–benefit analysis, including valuation of nonmarket goods, ecosystem complexity, the social rate of discount, irreversibilities, and efficiency versus equity, see Ref. 32.

Again, ours is a cost analysis, not a cost-benefit analysis. Of course, we have not forgotten that there are benefits of motor vehicle use, and certainly have not presumed that the benefits somehow are less important than the costs.[17,33] To the contrary, it is obvious that motor vehicle use is enormously beneficial, and that its total social benefit vastly exceeds its cost. Moreover, it is worth noting that in some places automobiles are more environmentally benign than the transportation modes (e.g., horse-drawn carriages) that they have replaced.[34] The problem is that, although it is possible to estimate the benefits of small changes in motor vehicle use, it is very difficult to estimate credibly the benefits of *all* motor vehicle use. The root of the problem is that we do not know what the total demand curve looks like near zero quantity: Trips by car for which there are no good substitutes must be extremely valuable, but precisely how valuable we don't know. Because this is a cost analysis only, we are unable to say much about net dollar benefits or cost-benefit ratios, or whether a particular transportation system or plan is worthwhile, or better or worse than another system or plan. For example, this analysis indicates that motor vehicle use might cost us more than people realize. That is, the total social cost could appreciably exceed the commonly recognized private cost. But even if this is so, it does not mean that motor vehicle use costs society more than it is worth, or that we should prefer any transportation option that might have near-zero external costs, or even any transportation option that might have lower total social costs. To make such choices, one must estimate the dollar value of all the benefits as well as the dollar value of all the costs, for all of the relevant policies or investment alternatives.

4.5 Minor Conceptual Issues on Social-Cost Analysis

There are other minor conceptual issues in social-cost analysis worth mentioning. One is that the cost/quantity function for increases in motor vehicle use might be different than for decreases. Another is that for some of the government services (say, police protection) that support motor vehicle use, long-run cost might be a nonlinear function of some measure of cost-related activity (say, crimes or arrests). In the extreme, cost might be a step-function of activity, such that over some range of activity, the cost of changes in activity might be zero. But one should be careful here, because many small changes in activity—each change by itself not large enough to reach the next cost step—might together create enough additional use to reach the next cost step. Put another way, the problem with assuming that any particular change does not have a cost is that, in the absence of information to the contrary, the starting point for any change is just as likely to be very close to the next cost step as very far, which means that it is just as likely that an infinitesimal change in use will occasion the entire cost of the step as a much bigger change in use will occasion no cost at all. To avoid this mistake, an analyst should treat a step-function as a continuous function, which is tantamount to using average cost as a proxy for marginal cost over the relevant range. This is an advantage of an average-cost analysis.

4.6 Classification of Components of the Total Social Cost

There are *many* components of the social cost of motor vehicles use, and one naturally has the urge to classify them. But should these components be classified or organized in any particular way? It seems sensible to organize cost components in consonance with how the cost estimates will be used. Thus, if one were interested *only* in estimating the total social cost of motor vehicle use and did not care at all about how the estimates might be used, then actually one would not need to categorize the components of the social cost. One would just estimate and perhaps add up every component of the social cost. This, however, would not be of much use to anybody.

As already discussed above, estimates of total social cost of motor vehicle use legitimately can be used for three purposes: (1) to evaluate the costs of transportation projects, policies, and long-range scenarios; (2) to establish efficient prices for and ensure efficient use of transportation services and commodities; and (3) to prioritize research and funding. Of these uses, only the second one, *efficiency of use*, comes with an established set of principles and conditions—namely, the conditions of efficient resource use—that can be used to categorize costs. Consequently, if one wishes one's social-cost estimates to be useful to policy makers who want improve the efficiency of the use of the transportation system, then one should categorize and analyze cost items with respect to the economic efficiency of their production or consumption. * We recognize, of course, that policy makers rarely focus on maximizing economic efficiency or social net welfare. Unquestionably, matters of distribution—who gets what, who wins, and who loses—loom larger in the political arena. However, efficiency is an interesting enough topic itself, and easily distinguished conceptually from equity.

In Table 1, we also use other organizing criteria, such as whether a cost is valued in dollars (this is discussed more below), and end up with six categories of costs. Of course, one could come up with other classifications, even using the same general organizing principles. One could, for example, merge or split some of my categories.

Classification with Respect to the Efficiency Condition That Marginal Social Value Equals Price Equals Marginal Social Cost

Resources are used efficiently when the marginal value to society (MSV) equals the market price (P) equals marginal cost to society (MSC). However, most real markets do not allocate resources efficiently, according to $MSV = P = MSC$, because at a minimum most production and consumption involves some sort of externality, and most prices are influenced by distortionary (nonoptimal) taxes. In fact, there are a variety of reasons that a market might not allocate resources optimally, or what is worse, why no private market might exist. *These reasons—the reasons for inefficiencies—are a natural organizing principle for a social-cost analysis, because there are prescriptions for every kind of inefficiency.* To organize costs with respect to efficiency or inefficiency of allocation is tantamount to organizing costs with respect to prescriptions for maximizing efficiency. This is useful to policy makers.

The conditions required for markets to exist and to allocate resources efficiently, and the theoretical consequences of not having these conditions, are well known. Here, we emphasize an important general point. It is generally true that, for society to use resources efficiently, each individual who makes a resource-use decision must count as a cost of that use everything that is an opportunity cost from the standpoint of society. It does not matter whether motor vehicle users *as a class* pay for a particular cost generated within the class; what matters is whether each individual decision maker recognizes and pays the relevant social marginal-cost prices. If the responsible individual decision maker does not account for the cost, it does not matter who actually pays for it—the resource (usually) is misallocated.

To account for a cost, a consumer must know its magnitude and be required or feel obliged to bear it. Generally, a *price* accomplishes both of these things: It tells the consumer what he must give up in order to consume the item. Although a market price on an item is sufficient to make a consumer account for the item in his decision making, in principle it is not necessary. What is necessary is that one way or another the consumer know and bear the cost. A cost can be borne abstractly, as, for example, a feeling of guilt. Thus, in principle, pollution could be satisfactorily accounted for in consumer decisions if everyone knew all the costs of pollution and cared enough to act as though they paid the costs in dollars.

This emphasis on price, and on individual resource-use decisions, keeps the analysis properly focused on economic efficiency. In an analysis of efficiency, one must not think of motor vehicle users as a class and imagine that the distinction between users and nonusers *as a class* is relevant. It is not. The class distinction may be relevant to questions of equity, but it certainly is not relevant to questions of efficiency. Indeed, thinking in terms of classes often will lead one to the wrong answer. For example, it might seem at first glance that because congestion costs are internal to motor vehicle users as a class, there is no imperative to do address them. However, when one person slows down another and does not account for the imposed delay, the resulting congestion, or delay, is an externality, and hence a source of economic inefficiency. In an analysis of efficiency, it does not matter that in this case motor vehicle users as a class might bear all of the consequences; the point is that if there is a delay externality, then the motor-vehicle users *themselves* are using their motor vehicles inefficiently, and can improve their total welfare if each person has to account for his or her effect on the travel time of others. To maximize the net social benefits of motor-vehicle use we must eliminate *all* externalities, not just those that affect the class of nonusers (however defined).

Methodological Organizing Criterion

We have included in Table 1, a classificatory criterion that has to do not with economic efficiency, but rather with methods of estimating costs: *monetary* versus *nonmonetary* costs. The distinction here is *not* between cost items that ought to be valued in dollars and costs that ought not, nor between efficiently and inefficiently priced items, but rather, between cost items that are traded in real markets and hence valued directly in dollars, and items that are not.

Although this distinction is not directly relevant to efficiency of resource use, it is relevant to the practical estimation of social cost. Abstractly, the social cost of any item X (tires, roads, disturbance by noise, suffering from asthma caused by air pollution) is equal to the quantity of X (number of tires, miles of roads, excess decibels of exposure, days of suffering asthma) multiplied by the unit cost of X ($/tire, $/road-mile, $/excess decibel, $/day of suffering). In Table 1, the distinction between monetary and nonmonetary costs pertains to the estimation of the $/unit part of the calculation of social costs. An item is classified as a monetary cost if we can observe or estimate its $/unit cost (or value) directly from market transactions. Thus, because we can observe the $/unit cost of tires, and the $/mile cost of building roads, tires and roads are classified as monetary costs. By contrast, we cannot observe directly the unit cost of noise or air pollution ($/decibel, or $/day of suffering), because noise disturbance and suffering per se are not traded and valued in markets.

Protective or ameliorative measures, such as ear plugs or asthma medicine, often are valued in markets. Ideally, one would distinguish these as monetary externalities. However, not only is it difficult in most cases to quantify the monetary-cost components of air-pollution and noise, it seems more natural to classify all of the costs of pollution in one place, as nonmonetary externalities. And in any event, failure to distinguish all monetary costs does not undermine the classification with respect to economic efficiency, because from the perspective of efficient resource allocation and proper pricing, there is little difference between a monetary externality and a nonmonetary externality.

Why, then, bother to distinguish monetary from nonmonetary externalities at all? One reason is that nonmonetary externalities usually are harder to estimate and more uncertain. A second reason is that some public-sector infrastructure and service costs can be considered to be monetary externalities, and hence to straddle the public-sector and the monetary-externality categories. If we do not distinguish monetary from nonmonetary externalities, then some of the public infrastructure and service costs, such as fire protection, will straddle the category that includes environmental externalities, such as global warming. This seems too much of a stretch; it is better to separate public-sector costs from environmental externalities by having an intermediate category called *monetary externalities*.

The distinction between monetary and nonmonetary externalities is methodologically important because it is much more difficult to estimate the $/unit cost of nonmonetary items. Although economists have a variety of techniques (e.g., hedonic-price analysis and stated-preference analysis) to estimate the $/unit costs of (or demand curves for) nonmonetary items, all of the techniques can be problematic. As a result, the social nonmonetary costs of motor vehicle use often are very uncertain—typically, much more uncertain than are the monetary costs. Of course, some monetary costs also are difficult to estimate and very uncertain. An example is the GNP loss due to a sudden change in the price of oil.

Other conceptual and methodological issues are explored in more detail in Report #2 of Delucchi et al. We turn now to the six general cost categories of Table 1.

5 COMPONENTS OF THE SOCIAL COST OF MOTOR VEHICLE USE

5.1 Column 1 of Table 1: Personal Nonmonetary Costs of Motor Vehicle Use

In our classification, personal nonmonetary costs are those unpriced costs of motor vehicle use that a person imposes on herself as a result of her decision to travel. The largest personal costs of motor vehicle use are personal travel time in uncongested conditions, and the risk of getting into an accident that (loosely speaking) involves nobody else.

Note the distinction between personal nonmonetary costs (column 1) and externalities of the same sort (column 6). Personal costs are caused and borne by the same party, whereas externalities are imposed by one party on another but not accounted for by the imposing party. The (expected value of the) risk that I will cause an accident and injure myself is a personal nonmonetary cost; the risk that someone else will injure me is an external risk, if the other person does not account for it. The congestion delay that others impose on me is an external cost; the *rest* of my travel time is a personal nonmonetary cost. These distinctions are relevant to policy making because personal costs are unpriced but efficiently allocated if consumers are informed and rational, whereas externalities are unpriced and inevitably a source of inefficiency. As discussed below and indicated in Table 2, the usual prescription for externalities is a Pigovian tax, whereas the "prescription" for a personal cost (whether caused by the affected party or not) is just that the affected party be fully aware of it. Thus, any individual should be charged for the accident or travel time costs he imposes on others, *and* be fully aware of the costs that he himself faces as a result of using a motor vehicle.

If an individual does not correctly assess the personal costs to himself, then he will consume more or less than he would have had he been fully informed and rational.[35,36] For example, there is evidence that most drivers overestimate their alertness and driving skill, and underestimate their chances of getting into an accident.[37] To the extent that they do, they underestimate the expected personal cost of driving and make more trips, or more risky trips, than they would if they were properly apprised of their abilities and chances.

Report #2 of Delucchi et al. contains further discussion of the classification and interpretation of personal nonmonetary costs.[1] In that report, we note that it is more sensible to classify the costs of drunk driving and motor vehicle crime not as external costs within a framework of economic efficiency, but as costs of immoral and illegal behavior, within a broader framework that classifies costs by nonefficiency as well as efficiency concerns.

Personal nonmonetary costs are estimated in Report #4 of Delucchi et al.[1]

5.2 Column 2 of Table 1: Motor Vehicle Goods and Services Produced and Priced in the Private Sector (Estimated Net of Producer Surplus and Taxes and Fees)

The economic cost of motor-vehicle goods and services supplied in private markets is the area under the private supply curve: the value of the resources that a private market allocates to supplying vehicles, fuel, parts, insurance, and so on.

However, we do not observe the supply curve itself, and so cannot estimate the area under the supply curve directly. Rather, we must estimate this area indirectly, starting from what we can observe: total price-times-quantity revenues. Thus, the private-sector resource cost under the supply curve is equal to price-times-quantity revenues minus producer surplus and taxes and fees. We

Table 2 Efficient Pricing of Motor Vehicle Goods and Services

Private-sector Costs	Bundled Private-Sector Costs	Public-Infrastructure and Services	Externalities
Factors affecting efficient marginal-cost pricing			
General taxes and subsidies; controls on quantity or price; nonoptimal standards; imperfect competition	High transaction costs of unbundling and establishing prices; tax and regulatory disincentives to charging for parking; perceived economic benefits of free parking and roads	Possible indivisibility in consumption (MC = 0; e.g., defense); decreasing long-run costs (e.g., some roads); government is concerned with generating revenue, encouraging or discouraging certain behaviors, distributing benefits, providing security and justice, and other things besides economic efficiency	Impossible, or too costly, or otherwise undesirable to assign and enforce property rights to the unpriced resources or effects (hence, no price)
Ideal prescriptions			
Set taxes to minimize deadweight losses (or use lump-sum transfers instead of taxes); set standards such that MCC = MDC; remove controls on price and quantity; break up monopolies and oligopolies; and so on.	If there are no external benefits to unbundling, and no distorting taxes, and if transaction costs cannot be lowered and private assessments are not wrong, then do nothing; otherwise, remove tax and regulatory disincentives to unbundling, and remove any institutional barriers to private ownership and operation of roads.	Turn ownership over to private sector, where possible and efficient; short-run marginal-cost pricing, where possible (highway use charges set equal to marginal wear and tear plus congestion costs; registration and license fees set at marginal administration costs; parking priced at marginal cost; etc.); lump-sum transfers to finance any "public good" portion of highway infrastructure and services.	If feasible, establish property rights; otherwise, if few are involved, then do collective bargaining; otherwise, levy a dynamic[a] tax, at the source, equal to marginal external costs [damage costs not accounted for], but do not compensate victims.

See also Refs. 35 and 36. MC = marginal cost; MCC = marginal control cost; MDC = marginal damage cost.

Note that the prescriptions generally all must be satisfied at once in order to achieve the most optimal resource use. The general theory of the "second best" tells us that, in the real world in which many of the conditions for optimality are not satisfied, it is not necessarily best to satisfy just one additional condition. For example, given nonoptimal emissions standards, emissions regulations, and fees and taxes on automobile producers, it is not necessarily true that it is most efficient to assess a Pigovian tax equal to the marginal cost of the residual emissions.

[a] In most cases, damage is a nonlinear function of output, with the result that the marginal damage rate (the slope of the total damage function) changes with the level of output. In these cases, the Pigovian tax will have to be iterated to stay equal to the marginal damage rate, because the initial application of the Pigovian tax will change the output and hence the marginal damage.

deduct our best guess at producer surplus because it is defined as revenue in excess of economic cost, and hence is a noncost wealth transfer from consumers to producers. (However, a net (equilibrium) transfer from U.S. consumers to foreign producers is a real cost to the United States.) We deduct taxes and fees assessed on producers and consumers because in no case are they marginal-cost prices that can be used in a price-times-revenue calculation of costs.

Recall that the point here is to estimate private-sector resource cost. The cost of the private-sector resources devoted to, say, making gasoline, does not include the federal and state gasoline tax, because that tax is a charge for the use of the roads, not part of the marginal-cost price of making gasoline. But why not then use the gasoline tax as an estimate of the cost of the roads, just as one uses price-times-quantity payments (less producer surplus) to estimate private-sector resource cost? There are two reasons. First, we have data on expenditures on road construction and maintenance anyway, and so do not need to use price-times-quantity to approximate cost. Second, even if we did want to use price-times-quantity to approximate the infrastructure cost, we would not treat the gasoline tax as a price, because it is not a marginal-cost price, but rather is a charge that bears no obvious resemblance to an efficient price.

Note that these considerations bear directly on comparisons of alternatives. For example, in comparing the cost of oil with the cost of alternative energy sources, it is misleading to count all price-times-quantity oil revenues as the cost, because the true private resource cost is much less than this, on account of the enormous producer surplus that accrues to some oil producers.

The prices and quantities that obtain in private markets rarely are optimal— that is, the actual prices (P) paid rarely satisfy $MSV = p = MSC$— not only because of distortionary taxes and fees, but because of imperfect competition, standards, and regulations that affect production and consumption, price controls, subsidies, quotas, externalities, and poor information. For example, the market for crude oil is not always competitive. The reason, of course, is that the Organization of Petroleum Exporting Countries (OPEC) sometimes manages to restrict oil output and thereby raise oil price above marginal cost. This is inefficient from the standpoint of the world because it cuts off production of oil that could be produced for less than the (formerly) prevailing market price and hence from a social-efficiency standpoint *should* be produced and consumed.

This also results in an increased transfer of wealth from consumers to producers (who are receiving a price above their marginal cost), and can be a real loss to heavy oil importers like the United States. Note, though, that this extra wealth transfer is not in addition to price-times-quantity payments; to the contrary, it already is part of price-times quantity payments. Rather, the extra wealth transfer is with respect to the total transfer in a competitive market.[38] The total resource cost of fuel use to the United States, competitive market or not, is equal to price-times-quantity payments less *domestic* producer surplus, which is a non-cost transfer from U.S. consumers to U.S. producers.

Standards and regulations also can be economically inefficient. For example, the cost of vehicles and fuels includes items, such as catalytic converters and airbags, used to meet government standards for emissions, safety, and fuel economy. Now, if the government standards are not the most efficient corrective, then the corresponding resources (for catalytic converters, air bags, etc.) are not efficiently allocated. Of course, it is well known that, transaction costs and uncertainty aside (and these admittedly are *big* asides), Pigovian taxes indeed are more efficient than are standards. However, Pigovian taxes can be more expensive to administer, less predictable, and more difficult to change on short notice, to the point that standards might be preferable in some and perhaps many situations.[39] It thus is not necessarily always the case that in the real world, standards and regulations are less efficient than Pigovian regulations.

I emphasize that the question here is not whether the resources used to meet government standards should be counted as a cost of motor-vehicle use — certainly they should be — but whether they are efficiently allocated. Catalytic converters are a cost of motor vehicle use today, and barring unforeseen changes in regulations, will continue to be a cost of motor-vehicle use, regardless of whether or not there would be catalytic converters in an economically efficient world. Furthermore, regardless of whether standards or taxes are used to address an externality, the relevant total cost is the resource cost of whatever control measures are used (including defensive behavior, broadly construed) *plus* the estimated cost of the residual (uncontrolled) effects, such as emissions.

Finally, consumers can be ignorant and irrational. For example, some and perhaps many people routinely underestimate the probability that they will be in an accident, and as a result undervalue safety equipment in motor vehicles.

In sum, then, it certainly is not true that all private markets are perfect and need no corrective measures. Rather, there are a variety of imperfections, in every sector of the economy, including the most competitive, unregulated private sectors. As a result, we face a range of analytical and policy issues pertaining to pricing, taxation, regulation, and so forth.

The costs of priced private-sector goods and services are estimated in Report #5 of Delucchi et al.[1]

5.3 Column 3 of Table 1: Motor Vehicle Goods and Services Bundled in the Private Sector

Some of the motor vehicle goods and services provided in the private sector are not priced explicitly, but are *bundled* in the prices of nontransportation goods and services. For example, "free" parking at a shopping mall is unpriced, but it is not costless; the cost is included — bundled — in the price of goods and services sold at the mall. Similarly, residential garages are not sold as separate commodities, but are included in the total price of a home. In the United States, nearly all parking, commercial and residential, is bundled. Some local roads also are bundled, usually with the cost of a home.

Parking

The typical motor vehicle is driven less than one hour every day. The rest of the time, it is parked. In the United States, a considerable amount of resources are devoted to providing parking for nearly 200 million vehicles parked for 23 hours a day. As estimated in Appendix A of Report #6 of Delucchi et al.,[1] parking spaces for vehicles consume on the order of 2,000 to 3,000 square miles of land. More importantly, most of the roughly $100 billion resource cost of parking is not priced as a separate charge for parking, but, rather, is bundled with other goods, such as items at a shopping center, or a family's home, and priced as a package.

There are several ways to classify and analyze parking: on street versus offstreet, commercial versus residential, publicly versus privately owned and operated, parking garage versus parking lot, and more. In our analysis, parking costs are classified and estimated as shown in Table 3.

Table 3 Classification of Parking Lots

Type of Parking Space	a. Classification If Parking Is Priced	b. Classification If Parking Is Unpriced (Bundled)
i. On-street parking		
Publicly owned	Public-sector cost (included with cost of public roads)	Public-sector cost (included with cost of public roads)
Privately owned	Priced private-sector cost (assume zero cost, or with private roads)	Bundled private-sector cost (private roads)
ii. Off-street loading ramp or commercial driveway		
Publicly owned	Not estimated	Not estimated
Privately owned	Not estimated	Not estimated
iii. Unimproved land*		
Publicly owned	Assume zero cost	Assume zero cost
Privately owned	Assume zero cost	Assume zero cost
iv. Offstreet residential		
Publicly owned	Public-sector cost (assume zero cost)	Public-sector cost (assume zero cost)
Privately owned	Priced private-sector cost	Bundled private-sector cost
v. Offstreet nonresidential		
Publicly owned	Public-sector cost	Public-sector cost
Privately owned	Priced private-sector cost	Bundled private-sector cost

*The cost of parking in, say, a dirt field is just the forgone stream of rent from alternative uses of the land. In areas where such parking occurs, this generally will be small; certainly, it will be small compared to the land, capital, and operating costs of improved parking spaces.

Bundled private-sector parking costs (i-b, iv-b, and v-b) are classified in column 3 of Table 1 and estimated in Report #6 of Delucchi et al.[1] In that report we develop our estimates in detail, with special attention to important and uncertain parameters, such as the number of offstreet, nonresidential parking spaces, the cost of parking spaces, the number of residential garages and parking spaces, the fraction of residential parking space actually used by cars, and maintenance and repair expenditures for garages. We also discuss the reasons for and efficiency implications of the practice of bundling parking.

Other Bundled Costs
Report #6 of Delucchi et al.[1] also presents rough estimates of the cost of local roads funded by private parties and included in the price of homes.

Although there are benefits to unbundling a commodity and pricing it explicitly, there also can be costs, and as a result it is not necessarily true that bundling is inefficient. For example, although priced parking generally is supplied and used more efficiently than is unpriced (bundled) parking, there is a cost to actually administering a pricing system, and this transaction cost may exceed the benefit of more efficient use of parking. One must do a complete social cost–benefit analysis, in which transaction costs are included, to determine if bundling is superior to pricing. If the decision to bundle is distorted by such things as minimum parking requirements and tax laws that do not count free parking for employees as a taxable benefit, the ideal solution is to eliminate the inefficient taxes and standards, and not necessarily to force parking costs to be unbundled. See Report #6 of Delucchi et al. and Gomez-Ibanez for further discussion.[1,40]

5.4 Column 4 of Table 1: Motor Vehicle Infrastructure and Services Provided by the Public Sector

The public sector provides a wide range of infrastructure and services in support of motor vehicle use. We use data on government expenditures for capital and operations and maintenance, and estimates of motor vehicle–related activity in various cost categories (police protection, fire protection, and so on), to estimate the long-run annualized capital cost and annual operating and maintenance cost of this motor vehicle–related infrastructure and service. We categorize these public-sector costs separately because governments, unlike private firms, do not charge efficient prices for their goods and services.

Note that some cost items straddle columns 4 and 5. In at least one respect, the distinction between column 4 and column 5 is somewhat arbitrary: Items in column 4 are priced but not priced efficiently (or as efficiently as is possible), whereas items in column 5 are not priced at all. The distinction is somewhat arbitrary because whether there is an inefficient charge or no charge at all, the result is similar: inefficient use of resources. (Of course, this statement does not apply to pure public goods, for which the optimal price is zero.) Nevertheless, for several reasons, it is useful and natural to distinguish improperly priced from unpriced items. In the first place, analyses of social cost often are framed around the distinction between private costs and external costs, wherein external costs are unpriced and completely unaccounted for by consumers. Thus, to identify pure externalities, one must distinguish unpriced from improperly priced items. Second, analysts and policy makers need to know which items are being charged for already, but incorrectly, versus which items are not being charged for at all, because generally it will be easier to correctly charge for the former group than the latter. Third, much of the motor vehicle–related infrastructure and service provided by the public sector is priced, but not efficiently. Thus, if one wants to identify public infrastructure and service costs charged at least partly to motor vehicle users—and it certainly seems natural to do so—one must distinguish improperly priced from unpriced costs.

This distinction does make for a messy classification, though, because it is difficult to decide which taxes or fees are payments for which public services. For example, as we argue in Report #17 of Delucchi et al., the portion of the motor-fuel tax that is officially dedicated to deficit reduction should be counted as a payment by motor-vehicle users for motor vehicle use, regardless of the actual legislative earmarking.[1] But to which publicly provided motor vehicle services does it apply? Fire protection? Highway construction only? Defense of oil interests? The answer is a matter of judgment, and as a result, whether a particular public service is priced inefficiently or instead is completely unpriced also is a matter of judgment. We have placed in column 4 those public infrastructure and service items that by law are funded at least partly by taxes fees on motor vehicle use. The rest of the items—those that are not definitely and universally understood to be funded by motor vehicle users—straddle columns 4 and 5.

Of course, whereas all government expenditures on highways and the highway patrol are a cost of motor vehicle use, only a portion of total government expenditure on local police, fire, jails, and so on, is a cost of motor vehicle use. We have estimated the portion of these expenditures that, in the long run anyway, is a cost of motor vehicle use. This sort of allocation is valid for expenditures (such as for police protection) that arguably are opportunity costs of motor vehicle use. (For example, using or having motor vehicle goods, services, and infrastructure has some effect on crime, which requires police protection.)

Another point: For at least three reasons, it is likely that expenditure data do not represent purely economic cost (area under the supply curve). First, even if competitive bidding forces each contractor to offer no more than his minimum willingness to supply, the amounts that the highway contractors themselves pay for materials and services (and which they incorporate into their bids) may include producer surplus. Second, as Lee notes, "it is possible to argue that kickbacks from corrupt contractors and [a portion of] politically inflated labor rates are transfers, not costs" (p. 19; bracketed comments mine).[41] Third, to the extent that highway expenditures are financed from incremental tax revenues, the economy suffers deadweight losses of consumer and surplus due to the contraction of consumption and production caused by price distortion by the incremental taxes.

Note that our estimates of total public-sector costs include the annualized cost of the capital stock. Because capital is foregone (liquidated, not replaced, or not expanded) only in the long run, and only as a result decisions by public officials, the costs estimated here are long-run costs of public decision making.

Government expenditures are estimated in Report #7 of Delucchi et al.[1]

5.5 Column 5 of Table 1: Monetary Externalities of Motor Vehicle Use

An external cost of motor vehicle use is a cost of motor-vehicle use that is imposed on person A by person B but not accounted for by person B. (In section 4.6 we give a more formal definition.) A *monetary* external cost is one that happens to be valued monetarily by markets, in spite of being unpriced from the perspective of the responsible motor vehicle user. The clearest example, shown in column 5 of Table 1, is accident costs that are paid for by those *not* responsible for the accident. These repair costs, inflicted by uninsured or underinsured motorists, clearly are unpriced in the first instance—that is, unpriced from the perspective of the uninsured motorist responsible for the accident—but nevertheless are valued explicitly in dollars in private markets. With respect to economic efficiency, the concern here is that the costs in this category are not priced at all, and hence are larger than is socially optimal.

The largest monetary externalities are those relating to accidents, travel delay, and the macroeconomic costs of oil use.

Monetary externalities are estimated in Report #8 of Delucchi et al.[1] Leiby and Jones et al. provide recent discussions and estimates of the macroeconomic costs of oil use.[25,42]

5.6 Column 6a of Table 1: Nonmonetary Externalities of Motor Vehicle Use

We follow Baumol and Oates (1988) and state that a nonmonetary externality is present when agent A chooses the value of [a] nonmonetary variable[s] in agent B's utility or production relationships without considering B's welfare.[39] Thus, by this definition, *externality* is synonymous not with *damage*, but with *unaccounted for cost*. A nonmonetary externality is one that is not valued directly by economic markets. Environmental pollution, traffic delay, and pain and suffering due to accidents are common examples of nonmonetary externalities.

Environmental costs include those related to air pollution, global warming, water pollution, and noise due to motor vehicles. To estimate these costs, one must model complex physical processes and biological responses, and then estimate the dollar value of the responses.

The economic problem created by externalities is the classic divergence between private cost and social cost, discussed above and illustrated inbreak Figure 1. As indicated in Table 2, the usual prescription for nonmonetary externalities is to assign property rights, bargain, or apply a dynamic Pigovian tax on the perpetrator or emissions source, with no direct compensation of the victim. The Pigovian tax must be levied on the immediate damaging activity, and not on some related activity. In the case of air pollution, the tax should be levied on the source of the emissions. For example, the environmental damages from pollution from petroleum refineries should be internalized by a tax on refinery emissions, not by a tax on the final uses of the fuel products of the refinery. This remains true even if there is a clear economic and physical linkage between the final use of the refinery products and the emissions from the refinery. Now, if there is such a linkage, we may say that refinery pollution is a cost of motor vehicle use—because motor fuel use does, through a chain of events, give rise to the environmental costs of the refinery—and one way or another, whether via the Pigovian tax or a separate calculation of marginal damages, we must count the refinery pollution as a cost of motor vehicle use. However, linkage or no, we should levy the pollution tax at the refinery stacks. The definition, treatment, and estimation of external costs in the U. S. is discussed in more detail Report #9 of Delucchi et al. and in Delucchi.[1,43] For a discussion of estimates of the environmental externalities of transport in Europe, see Ref. 29.

In this chapter, we have distinguished nonmonetary costs inflicted—even if only indirectly, by motor vehicle user A on party B and not accounted for by A—from personal nonmonetary costs, which are inflicted by a motor vehicle user on herself. We also might have distinguished a third kind of nonmonetary or environmental-damage cost: that inflicted by motor vehicle user A on party B but accounted for by A as a marginal cost of motor vehicle use. When an externality is properly taxed, it becomes this third type of cost. (One perhaps could argue that once a [formerly] nonmonetary cost is properly taxed, it becomes a monetary cost, but this is merely semantics.) Thus, the third category would consist of true Pigovian taxes.

However, there are at most only three quasi-Pigovian taxes related to motor vehicle use: (1) the portion of the oil-spill environmental excise tax that covers costs other than clean-up costs; (2) the tax, which Barthold says is *Pigovian*, on ozone-depleting chemicals; and (3) the gas-guzzler tax, which arguably is partly a tax on energy-security costs.[44] However, the oil-spill tax and the gas-guzzler tax probably are not equal to marginal expected damages, and hence probably are not true Pigovian taxes, and the tax on ozone-depleting chemicals now is largely irrelevant because new automobiles use a more ozone-friendly refrigerant that is not subject to the tax. For these reasons, we have not created a separate category called "properly taxed, efficiently allocated environmental damages."

Note that, if one were tallying the marginal social cost and found that there were optimal Pigovian taxes, one would count either the tax or the value of the actual marginal damage, but not both, because if the tax had been calculated correctly it would equal the damage. Note, too, that the cost of pollution control equipment cannot be construed as a Pigovian tax: The economic cost of pollution-control equipment is the value of the resources used to make and operate control equipment, whereas a correct Pigovian tax is equal to the marginal cost of the remaining (post-control) pollution. In a social-cost analysis, control costs and post-control damage costs are additive, not equivalent.

Suppose that we wish to estimate the social cost (private cost plus external cost) of using motor gasoline. We know that there is a relationship between the amount of motor fuel consumed and the amount that refineries produce, and a relationship between the amount of fuel that refineries produce and the amount of pollutants they emit. We therefore may count as a cost of using motor gasoline the value of the environmental damages from emissions from petroleum refineries making gasoline. In a world without true marginal-cost Pigovian taxes—that is, in the real world of today and tomorrow—we can make an independent estimate of the value of the environmental damages from making motor gasoline, and add to it the refineries' actual private cost (exclusive of taxes) of making gasoline, as part of our estimate of the social cost of motor gasoline. This is what we do here. But what if the emissions from refineries actually were assessed a Pigovian charge equal to the marginal damage that they caused? In that case, the damage cost would be internalized at the refineries (which, as pointed out above, is where it should be internalized), and the refineries' private cost would include the cost of environmental damage. To add to this private cost an independent estimate of the environmental damages in this case would double-count the damages.

Nonmonetary externalities are estimated in Report #9 of Delucchi et al. and in Delucchi.[1,43]

5.7 Column 6b of Table 1: Nonmonetary Costs of Motor Vehicle Infrastructure

Note that we have classified the nonmonetary social and environmental impacts of the motor vehicle infrastructure in part b of column 6, separate from the nonmonetary externalities of motor vehicle use. Although these infrastructure costs ultimately are a long-run cost of total motor vehicle use, they are not a cost of marginal or incremental motor vehicle use, because they do not vary with each mile or trip. Hence, infrastructure costs are not externalities of motor vehicle use, according to our definition of *externality*, and for this reason are categorized separately from external costs. Note, too, that we we have not actually estimated any of these environmental costs of infrastructure. (One should not presume, though, that omitted costs necessarily are trivial.)

5.8 Summary Observations Regarding Table 1

Divergence between Price and Marginal Social Cost Increases from Left to Right

One perhaps can argue that, in general terms, the "typical" divergence between the marginal social cost and the actual price (or the marginal social value) in each column of Table 1 increases as one moves from column 1 to column 6. For the items in the first column, there is little or no divergence between marginal social cost and marginal social value; for those in the last column, the price is zero but the marginal social cost can be considerable.

Long Run versus Short Run, and Direct versus Indirect Costs

In order to keep Table 1 manageable, we have not distinguished in the table between costs incurred immediately as a result of motor-vehicle use (one might call these "direct short-run" costs), and costs incurred in the long run, or only indirectly, as a result of motor-vehicle use. However, these distinctions can be important.

Motor vehicle use does not give rise to costs *automatically*, according to immutable laws of physics or to the logic of mathematics, but, rather, is linked to costs—to particular effects, or changes in actual resource consumption—by economic, political, technological, and natural processes. Some links are direct and almost immediate. For example, motor vehicle use is linked directly by combustion processes to motor vehicle emissions of CO, emissions of CO, in turn, are linked directly by atmospheric processes to ambient levels of CO, and ambient levels of CO are linked statistically, by behavioral and biological processes, to headaches. In this case, the linkage between use and cost (headaches) is largely physical, and almost immediate.

But linkages can be much more attenuated than this. For example, the linkage between motor vehicle use and a change in refinery emissions is more complicated than the linkage between motor vehicle use and a change in motor vehicle tailpipe emissions, because there are intervening economic as well as physical processes. In theory, a change in motor vehicle use will change quantity and hence price in the market for gasoline. This, in turn, will affect price in the market for crude oil, which will affect price in the market for other petroleum products (such as heating oil). In theory, refinery owners will adjust to the price changes by changing the mix and amount of refinery products. This economically induced change in output will be linked physically to changes in refinery emissions, which will be linked to ambient pollution and then to health effects. All of this is a theoretical simplification. In reality, political factors and economic variables other than price will be important, too.

Table 4 Description of Our Ratings of the Quality and Complexity of Our Analysis

Quality of Our Analysis	Rating
Detailed and largely original analysis, with extensive calculations based mainly on primary data. Primary data include: original censuses and surveys of population, economic activity, energy use, travel and more; financial statistics collected by government agencies such as state MV departments; measured environmental data, such as of ambient air quality and visibility; surveys and inventories of physical infrastructure, such as housing stock and roads; and the results of empirical statistical analyses, such as epidemiological analyses of air pollution and health.	A1
Detailed and original analysis based mainly on primary data, but less involved than level A1 analysis (see A1 for examples of primary data).	A2
Straightforward analysis based partly or mainly on primary data, with few and relatively simple calculations. Less involved than A2 analysis.	A3
Direct use of a few primary data, with no significant analysis, calculations, or adjustments. A simple citation of primary data.	A4
Review and analysis of existing estimates of the whole cost or its major components. The difference between B work and A work is that A work is based mainly on primary data, such as from government surveys or data series or physical measurements (see above), whereas B work is more dependent on the secondary literature (i.e., on someone else's original analysis of some major components of the social cost). However, the analysis in B work can be more extensive than that in A3 and certainly A4 work.	B
Review of a few existing estimates, with little or no analysis. This is essentially a literature review.	C
Estimate or simple, illustrative calculation based ultimately on supposition or judgment. Whereas C work cites a substantive analysis or estimate of the cost under consideration, D work is based on judgment without reference to any direct estimate of the cost or its major components.	D

Table 5 Personal Nonmonetary Costs of Motor Vehicle Use, 1991 (10^9 1991\$)

Cost Item	Low	High	Q^a
Travel time, excluding travel delay imposed by others, that displaces unpaid activities	406.8	629.0	A2
Accidental pain, suffering, death, and lost nonmarket productivity inflicted on oneself	70.2	227.0	A2/B
Personal time spent working on motor vehicles and garages, and refueling motor vehicles	49.5	109.6	A3
Personal time spent buying and selling and disposing of vehicles, excluding dealer costs	0.8	2.6	A3
Motor vehicle noise inflicted on oneself	Included with external noise costs		
Motor vehicle air pollution inflicted on oneself	Included with external pollution costs		
Total	527.3	968.2	

See Report #4 of Delucchi et al. for details. See Ref. 1.

[a] Q = Quality of the estimate (see Table 4).

Table 6 Motor Vehicle Goods and Services Priced in the Private Sector (Cost Estimated Net of Producer Surplus and Taxes and Fees), 1991 (10^9 1991$)

Cost Item	Low	High	Q^a
Usually Included in GNP-type Accounts			
Annualized cost of the entire motor vehicle car and truck fleet, excluding sales taxes[b]	269.2	350.2	A3
Cost of transactions for used cars	12.7	12.7	A3
Parts, supplies, maintenance, repair, cleaning, storage, renting, towing, etc.[b]	159.9	188.1	A3
Motor fuel and lubricating oil, excluding excise and sales taxes and fuel costs attributable to travel delay	74.9	82.2	A2
Motor vehicle insurance: administrative and management costs	36.7	36.7	A4
Priced private commercial and residential parking, excluding parking taxes	3.2	3.2	A3
Usually Not Included in GNP-type Accounts			
Travel time, excluding travel delay imposed by others, that displaces paid work	190.1	229.1	A2
Overhead expenses of business, commercial, and government fleets	90.3	112.9	A3
Private monetary costs of motor vehicle accidents, excluding user payments[c]	65.7	65.6	A2/B
Motor vehicle user payments for the cost of motor vehicle accidents inflicted on others	55.7	58.8	A4/D
Deduction for property damage, and motor vehicle insurance administration costs counted elsewhere (as private monetary costs here, or as external monetary costs)	(65.2)	(74.8)	A2/B
Deduction for embedded taxes included in the price-times-quantity estimates above	(59.8)	(57.6)	A2/A3
Deduction for bundled parking costs included in cost of any industries above, but counted separately here as a bundled parking cost	(6.4)	(26.6)	D
Total	826.9	980.4	

See Report #5 of Delucchi et al. for details. See Ref. 1.
[a] Q = Quality of the estimate (see Table 4).
[b] These estimates include costs related to motor vehicle accidents. Because these costs also are counted in the line "private monetary costs of motor vehicle accidents," they are deducted in a separate line ("Deduction for property damage...") to avoid double counting.
[c] The estimate under "Low" might be higher than the estimate under "High" because a total estimated accident cost is allocated to the different cost categories on the basis of low and high externality fractions, whereby "Low" means low external cost—and hence high private or personal cost—and "High" means high external cost.

Table 7 Motor Vehicle Goods and Services Bundled in the Private Sector, 1991 (Billion 1991$)

Cost Item	Low	High	Q^a
Annualized cost of nonresidential offstreet parking included in the price of goods or services or offered as an employee benefit	48.5	162.2	A2
Annualized cost of home garages, carports, and other residential parking included in the price of housing	15.4	40.6	A2
Annualized cost of roads provided or paid for by the private sector and recovered in the price of structures, goods, or services	11.8	75.9	A3, D^b
Total	75.7	278.7	

See Report #6 of Delucchi et al. for details, Ref 1.
^{a}Q = Quality of the estimate (see Table 4).
^{b}A simple calculation involves some solid numbers and some guesswork.

The linkages between motor vehicle use and cost can be even more tenuous: They can depend not only price changes, which at least in economic theory are *mechanisms*, but on the decisions of public policy makers as well. Consider the links between motor-vehicle use and defense expenditures in the Middle East. First, the change in motor-fuel use will change demand for oil, but not barrel for barrel, because prices of and hence demand for other petroleum products will change. The change in demand for oil might change demand for oil imported from the Middle East, depending on the price of domestic versus imported oil, sunk costs, contractual arrangements, political conditions, and other factors. Congress then *might* notice any change in oil imports from the Middle East, and then *might* decide that it means that the United States cares less about the region and need not devote as many resources to policing it. Such government decisions make the link between motor vehicle use and military expenditures especially hard to represent formally.

Although Table 1 does not make these distinctions, they nevertheless are important because the more tenuously linked costs are harder to estimate, often are lagged considerably with respect to the causal changes in motor vehicle use, and often depend greatly on the specific characteristics and amount of the change in motor-vehicle use. The upshot is that it is *especially* dubious to use willy-nilly, in any context, our estimates of the total or average cost of the more tenuously linked costs.

5.9 Quality of the Estimates

Table 1 lists nearly 50 individual components of the total social cost of motor vehicle use. For some of these cost components, we were able to develop original, reasonably detailed estimates. However, in many other cases we simply took estimates from the literature or made educated guesses. Thus, there is quite a wide range in the quality of our estimates. In order to provide an overview of

Table 8 Motor Vehicle Infrastructure and Services and Services Provided by the Public Sector, 1991 and 2002 (10^9 $)

Cost Item	10%Δ MVU (1991) Low	10%Δ MVU (1991) High	100%Δ MVU (1991) Low	100%Δ MVU (1991) High	100%Δ MVU (2002) Low	100%Δ MVU (2002) High	Q^a
A1. Direct expenditures (FHWA)^b							
Annualized cost of highways (FHWA)	9.0	18.5	90.4	184.9	159.9	335.7	A2
Highway law enforcement and safety	0.45	0.70	7.4	8.7	12.6	15.8	A3
A2. Other direct expenditures^c							
Collection expenses, LUST, extra m&r	0.46	0.46	4.7	4.7	8.3	8.3	A3
Annualized cost of municipal and institutional offstreet parking	n.e.	n.e.	11.9	19.8	17.5	29.0	A2/3
Deduction for embedded private investment in roads	(0.30)	(0.75)	(3.0)	(7.5)	(6.6)	(16.7)	C
B. Indirect expenditures							
Other police-protection costs (not estimated by FHWA) related to MV use	0.10	0.47	0.8	4.1	1.9	9.3	A2
Fire-protection costs related to MV use	0.07	0.27	0.7	2.8	1.4	5.5	A2
Emergency-service costs of MV accidents included in police and fire costs	(0.15)	(0.16)	(1.1)	(1.1)	(1.4)	(1.4)	A2/B
Judicial and legal-system costs	0.46	0.59	4.8	6.2	8.9	11.6	A2
Legal costs of MV accidents included under judicial and legal-system costs	(0.09)	(0.12)	(0.9)	(0.9)	(1.2)	(1.2)	A2
Jail, prison, probation, and parole costs related to MV use	0.39	0.61	3.9	6.2	7.0	9.4	A2
Regulation and control of air, water, and solid-waste pollution related to MV use	0.17	0.56	2.1	5.9	7.1	15.4	A2
Energy and technology R & D	n.e.	n.e.	0.3	0.5	0.3	0.8	A3
MV-related costs of other agencies	n.e.	n.e.	0.1	0.1	0.1	0.1	D
Military expenditures related to the use of Persian-Gulf oil by MVs	n.e.	n.e.	0.8	8.5	0.8	11.2	B, D^d
Annualized cost of the SPR	0.00	0.06	0.1	0.7	0.0	0.9	A2
Total	n.e.	n.e.	122.9	243.2	216.5	433.6	

See Report #7 of Delucchi et al. for details, Ref. 1.
Δ MVU = change in motor-vehicle use; MV = motor vehicle; O & M = operation & management; FWHA = Federal Highway Administration.
^{a}Q = Quality of the baseline year—1991 estimate (see Table 3).
^{b}With minor exceptions, these are based on Federal Highway Administration (FHWA) estimates of government expenditures for highways. See Ref. 49.[49] The A1 estimates shown here *exclude* user tax-and-fee collection expenses, LUST-fund costs, and extra maintenance and repair (m&r) costs, but *include* the embedded private-sector investment in roads, because the FHWA expenditure estimates exclude collection, LUST, and extra m&r costs, but include embedded private costs. In part A2 of this table the excluded collection, LUST, and extra m&r costs are added back in, and the included embedded private costs are deducted.
^{c}See note b.
^{d}A review and analysis of the literature with a good deal of supposition. See Report #15 of Delucchi et al. for details.[1]

Table 9 Monetary Externalities of Motor Vehicle Use, 1991 (10^9 1991\$)

Cost Item	Low	High	Q^a
Monetary costs of travel delay imposed by others: forgone paid work	9.1	30.5	A2
Monetary costs of travel delay imposed by others: extra consumption of fuel	2.3	5.7	A2
Accident costs not accounted for by economically responsible party: property damage, medical, productivity, legal and administrative costs	26.0	28.0	A2/B
Macroeconomic adjustment costs related to oil-price shocks	1.8	31.5	B [A1]
Pecuniary externality: increased payments to foreign countries for oil used in non-MV sectors, due to ordinary price effect of using petroleum for motor vehicles	3.8	8.0	A3
Monetary, non-public-sector costs of net crimes related to using or having MV goods, services, or infrastructure	0.1	0.4	A3
Monetary costs of injuries and deaths caused by fires related to MV use	0.0	0.1	A3
Total	43.1	104.2	

See Report #8 of Delucchi et al. for details, Ref. 1.
^{a}Q = Quality of the estimate (see Table 3). Ratings in brackets refer to the quality of the analysis in the literature reviewed.

the quality of our estimates and help readers understand initially which estimates are sound and which are little better than guesswork, we have rated each of our estimates. The rating system is delineated in Table 4, and the ratings are presented in Tables 5 to 10.

6 RESULTS OF THE ANALYSIS

The results of this analysis are shown by individual cost item in Tables 5 to 10, and summarized by aggregate cost category in Tables 11 The cost items correspond to those in Table 1. We show the aggregated totals here in order to provide a sense of magnitudes, not because such aggregated totals are themselves useful. Indeed, as discussed next, one must be careful to avoid misusing estimates of the total social cost of motor vehicle use.

6.1 Allocation of Costs to Individual Vehicle Categories

All of the costs shown in Tables 5 to 10 pertain to all motor vehicles: all autos, trucks, and buses. Although it can be interesting to estimate the cost of all motor vehicle use, it typically will be more useful to estimate the cost of different classes of vehicles or of different fuel types, because analysts, policy makers, and regulators typically are interested in specific classes of vehicles, and specific

Table 10a Nonmonetary Externalities of Motor Vehicle Use, 1990–1991 (10^9 1991\$)

Cost Item	Low	High	Q^a
Accidental pain, suffering, death, and lost nonmarket productivity inflicted on oneself	9.5	97.7	A2/B
Travel delay, imposed by others, that displaces unpaid activities	22.5	99.3	A2
Air pollution: human mortality and morbidity due to particulate emissions^b from vehicles	16.7	266.4	A1
Air pollution: human mortality and morbidity due to all other pollutants from vehicles	2.3	17.1	A1
Air pollution: human mortality and morbidity, due to all pollutants from upstream processes	2.3	13.0	A1
Air pollution: human mortality and morbidity, due to road dust	3.0	153.5	A1
Air pollution: loss of visibility, due to all pollutants attributable to motor vehicles	5.1	36.9	A1
Air pollution: damage to agricultural crops, due to ozone attributable to motor vehicles	3.3	5.7	A1
Air pollution: damages to materials, due to all pollutants attributable to motor vehicles	0.4	8.0	B [A1]
Air pollution: damage to forests, due to all pollutants attributable to motor vehicles	0.2	2.0	B [A2]
Climate change due to lifecycle emissions of greenhouse gases (U.S. damages only)	0.0	3.5	A1, B [A1]^c
Noise from motor vehicles	0.5	15.0	A1
Water pollution: health and environmental effects of leaking motor-fuel storage tanks	0.1	0.5	D
Water pollution: environmental and economic impacts of large oil spills	0.2	0.5	C [A1]
Water pollution: urban runoff polluted by oil from motor vehicles, and pollution from highway deicing	0.7	1.7	D^d
Nonmonetary costs of net crimes related to using or having motor vehicle goods, services, or infrastructure	0.7	2.8	A3
Nonmonetary costs of fires related to using or having motor vehicle goods, services, or infrastructure	0.0	0.2	A3
Air pollution: damages to natural ecosystems other than forests, due to all pollutants attributable to motor vehicles	n.e.	n.e.	n.a.
Environmental and esthetic impacts of motor vehicle waste	n.e.	n.e.	n.a.
Vibration damages from motor vehicles	n.e.	n.e.	n.a.
Fear and avoidance of motor vehicles and crimes related to motor vehicle use	n.e.	n.e.	n.a.
Total	68.0	729.6	

Table 10b Nonmonetary Environmental and Social Costs of the Motor Vehicle Infrastructure

Cost Item	Low	High	Q^a
Land-use damage: habitat destruction and species loss due to highway and motor vehicle infrastructure	n.e.	n.e.	n.a.
The socially divisive effect of roads as physical barriers in communities	n.e.	n.e.	n.a.
The esthetics of highways and service establishments	n.e.	n.e.	n.a.
Total	n.e.	n.e.	

See Report #9 of Delucchi et al. for details, Ref. 1. n.e. = not estimated; n.a. = not applicable. Note that all air pollution estimates include costs of air pollution inflicted by drivers on themselves, technically part of personal nonmonetary costs.

^{a}Q = Quality of the estimate (see Table 3). Ratings in brackets refer to the qualityy of the analysis in the literature reviewed.
^{b}Includes secondary PM, formed from direct emissions of SO_x, NO_x, and NH_3.
^{c}The estimate of lifecycle emissions of greenhouse gases is original and detailed (A1), whereas the estimate of the $/ton cost of emissions is based on a review of literature (B) that reports results from detailed model calculations ([A1]).
^{d}This is my estimate of the cost as of 1997. The cost probably was higher in 1991, because the leakage-prevention and clean-up programs were not in place everywhere. We speculate that the external costs in 1991 were three times the costs today.

fuels, rather than all motor vehicles as a group (e.g., pollution regulations are set for individual classes of vehicles, not for all motor vehicles as a class).

For some cost items, such as the some of the costs of air pollution, we have estimated marginal costs by individual vehicle class (see Report #9 of Delucchi et al., and Delucchi).[1,43] In most cases, though, we have not actually estimated costs by vehicle class. However, we have developed simple *cost-allocation factors*, which can be used to apportion or disaggregate some total costs to specific vehicle and fuel classes. These factors are developed in Report #10 of Delucchi et al.[1]

6.2 How the Results of This Analysis Should Not Be Used

Earlier in this chapter, we explain the proper uses of a social-cost analysis. In this section, we caution against several common misuses of estimates of the total social cost.

First, one should resist the temptation to add up all of the unpriced costs, and express the total per gallon of gasoline, as if the optimal strategy to remedy every inefficiency were simply to raise the gasoline tax. Rather, as indicated in Table 2, the various kinds of inefficiencies, or market failures or imperfections, ideally require various kinds of remedies. In fact, it turns out that there is not a

Table 11 Summary of the Costs of Motor Vehicle Use

	Total Cost (10^9 $)		Percentage of Total	
	Low	*High*	*Low*	*High*
1. Personal nonmonetary costs of motor-vehicle use	$527	$968	32%	29%
2. Motor vehicle goods and services produced and priced in the private sector (estimated net of producer surplus, taxes, fees)	$827	$980	50%	30%
3. Motor vehicle goods and services bundled in the private sector	$76	$279	5%	8%
4. Motor vehicle infrastructure and services provided by the public sector[a]	$123	$243	7%	7%
5. Monetary externalities of motor vehicle use	$43	$104	3%	3%
6. Nonmonetary externalities of motor vehicle use	$68	$730	4%	22%
Grand total social cost of highway transportation	$1,664	$3,304	100%	100%
Subtotal: monetary cost only (2 + 3 + 4 + 5)	$1,069	$1,606	64%	49%

For details, see other summary tables in this chapter, the text in this chapter, and Delucchi et al., Ref. 1.

[a] Includes items in Table 1 that straddle columns 4 and 5.

single external cost, with the possible exception of CO_2 emissions from vehicles, that in principle is most efficiently (or ideally) addressed by a gasoline tax. However, a gasoline tax as a less-than-ideal measure to address the external costs of motor vehicle use still can improve social welfare. See Ref. 45 for an analysis of the components of an optimal second-best gasoline tax in the United States.

In the first place, some sources of inefficiency, such as imperfect competition and distortionary income tax policy, are not externalities, and hence should be addressed not by Pigovian taxation, but by ensuring that the markets are competitive and only minimally distorted by taxation. Similarly, it is not theoretically ideal (in a first-best world), to force privately provided free parking to be priced; rather, one should amend any tax and regulatory policies that distort the pricing and bundling decisions of private suppliers.

Even where Pigovian taxation is called for, a tax on gasoline is not the ideal application. For example, an optimal air pollution tax would be a function of the amount and kind of emissions, the ambient conditions, and the size of the exposed population; it would not be simply proportional to gasoline consumption.

Similarly, an optimal congestion charge would be a dynamic function of traffic conditions. Costs that arise from the use of particular sources of oil, such as oil imported from the Middle East, should be addressed at the source, not at the level of all gasoline end use. And in any case, it is not even necessarily true, in the real and far-from-best world of regulations, standards, taxes, imperfect taxes, poor information, imperfect competition, and so on, that the optimal emissions tax is equal to the cost of the marginal residual emissions. Against this, however, Freeman notes that even if the emissions standards results in lower emissions than is consistent with economic efficiency, there still should be a tax on miles equal to the residual marginal damages.[46,47]

Second, we caution that it might be misleading to compare the total social cost of motor vehicle use with the gross national product (GNP) of the United States, because the GNP accounting is quite different from and generally more restricted than our social-cost accounting. See Han and Fang for a discussion of different ways of measuring the economic importance of transportation.[48] For example, the GNP does not include any non-market items, which constitute a substantial portion of the social cost estimated here.

Third, one should properly represent and interpret the considerable uncertainty in any estimate of social cost. Uncertainty can be represented by low-high ranges, scenario analyses, probability distributions, and other techniques. Our analysis presents low and high estimates of cost. Yet, strictly speaking, these estimates are not lower and upper bounds, even where the high is much higher than the low, because we did not estimate every conceivable component or effect of every cost, and did not always accommodate the entire span of data or opinions in the literature. Moreover, we do not know how probable the higher and lower values are, or even if the higher is more probable than the lower; in fact, we do not know anything about the probability distribution of the estimated total cost. We can not even offer a "best" guess between our low and high estimates.

Fourth, it is not *economically* meaningful to compare estimates of user tax and fee payments for public motor vehicle goods and services with estimates of government expenditures for same. Most emphatically, it simply is not true that, in order to have the economically optimal amount and use of public motor vehicle goods and services, the revenues collected from the current system of user charges must equal government expenditures. It is not true because the current taxes and fees look nothing like efficient marginal-cost prices, and because in any case it is not a necessary or sufficient condition of economic efficiency that the government collect from users of the highway infrastructure revenues equal to expenditures. Comparisons between current user payments and government expenditures are relevant only to concerns about equity (see Report #17 of Delucchi et al. for further discussion).[1]

Finally, given that ours is an analysis of the *total* social cost of motor vehicle use, whereas any particular policy or investment decision will involve costs incremental or decremental to the total, one should not use our average-cost estimates in marginal analyses, unless one believes that the total-cost function is approximately linear and hence that any marginal-cost rate is close to the average rate. Certainly, our results will become less and less applicable as one considers times and places increasingly different from the United States in 1990 and 1991. However, we note that, even if our results per se are irrelevant, our data, methods, concepts, and cost models might be useful in an analysis of specific pricing policies or investments.

7 SUMMARY

We have classified and estimated the social costs of motor vehicle use in the United States, on the basis of 1990 to 1991 data. Our analysis is meant to inform general decisions about pricing, investment, and research. It provides a conceptual framework for analyzing social costs, develops analytical methods and data sources, and presents some detailed first-cut estimates of some of the costs.

By now it should be clear that a social-cost analysis cannot tell us precisely what we should do to improve our transportation system. There are several kinds of inefficiencies in the motor vehicle system, and hence several kinds of appropriate correctives. Many of our estimates are simply too generic or uncertain to be of much use—as hard numbers—to policy makers and analysts faced with specific problems. Moreover, society cares at least as much about equity, opportunity, and justice as it does about economic efficiency. At the end of the day, a total social-cost analysis contributes modestly to but one of several societal objectives for transportation.

REFERENCES

1. M. A. Delucchi et al., *The Annualized Social Cost of Motor-Vehicle Use in the U. S.*, UCD-ITS-RR-96-3, Institute of Transportation Studies, University of California, Davis (1996-2007). A series of 21 research reports, most of them available at www.its.ucdavis.edu/people/faculty/delucchi.
2. K. Ozbay, B. Bartin and J. Berechman, "Estimation of the Full Marginal Costs of Highway Transportation in New Jersey," *Journal of Transportation and Statistics*, **4**(1): 81–103 (2001).
3. J. Gifford, "Toward the 21st Century," *The Wilson Quarterly*, **17**(1) 40–47 (Winter 1993).
4. F. Popoff and D. Buzzelli, "Full-Cost Accounting," *Chemical and Engineering News*, January 11, 1993, pp. 8–10.
5. P. Stopher, "Panacea—or Solution in Search of a Problem," *TR News*, **165** (March–April, 6–7, 1993).
6. L. S. Thompson, "Trapped in the Forecast, an Economic Field of Dreams," *TR News*, **165** (March–April, 3–4, 1993).

7. J. J. MacKenzie, R. C. Dower and D. T. Chen, *The Going Rate: What It Really Costs to Drive*, World Resources Institute, Washington, D. C., 1992.
8. P. Miller and J. Moffet, *The Price of Mobility: Uncovering the Hidden Costs of Transportation,* Natural Resources Defense Council, San Francisco, California, 1993.
9. C. E. Behrens, J. E. Blodgett, M. R. Lee, J. L. Moore, L. Parker, *External Costs of Oil Used in Transportation*, 92-574 ENR, Congressional Research Service, Environment and Natural Resources Policy Division, Washington, D. C., June 1992.
10. California Energy Commission, *1993–1994 California Transportation Energy Analysis Report*, P300-94-002, Draft Staff Report, Sacramento, California, February 1994.
11. Apogee Research Inc., *The Costs of Transportation: A Review of the Literature*, Final Report, prepared for the Conservation Law Foundation, August 1993.
12. COWIconsult, *Monetary Valuation of Transport Environmental Impact—The Case of Air Pollution*, Danish Ministry of Energy, The Energy Research Programme for Transport, Lyngby, Denmark, December 1991.
13. KPMG Peat Marwick Stevenson & Kellog, Management Consultants, *The Cost of Transporting People in the Lower British Columbia Mainland*, prepared for Transport 2021, Burnaby, British Columbia, Canada, March 1993.
14. B. Ketcham and C. Komonaff, *Win-Win Transportation: A No-Losers Approach to Financing Transport in New York City and the Region*, draft report, Transportation Alternatives, New York, New York, July 9, 1992.
15. T. Litman, *Transportation Cost Analysis: Techniques, Estimates, and Implications*, Draft2, Victoria, British Columbia, Canada, May 26, 1994.
16. National Research Council, *Automotive Fuel Economy: How Far Should We Go?* National Academy Press, Washington, D. C., 1992.
17. K. Green, *Defending Automobility: A Critical Examination of the Environmental and Social Costs of Auto Use*, Policy Study No. 198, Reason Foundation, Los Angeles, California, December 1995.
18. E. W. Beshers, *External Costs of Automobile Travel and Appropriate Policy Responses*, Highway Users Federation, Washington, D. C., March 9, 1994.
19. R. S. Dougher, *Estimates of Annual Road User Payments versus Annual Road Expenditures*, Research Study #078, American Petroleum Institute, Washington, D. C., March 1995.
20. H. Morris and J. DeCicco, "Extent to Which User Fees Cover Road Expenditures in the United States," *Transportation Research Record*, **1576**: 56–62 (1997).
21. Federal Highway Administration, Federal Transit Administration, and Federal Railroad Administration, *1997 Federal Highway Cost Allocation Study Summary Report*, HPP-10/9-97(3M)E, U.S. Department of Transportation, Washington, D. C., August 1997.
22. K. A. Small and C. Kazimi, "On the Costs of Air Pollution from Motor Vehicles," *Journal of Transport Economics and Policy*, **29**: 7–32 (1995).
23. A. J. Krupnick, R. D. Rowe and C. M. Lang, "Transportation and Air Pollution: The Environmental Damages," in *Measuring the Full Social Costs and Benefits of Transportation*, D. L. Greene, D. Jones, and M. A. Delucchi (eds.), Springer-Verlag, Heidelberg, Germany, 1997, pp. 337–370.
24. T. R. Miller, "Societal Costs of Transportation Crashes," in *Measuring the Full Social Costs and Benefits of Transportation*, D. L. Greene, D. W. Jones, and M. A. Delucchi (eds.), Springer-Verlag, Heidelberg, Germany, (1997) pp. 281–314.

References

25. P. N. Leiby, *Estimating the Energy Security Benefits of Reduced U. S. Oil Imports*, ORNL/TM-2007/028, Oak Ridge National Laboratory, Oak Ridge, Tennessee, February 28, 2007. www.epa.gov/otaq/renewablefuels/ornl-tm-2007-028.pdf.
26. D. J. Forkenbrock, "External Costs of Intercity Truck Freight Transportation," *Transportation Research A* **33**: 505–526 (1999).
27. M. Janic, "Modelling the Full Costs of an Intermodal and Road Freight Transport Network," *Transportation Research D*, **12**: 33–44 (2007).
28. I. Mayeres, S. Ochelen, and S. Proost, "The Marginal External Costs of Urban Transport," *Transportation Research D* **1D**: 111–130 (1996).
29. P. Bickel, R. Friedrich, H. Link, L. Stewart and C. Nash, "Introducing Environmental Externalities into Transport Pricing: Measurement and Implications," *Transport Reviews* **26**: 389–415 (2006).
30. T. Keeler, et al., *The Full Costs of Urban Transport*, part III, "Automobile Costs and Final Intermodal Comparisons," monograph 21, Institute of Urban and Regional Development, University of California, Berkeley, July 1975.
31. E. J. Mishan, *Cost-Benefit Analysis: An Introduction*, Praeger, New York, 1976.
32. N. Hanley, "Are There Environmental Limits to Cost-Benefit Analysis?," *Environmental and Resource Economics*, **2**: 33–39 (1992).
33. *Science News*, **143**(23), Letters to the editor, June 5, 1993, p. 355.
34. K. Button, *Transport, the Environment, and Economic Policy*, Edward Elgar Publishing Company, Vermont, 1993.
35. Congressional Budget Office, *Paying for Highways, Airways, and Waterways: How Can Users Be Charged?*, U. S. Government Printing Office, Washington, D. C., May 1992.
36. D. Gillen, "Efficient Use and Provision of Transportation Infrastructure with Imperfect Pricing: Second Best Rules," in *Measuring the Full Social Costs and Benefits of Transportation*, D. L. Greene, D. W. Jones, and M. A. Delucchi (eds.), Springer-Verlag, Heidelberg, Germany, 1997, pp. 193–218.
37. D. M. DeJoy, "The Optimism Bias and Traffic Accident Risk Perception," *Accident Analysis and Prevention* **21**: 333–340 (1989).
38. D. L. Greene and P. N. Leiby, *The Social Costs to the U. S. of Monopolization of The World Oil Market, 1972–1991*, ORNL-6744, Oak Ridge National Laboratory, Oak Ridge, Tennessee, March 1993.
39. W. J. Baumol and W. E. Oates, *The Theory of Environmental Policy*, 2nd ed., Cambridge University Press, New York, 1988.
40. J. Gomez-Ibanez, "Estimating Whether Transport Users Pay Their Way: The State of the Art," in *Measuring the Full Social Costs and Benefits of Transportation*, D. L. Greene, D. W. Jones, and M. A. Delucchi (eds.), Springer-Verlag, Heidelberg, Germany, 1997, pp. 150–172.
41. D. B. Lee, *A Market-Oriented Transportation and Land-Use System: How Different Would It Be?*, Research and Special Programs Administration, U. S. Department of Transportation, Cambridge, Massachusetts, September (1992). Also published in *Privatization and Deregulation in Passenger Transport: Selected Proceedings of the 2nd International Conference*, Espoo, Finland, Viatek, Ltd., June 1992, pp. 219–238.
42. D. W. Jones, P. N. Leiby and I. K. Paik, "Oil Price Shocks and the Macroeconomy: What Has Been Learned since 1996," *The Energy Journal*, **25**(2): 1–32 (2004).

43. M. A. Delucchi, "Environmental Externalities of Motor-Vehicle Use in the U. S.," *Journal of Transport Economics and Policy* **34**: 135–168, (May 2000).
44. T. A. Barthold, "Issues in the Design of Environmental Excise Taxes," *Journal of Economic Perspectives* **8**: 133–151 (1994).
45. I. W. H. Parry and K. A. Small, "Does Britain or the United States Have the Right Gasoline Tax?" *American Economic Review*, **95**: 1276–1289 (2005).
46. A. M. Freeman III, "Externalities, Prices and Taxes: Second Best Issues in Transportation," in *Measuring the Full Social Costs and Benefits of Transportation*, D. L. Greene, D. W. Jones, and M. A. Delucchi (eds.), Springer-Verlag, Heidelberg, Germany, 1997, pp. 173–192.
47. D. Burtraw, W. Harrington, A. M. Freeman III and A. J. Krupnick, *Some Simple Analytics of Social Costing in a Regulated Industry*, Discussion Paper QE93-13, Resources for the Future, Washington, D. C., April (1993).
48. X. Han and B. Fang, "Four Measures of Transportation's Economic Importance," *Journal of Transportation and Statistics*, **3**(1): 15–30 (2000).
49. Federal Highway Administration, *Highway Statistics 1992*, FHWA-PL-93-023, U.S. Department of Transportation, Washington, D. C., 1993. Annual data available online at www.fhwa.dot.gov/policy/ohpi/hss/index.htm.

CHAPTER 5

TRAFFIC CONGESTION MANAGEMENT

Nagui M. Rouphail
Institute for Transportation Research and Education
North Carolina State University
Raleigh, North Carolina

1	**INTRODUCTION AND BACKGROUND**	97	5.1 Travel Demand Management (TDM)	105
2	**SCOPE OF THE CHAPTER**	98	5.2 Traffic Operational Strategies (TOS)	108
3	**ORGANIZATION OF THE CHAPTER**	99	6 **ASSESSING EMISSION IMPACTS OF TRAFFIC CONGESTION MANAGEMENT**	113
4	**FUNDAMENTALS OF VEHICLE EMISSION ESTIMATION**	99	6.1 Direct Measurement Methods	114
	4.1 VMT	99	6.2 Mobile Emission Estimation Models	118
	4.2 VMT-S	100	6.3 Transportation Models	122
	4.3 Mode	100	7 **SUMMARY**	124
	4.4 VSP	101		
	4.5 Hybrid VSP-S	103		
5	**INVENTORY OF TRAFFIC CONGESTION MANAGEMENT METHODS**	105		

1 INTRODUCTION AND BACKGROUND

In striving toward an environmentally conscious transportation system, the field of traffic congestion management offers a diverse toolbox of solutions to traffic and transportation engineers, all of which are aimed at mitigating the effect of congestion on emissions and air quality. Traffic congestion continues to be on an upward trajectory in the United States. According to the 2005 Urban Mobility Report by the Texas Transportation Institute, urban daily vehicle miles of travel (VMT) more than doubled in the period 1982 to 2003, resulting in a near threefold increase in peak hour delays experienced by motorists in the same time period (from 16 to 47 hours of annual delay per peak traveler).[1] That same study estimated that traffic operational improvements (not including public transportation solutions) could reduce the overall hours of delays by nearly 10 percent and annual congestion costs by nearly 9 percent.

The evidence connecting transportation sector use in the United States to prevailing emissions levels for various pollutants is indeed compelling: Direct emissions from surface transportation account for nearly 45 percent of total national annual emissions of nitrogen oxides (NO_x), 59 percent of CO, 38 percent of volatile organic compounds (VOC), and 50 percent of particulate matter less than 10 microns in diameter (PM_{10}).[2] In addition, electrified surface transportation systems, such as light rail or electric vehicles, generate indirect emissions at the power plants. NO_x, CO, and VOC are precursors to ozone formation in the troposphere layer (the lowest portion of Earth's atmosphere). This layer contains approximately 75 percent of the mass of the atmosphere and almost all the water vapors and aerosols. Tropospheric ozone is a greenhouse gas that contributes significantly to climate change. Over 158 million Americans live in regions exceeding the National Ambient Air Quality Standard (NAAQS) for ozone.[3] Thus, traffic congestion management strategies, whose focus is on improving the operation of the surface transportation system, offer a realistic and substantive means to reducing emissions and improving air quality in the short- and long-term horizons.

2 SCOPE OF THE CHAPTER

Congestion management programs that are aimed at reducing overall vehicle emissions on a corridor or network generally fall into two classes: (1) those intended to *restrict the demand or travel intensity* of the transportation system and (2) those aimed at *improving the operational performance* of the transportation system. Some strategies, such as improved transit operations and ridership could fall into both categories.

In this chapter, the first class will be referred to as *travel demand management (TDM) strategies* and the second as *traffic operational strategies (TOS)*. Absent from this classification are programs aimed at improving vehicle fleet characteristics as they relate to fuel efficiency, the use of alternative fuels and the associated emission rates, and those targeted at specialized vehicle fleets, such as idling reduction programs for trucks and other heavy vehicles. Thus, the focus here is primarily on the aggregate traffic performance of the average commuter fleet.

Finally, while there are obvious linkages between vehicle emissions and regional air quality measures, those linkages are not explored in this chapter. Air quality indicators generally incorporate multiple sources of emissions, including industrial, residential, and commercial uses, in addition to mobile sources, and are also affected by the regional topography and atmospheric conditions. Therefore, the presentation in this chapter focuses on the emissions consequences of traffic management strategies, with the understanding that these strategies relate to only a portion of the total emissions inventory in a region. The relative magnitude of this portion will also vary according to the pollutant considered, as indicated in the national inventory statistics presented in section 1.

3 ORGANIZATION OF THE CHAPTER

This chapter is organized around three principal themes. The first theme (section 4) covers the *fundamental measures of emissions* and how they relate to vehicular activity at the individual and traffic stream level. The second theme (section 5) provides a *survey of traffic congestion management methods* aimed at reducing vehicle emissions, focusing on both demand and supply management. The last theme (section 6) is related to emission-based *assessment methods of the effectiveness of congestion management techniques*, which form the empirical and modeling tools available to engineers and researchers to estimate or predict emission consequences of various traffic control actions or policies.

4 FUNDAMENTALS OF VEHICLE EMISSION ESTIMATION

Two fundamental measures underlie the estimation of emissions (E) as they relate to traffic congestion management. These are measures of the vehicle activity (A) and emission factors (EF), where $E = A \times EF$. In other words, the EF represents the generated emissions per unit vehicle activity.

Vehicle activity can be defined at various levels or scales of aggregation, each with its own purpose, and associated emission factor. For traffic congestion management purposes, five activity measures are discussed in this section:

1. Vehicle miles of travel (VMT)
2. Vehicle miles of travel at specified speeds or speed ranges (VMT-S)
3. Travel time distribution by driving mode (MODE)
4. Travel time distribution by vehicle specific power (VSP) mode
5. Travel time distribution using a hybrid VSP-speed (VSP-S) approach

4.1 VMT

At a very coarse level, one can define the vehicle activity to be the number of annual vehicle miles traveled by the average gasoline-powered, light duty vehicle in the United States. According to the 2005 National Transportation Statistics such a vehicle has hydrocarbon (HC), carbon monoxide (CO), and nitrogen oxides (NO_X) emission factors that are equivalent to 1.25, 12.57, and 0.92 grams/mile.[4] Therefore, if such a vehicle traveled 10,000 miles per year, its contribution to the annual emission inventory would be 12.5, 125.7, and 9.2 kg/year for HC, CO, and NO_X emissions, respectively.

Table 1 Illustrative Vehicle Speed Effect on CO Emissions

Speed range (mph)	<=10	10–30	30–50	>50
Annual mileage at given speed range	2000	3000	3000	2000
Corresponding EF (gm/mile)	18	9	10	11
Annual emissions in speed range (Kg.)	36	27	30	22

4.2 VMT-S

Although the previous approach could be useful in providing gross estimates of the effect of VMT changes on regional emission levels, it is fundamentally inadequate for characterizing the effectiveness of congestion management strategies. This is because the *nature* of the vehicle activity, not just its *magnitude*, can have a profound effect on the corresponding emission factor. To illustrate this effect, consider the case where additional information is known about the mileage distribution of the subject vehicle travel speed throughout the year, as summarized in Table 1 (considering CO emissions, and for illustrative purpose only). The implication here is that the emission factor varies with the vehicle speed. Activity in this context is defined as the annual miles traveled in a given speed range.

On the basis of the depicted values, the annual CO emissions in this case would add up to 115 kg/year (contrasted with the 125.7 kg/year value computed earlier). Therefore, a change in the speed distribution—for example, due to changes in the prevailing congestion level—will be automatically reflected in the estimation of total annual emissions produced by the vehicle.

4.3 Mode

It is possible to further pursue this line of increased detail by refining the definition of vehicle activity to include the effects of short-term (also known as *microscale*) events that are associated with changes in the emission factor. A *microscale* event represents a different load requirement on the vehicle engine and, therefore, generates different fuel and emission rates. One way to represent these events is to categorize the trip time (or trip distance) into driving modes. A simplified modal emission model by Frey et al. considers four such modes: acceleration, deceleration, idle, and cruise.[5] Each mode is then associated with a unique emission factor (per unit time or distance), and activity is defined as the trip travel time spent in each of the four modes.

An example of modal emission factors computed from on-board vehicle measurements in the Frey study is illustrated in Figure 1, which depicts the average and 95th percentile confidence intervals for four different pollutants. It is obvious that travel during the acceleration mode has a disproportionately higher emission factor for all pollutants considered, while the idle mode has the lowest emission factor. On average, the emission factors in Figure 1 for CO are 48, 11, 6, and

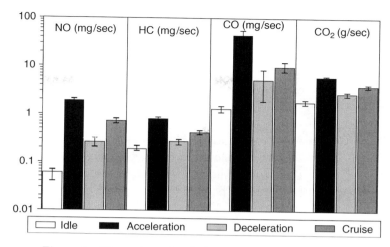

Figure 1 Illustration of emission factors by driving mode.

1.2 grams/sec for the acceleration, cruise, deceleration, and idle modes, respectively. Therefore, a commuter trip that includes many acceleration and deceleration cycles is likely to generate higher emissions than one in which speed does not vary substantially.

An application of the driving mode approach is shown in Table 2, which represents driving conditions in a congested environment for an urban commuter. In this case, travel time by mode in lieu of distance is used to describe vehicle activity. The results indicate that while acceleration constituted only about 17 percent of the vehicle modal activity, it accounted for more than 55 percent of the overall annual CO emissions.

4.4 VSP

Three obvious deficiencies in the MODE vehicle activity descriptor involve (1) the lack of accounting for the various levels of accelerations (the mode where the highest emission rates occur), (2) the fact that the cruise mode has a fixed emission factor independent of speed, and (3) that this approach does not account for the roadway infrastructure effects on power demand. These limitations can be partially overcome through the use of a *vehicle-specific power*

Table 2 Driving Mode Effect on CO Emissions

Driving Mode	Accel.	Decel.	Cruise	Idle
Travel time by mode (hrs/year)	60	40	150	100
Corresponding EF (mg/sec)	40	7	10	2
Annual Emissions by mode (kg)	86.4	10.1	54	7.2

term, or VSP. This parameter is highly correlated with emissions, and appears to form the basis for the development of the next generation emission models at the U.S. Environmental Protection Agency.[6,7] Instantaneous, second-by-second VSP values for a generic light duty vehicle is expressed as follows:

$$\text{VSP} = v \times [1.1\ a + 9.81 \times (\sin(a\tan(\varphi))) + 0.132] + 0.000302 v^3 \quad (1)$$

where

VSP = vehicle specific power (kW/ton)
v = instantaneous vehicle speed (in m/s)
a = instantaneous acceleration or deceleration rate (in m/s^2)
$a\tan$ = inverse tangent function, equivalent to $\tan^{-1}(\varphi)$
φ = road grade (dimensionless fraction)

It has been shown by Frey et al.[7] that it is possible to discretize the VSP values computed from equation 1 into a limited number of "bins," each having a unique average emission factor, which is also significantly different from all other bins. The bin assignments for a light duty vehicle are depicted in Table 3.

An example of CO emission factors for a generic light-duty vehicle associated with the VSP bins defined in Table 3 are shown in Figure 2. Note that modes 1 and 2 represent deceleration models (VSP < 0), while mode 3 includes the idle mode (VSP = 0). All the remaining modes represent combinations of speed and (positive) accelerations.

One obvious difficulty in applying a modal activity model (whether the four-mode or VSP) to assess the impact of congestion management strategies on emissions is the requirement for high-resolution vehicle travel data (e.g., second-by-second), which are not typically available to transportation engineers and transit planners. On the rare occasions where a passenger car or transit vehicle fleet is equipped with an automatic vehicle location (AVL) system, speed profiles can be captured and geo-referenced to a point on the roadway, which enables the implementation of modal or VSP models to estimate instantaneous emissions in real time. In practice, however, an instrumented roadway infrastructure can

Table 3 VSP Modal Definitions

VSP Mode	Bin Range	VSP Mode	Bin Range
1	VSP < −2	2	−2 ≤ VSP < 0
3	0 ≤ VSP < 1	4	1 ≤ VSP < 4
5	4 ≤ VSP < 7	6	7 ≤ VSP < 10
7	10 ≤ VSP < 13	8	13 ≤ VSP < 16
9	16 ≤ VSP < 19	10	19 ≤ VSP < 23
11	23 ≤ VSP < 28	12	28 ≤ VSP < 33
13	33 ≤ VSP < 39	14	VSP ≥ 39

Note: From Ref. 7.

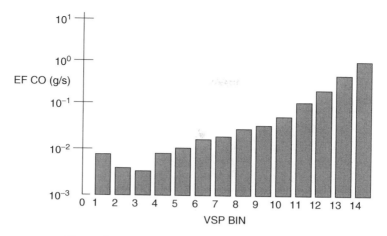

Figure 2 CO emission factors versus VSP bins.

at most produce measures of traffic flow and average speeds on those links that happen to be equipped with vehicle detectors at discrete points on the road. In this case, a hybrid approach that relates VSP to average speed on a roadway link offers a good compromise.

4.5 Hybrid VSP-S

This approach is derived from observational studies that are able to relate microscale events that occur on a traffic link, such as second-by-second VSP, to aggregate link performance measures, namely average traffic speed. It is based on taking concurrent measures of both parameters using portable emission measurement systems or PEMS.[8] The concept is straightforward. For each range of average link speed, there appears to be a corresponding, and relatively stable, distribution of VSP values that can then be used to estimate emission factors.[9] An example of this approach is shown in Figure 3. The top figure shows multiple speed profiles observed on a single roadway link. These profiles have one thing in common: the *average link speed* from all profiles varies from 8 to 12 mph. The bottom figure shows the corresponding travel time distribution of VSPs associated with that speed range. Remarkably, these distributions appear to be fairly uniform in shape, with a mode around VSP bin #3 (not surprising, given the low average speed on the link of 10 mph). The hybrid approach thus entails several steps, namely average link speed prediction, selection of the corresponding distribution of VSP, and application of the appropriate emission factor for each VSP bin (from Figure 2), to the link travel time. Mathematically, the emission estimation method is expressed as follows:

$$E_j = \sum_{t=1}^{T_j} a_{itj} \times EF_i, i = 1, \ldots, 14 \text{ and } j = 1, 2, \ldots, J \qquad (2)$$

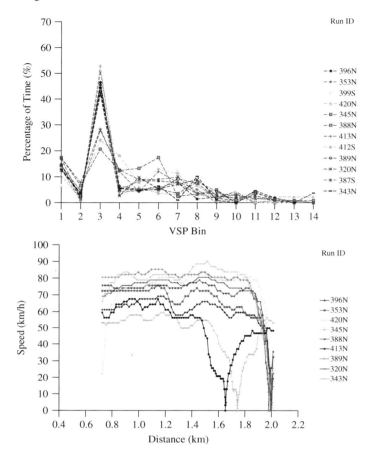

Figure 3 Speed profiles and VSP distribution (speed range 8–12 mph).

where

E_j = emissions *per vehicle* traveling on link (j), in grams/vehicle
a_{itj} = 1 if VSP mode (i) occurs in time step (t) on link (j), zero otherwise
EF_i = emission factor associated with VSP mode (i) (e.g., Figure 2 for CO)
T_j = Average travel time on link j (computed as link (j) length over average link travel speed)

By multiplying the per-vehicle emission rate with the link traffic flow (typically on an hourly basis), the above equation produced the overall link emission over the hour for the subject link. Aggregating over all the network links will generate networkwide emissions. Further disaggregation of eq. 2 by vehicle type can be carried out to distinguish between emission factors for various vehicle classes (cars, buses, etc.).

5 INVENTORY OF TRAFFIC CONGESTION MANAGEMENT METHODS

5.1 Travel Demand Management (TDM)

Travel demand management involves a family of strategies that are aimed at altering the demand pattern of travelers' activities. This requires the reader to distinguish between travel demand that is expressed in *person-miles* of travel, and one that is expressed in *motorized vehicle-miles* of travel. An implicit assumption in the following discussion is that the demand for person-miles of travel remains unchanged in order to satisfy the individual's objectives to have access to employment and to exercise other desirable economic and social activities. Demand management therefore focuses on altering the motorized vehicle-miles of travel and associated emissions, in order to enable the accomplishment of one or more of the following objectives:

- An overall reduction in demand for vehicular travel activities (VMT)
- A spatial shift in the current vehicular travel demand to less congested facilities
- A temporal shift in the current vehicular travel demand to less congested periods of the day
- A modal shift in the current vehicular travel demand to less-polluting motorized or nonmotorized alternatives

A sampling of some commonly used TDM strategies is described next.[10] It should be noted that several of the strategies can affect more than one of the stated objectives.

- *Congestion and parking pricing.* These strategies are aimed at altering traveler behavior by increasing the out-of-pocket cost for users of the transportation system during peak demand periods. Some of the common technologies include toll roads, high occupancy toll or HOT lanes (these are physically separated lanes that provide reduced cost of access to vehicles meeting a prespecified passenger occupancy and may provide more costly access to other vehicles not meeting the occupancy threshold), cordon pricing as exemplified by the Central London congestion pricing scheme,[11] peak-period parking costs, or restrictions are the most common. Their intended primary effect is to produce a temporal reduction in peak period demand (Objective 3) for the high pollutant travel modes. When combined with preferential pricing policies toward the use of public transportation, these strategies could also affect a modal shift (Objective 4).
- *Flexible work hours.* These employer-based programs are aimed at reducing peak travel (Objective 3) by allowing employees some flexibility in the start and end hours of their workday, and in some cases compress the weekly work schedule into four working days.

- *Park-and-ride facilities.* Park-and-ride facilities are targeted towards inducing a modal shift (Objective 4) from private to public transportation modes (mostly bus, light rail, or heavy rail modes). In addition to providing direct access to those modes at the trip origin point, free or low-cost parking is often provided for commuters at those facilities.
- *High occupancy vehicle (HOV) facilities.* HOV facilities apply to both carpool and (normally) bus transit vehicles. The facilities are exclusive to the users of these modes and, therefore, offer an improved level of service as an incentive for the modal shift (Objective 4). It is important to note that providing easy access to and egress from those facilities is critical; otherwise, their use becomes of limited value to the traveler. This could be in the form of separate entry and exit ramps on freeways, or by providing preferential access at the entry ramps to HOV facilities on freeways.
- *Ridesharing and vanpool programs.* These programs are aimed at producing a direct reduction in VMT (Objective 1) by promoting and facilitating ride sharing and (normally employer-based) vanpool programs. These may include incentives such as free parking, free fuel, and the use of HOV facilities among others. Membership in such programs would be on a subscription basis, although recent advances in computer-aided dispatching could enable real-time ride matching and vanpooling.
- *Nonmotorized alternatives.* These programs promote nonpolluting walking and bicycling modes (Objective 4) through a variety of infrastructure improvements. Examples include dedicated bicycle lanes, wide sidewalks, grade-separated pedestrian crossings at busy intersections, and a variety of priority treatments at signalized intersections. To accommodate longer trips the provision of bicycle storage space on transit buses and rail cars, and exclusive bicycle park-and-ride facilities at train stations are often considered.
- *Transit alternatives.* These programs are aimed primarily at shifting single occupancy vehicle (SOV) drivers to public transportation modes, which are highly effective in reducing pollutants per person-mile of travel (Objective 4). Improvements are generally implemented in four areas: (1) service, including added transit lines, shorter headways and higher reliability; (2) amenities at bus and transit stops, including covered and climatized shelters; (3) pricing, including reduced or fare-free rides for the general or special populations; and (4) travel information systems disseminated through a variety of media, including the Internet, 511 call lines, cable television, and strategically located traveler information kiosks.
- *Land-use strategies.* These programs represent a significant departure from the predominant suburban land-use development patterns with their reliance

on the auto mode for virtually all trip purposes, to one that is transit, pedestrian, and bicycle friendly. Known also as *smart growth*, or *neotraditional* patterns, these strategies promote mixed-use development that can negate or a least curtail the need for auto travel. By definition, such strategies are not expected to have a short- or even medium-term impact on trip making behavior, but in conjunction with improved vehicle emission technology, have the most significant potential for reducing emissions on a large scale through major reductions in auto VMT (Objective 1). The work by Johnston et al. and Rodier et al. quantifies the effect of land-use changes on vehicle emissions.[12,13]

- *Highway information services.* These strategies fall under the umbrella of *advanced traveler information systems (ATIS)*. Their impact is ubiquitous in that they can affect the spatial, temporal, modal, and number of trips that are taken by the traveler (Objectives 1 to 4). Yet, the effect of ATIS on VMT of travel, much less emissions and air quality, remains somewhat elusive, due to the difficulty of acquiring the appropriate empirical data. Travel information can be imparted pre-trip (at home), which could alter the traveler's departure time, route, and mode taken or forgo the trip altogether; it could be given en route via changeable message signs (CMS) or from in-vehicle navigation and routing systems, which will tend to alter the spatial pattern of the trips.

Table 4 provides a summary of the general impacts of the stated travel demand management strategies on VMT-based activities.

Table 4 Mapping TDM Strategies to VMT Changes and Time Horizon

TDM Strategy	Primary Objective 1 = Overall VMT reduction; 2 = spatial shift; 3 = temporal shift; 4 = modal shift				Implementation Time Horizon 1 = short; 2 = med; 3 = long;
	1	2	3	4	
Congestion and parking pricing	✓				2
Flexible work hours			✓		1
Park-and-ride facilities				✓	2
High occupancy vehicle facilities				✓	1
Ride sharing and vanpool programs				✓	1
Nonmotorized alternatives				✓	2
Transit alternatives				✓	2
Land-use strategies	✓				3
Highway information services	✓	✓	✓	✓	1

5.2 Traffic Operational Strategies (TOS)

A common characteristic across TDM strategies is that they all tend to target trip-making behavior *before* the trip is actually made. Traffic operational strategies, by contrast, are geared toward reducing emissions under prevailing traffic conditions. In most cases, congestion mitigation measures will also have a positive impact on emission reductions. This section, therefore, focuses on strategies that have an effect on the individual and traffic speed profiles of vehicles as they travel through the network. It has been established in section 3 that significant speed changes over a trip will result in increased emissions. A simple example by Unal illustrates this point vividly.[14] Figure 4 shows speed profiles for two trips that have identical average travel speeds (19 mph), using the same vehicle on the same route. Trip 2, however, appears to have much larger speed variance than trip 1. Table 5 depicts the pollutant emissions measured for both trips using a portable emission measurement system (PEMS). It is evident that a speed profile that is not *smooth* will generate much higher emissions. In the case presented, CO emissions for trip 2 are three times as high as for trip 1; NO emissions for trip 2 are close to five times as high as those for trip 1. The same pattern applies to the other pollutant, although less dramatically. In the following presentation, while the focus may be on solutions that increase travel speed, it is critical to note that speed variance is the stronger indicator of emission impacts.

Traffic operational strategies are typically identified by the facility type where they are applied, namely freeways or surface streets (which includes arterials, collectors, and local roads). An important consideration of the effect of TOS is the possibility of induced traffic demand as an outcome of congestion mitigation. In other words, while TOS are intended to reduce the unit emissions by smoothing traffic flow, as evident from the preceding charts, they could have the unintended

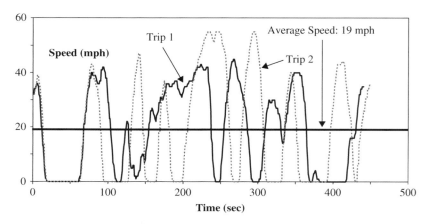

Figure 4 Speed profiles for two trips at an average speed of 19 mph.

Table 5 Measured Trip Emission Rates for Trips Shown in Figure 4

Trip	Pollutant Emission Rate			
	CO (mg/sec)	NO (mg/sec)	HC (mg/sec)	CO_2 (g/sec)
1	6.1	0.41	0.91	2.3
2	19	1.9	1.2	3.1

consequence of increasing the vehicle miles of travel, which could partially offset their benefits. Unfortunately, there are no standard methods to estimate the elasticity (or sensitivity) of travel demand to short TOS-type improvements, but the engineer should, nevertheless, be cognizant of such unintended consequences. Two comprehensive sources of information on the emission effects of traffic flow improvements include the USDOT Intelligent Transportation Systems Benefits Page and a National Cooperative Highway Research Project report by Dowling et al.[15,16]

Freeway Operational Strategies

- *Ramp metering.* Ramp metering has the beneficial effect of reducing traffic turbulence in the ramp entrance zone, particularly when the freeway mainline is operating close to capacity. Evidence shows that with ramp metering, average speeds increase, and delays are reduced when proper metering rates are applied.[17] Metering can be implemented in an isolated fashion at selected ramps, or via a systemwide coordinated metering plan that covers entire corridors. Possible drawbacks of metering include the possibility of long queues at ramps that may spill back onto the surface street, and inducing large acceleration rates for entering traffic(see Figure 1).

- *Incident management.* Incident management systems mitigate the effects of nonrecurring congestion due to accidents, stalled vehicles, unanticipated road work, and so on. Evidence show that between 52 to 58 percent of all congestion delay may be due to nonrecurring events, or incidents.[1] By reducing the response and clearance times for such incidents, delays and travel times can be markedly reduced. There are also safety benefits that can accrue from a reduction in secondary (mostly rear-end) accidents that occur as a result of the large speed differentials between approaching traffic and traffic within the vicinity of the incident.

- *Bottleneck mitigation.* Freeway bottlenecks are leading causes of recurring traffic congestion on urban freeways, resulting in local stop-and-go operations in their vicinity. They are typically activated when traffic demand surges over capacity downstream of a congested on-ramp or when geometric restrictions reduce the normal freeway capacity, as in the case of a lane drop or a weaving section. Besides major (and costly) capacity additions, there are low-cost solutions that may mitigate their effect, including short auxiliary lanes, the use of hard shoulders during the peak hours, and implementation of truck restrictions. A unique strategy that enables the use of a wide shoulder as a narrow lane for queue storage during peak flow periods only (called the plus-lane) has been applied in Utrecht, The Netherlands, and is illustrated in Figure 5.[18]

- *Lane control.* Lane control strategies are more commonly used in European cities to control speed and restrict lane use to specific time periods or vehicle classes. From an emissions control perspective, lane control promotes gradual speed reduction ahead of a congested region through the use of a series of advisory or regulatory speed limit signals erected on overhead gantries. This control will tend to smooth the traffic speed profile and minimize high acceleration and deceleration events. An example application of speed control ahead of a lane drop in Stockholm, Sweden, is illustrated in Figure 6.

In addition, many urban areas in the United States use a combination of reversible and HOV lanes to provide added capacity in the peak flow direction to meet the prevailing directional demand. More recently, entire

Figure 5 Bottleneck mitigation using plus-lane concept.

Figure 6 Lane and speed control system. Courtesy N. Rouphail (2001).

directional freeway facilities have operated in the reverse direction to evacuate traffic from hurricane-threatened coastal areas.[19]

- *High-speed tolling.* On toll facilities operating with manual toll collection, approaching vehicles will experience significant speed change cycles both upstream (deceleration) and downstream (acceleration) of the toll booth. The zone just downstream of the tool plaza, where large accelerations are prevalent, is considered to be a high CO emission hot spot.[20] The use of electronic, high-speed tolling negates the requirement for a vehicle to come to a full stop and then accelerate to freeway speeds, and, therefore, reduces the number and severity of speed change cycles in the vicinity of toll facilities. Such systems use electronic tag readers on the moving vehicle to record its arrival and to charge the user's account.

Surface Street Operational Strategies

- *Signal timing and coordination.* Improved traffic signal timing and coordination is considered to be one of the most cost-effective measures for congestion mitigation and emission reductions on surface streets. According to the Institute for Transportation Engineers, there are over 300,000 traffic signals in the United States, with the majority needing upgrading of equipment or adjusting of the timing or coordination, while servicing nearly two-thirds of the vehicle-miles traveled.[21] Comprehensive signal retiming programs have documented benefits of a 7 to 13 percent reduction in overall travel time, a 15 to 37 percent reduction in delays, and a 6 to 9 percent fuel savings. Signal coordination, in particular, has a significant impact on reducing vehicular stop-and-go cycles. Section 6.1 discusses the results of a study on the effect of improved timing and coordination on real-world measured vehicle emissions.

- *Tidal or reversible flow lanes.* Similar to reversible lanes on freeways, tidal flow lanes on arterials are used to provide flexible capacity in one direction to accommodate peak flow traffic. These are also used to cater for special event traffic that requires the provision of capacity in different directions before, during, and after the event (sporting events, concerts, major evacuation, etc.).

- *Roundabouts.* This form of intersection traffic control is becoming very popular in the United States, with over 700 modern roundabouts installed.[22] From a vehicle emission perspective, roundabouts, especially those that are not highly congested, promote a smoother speed profile in the vicinity of the intersection, by not requiring traffic to stop at the circle, and by reducing overall speeds on the approach and in the circle. Evidence, both from empirical and modeling studies, indicates positive environmental effects. As an example, a recent study in Sweden shows that at a small roundabout that replaced a signalized intersection, CO emissions decreased by 29 percent, NO_x emissions by 21 percent, and fuel consumption by 28 percent.[23] At roundabouts, replacing yield-controlled intersections, CO emissions increased on average by 4 percent, NO_x emissions by 6 percent and fuel consumption by 3 percent. An area that requires further investigation is the effect of exiting traffic on emissions, where drivers tend to accelerate from the circulating lane design speed to that of the open road ahead.

- *Continuous flow intersections.* Continuous flow intersections (CFIs) represent an innovative intersection design scheme that physically separates and services left-turn movements upstream of the main intersection, which are then rerouted to a lane to the left of the opposing flow. This design enables the left turns to merge without stopping into the cross-street traffic. According to recent studies, there are only two such installations in the United States, in New York and Maryland, with the majority of the installations in Mexico. Similar to other TOS strategies, CFIs have been shown to reduce delays and speed change cycles at intersections and, thus, would result in lower fuel and emission use. An illustration of an operational CFI in Mexico City is shown in Figure 7.

- *Bus signal priority.* This strategy entails the provision of traffic signal control priority for buses on arterials. This can be accomplished in several ways, depending on the signal indication when the bus arrives at the intersection: an early green start, an extension of the current green

Figure 7 Continuous flow intersection operation. Courtesy Francisco Miero.

(if needed), or an off-sequence special green phase. The purpose of these strategies is to reduce transit travel time, improve its reliability and, in the long term, help increase transit ridership (essentially a TDM strategy). The emissions consequences of implementing bus priority schemes are unclear and very much depend on the site conditions. Areas of concerns are the effect of the bus signal priority or preemption on disrupting the existing signal coordination plan, and on side street traffic that may be penalized as a result of the additional main street green times. Advanced algorithms that account for schedule delay (using automated vehicle location systems) and real-time passenger counts in deciding whether preemption is desirable may alleviate some of these concerns.

A comparison of TDM and TOS strategies indicates that the first has a diverse set of objectives, ranging from reducing travel demand, to shifting it in time, in space, or to other environmentally friendly modes. Traffic operational strategies, by contrast, tend to hone on the single objective of smoothing traffic flow by reducing stop-and-go cycles, maintain near-uniform speed profiles, and minimizing the episodes of high emissions associated with large acceleration and deceleration events.

6 ASSESSING EMISSION IMPACTS OF TRAFFIC CONGESTION MANAGEMENT

There are numerous methods available to transportation and environmental analysts to evaluate the effectiveness of traffic congestion management methods. These can be classified into two general categories: on-road direct measurements and emission estimation models. The latter can be further subdivided into emission factors and traffic models. Some methods combine elements of both types of models.

6.1 Direct Measurement Methods

This section covers two direct measurement methods for vehicle emissions.

Portable Emission Measurement Systems (PEMS)

Vehicle emissions can be directly measured under various traffic conditions using PEMS.[8] This method is rapidly gaining wide acceptance by the user community and regulatory agencies such as the U.S. EPA. These devices are installed in a vehicle in a matter of minutes and enable the analyst to measure tailpipe emissions of several pollutants, including CO, NO, HC, CO_2, and O_2, in addition to fuel consumption, on a second-by-second basis under real-world traffic conditions. PEMS is connected and synchronized with the vehicle engine data scanner, which includes, among other variables, a second-by-second speed profile. It draws AC power from the cigarette lighter outlet. Figure 8 depicts a PEMS device, the engine data scanner, its positioning in the vehicle, and the connection to the tailpipe sampling probe.

An example assessment study using PEMS was reported by Unal et al.[24] In this study, the effectiveness of a traffic signal timing and coordination scheme for an arterial corridor was carried out. The study involved multiple runs of two PEMS-instrumented vehicles using the same drivers and data collection periods before and after the signal system was coordinated. Figure 9 shows speed and CO traces for a representative single run in the before and after case.

It is evident that this strategy was effective in reducing the corridor travel time, as well as smoothing the speed profile by reducing the number of stops and CO emissions. Summary statistics of traffic and emission changes after the signals were retimed are reported in Table 6 for both directions of traffic, two peak periods, and two vehicles. The results confirm the cost-effectiveness of signal control strategies as an emission mitigation measure.

The same referenced study also compared emissions for uncongested and congested traffic conditions on another corridor, by contrasting trip emissions in the peak and off-peak directions.[24] The results were even more dramatic, with a decrease in trip emissions in the range of 40 to 60 percent, depending on the time period and the pollutant considered.

Remote Sensing Device (RSD)

Remote sensing technology utilizes infrared spectroscopy to flag high-emitting vehicles as they pass a point on the highway. It is motivated by statistics that show that only about 10 percent of the on-road vehicles are responsible for over 50 percent of the CO, NO_x, and HC emissions. The device consists of

(a)

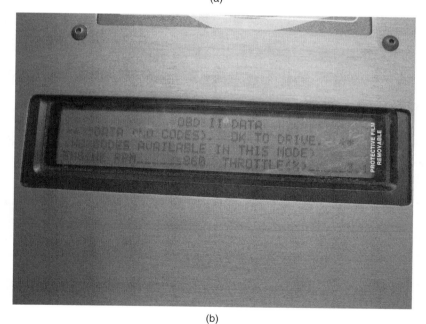

(b)

Figure 8 Example of a PEMS and its installation in a vehicle.

(c)

(d)

Figure 8 (*continued*)

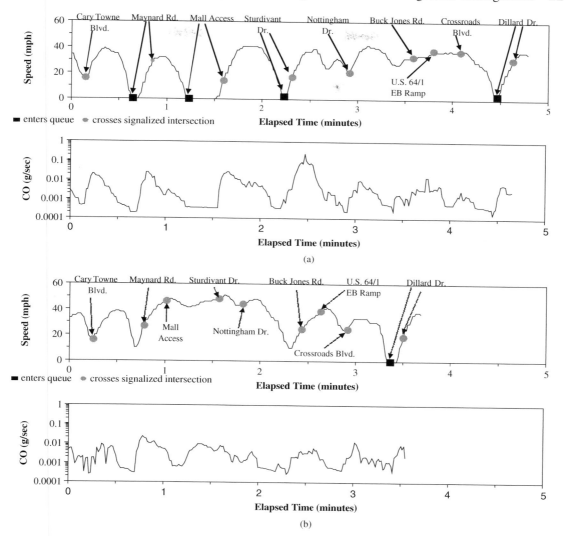

Figure 9 Representative speed profiles before and after signal coordination. (Source: From Ref. 24.)

an infrared source that emits a light beam across the highway just before and during the vehicle crossing the measurement point, as shown in the schematic Figure 10.[25] The system determines the ratio of CO/CO_2 in the vehicle plume, and if the level of carbon monoxide (CO) is above a certain threshold, it is tagged as a high-emitting vehicle, typically using video detection technology. The device could also be used to assess the emission consequences of traffic control strategies. However, it has many limitations with regards to lane (one lane at a time) and spatial coverage (single point on the road), and as such could miss many of the high-acceleration events that are concomitant with episodes of high

Table 6 Summary Traffic Performance and Emission Changes after Implementing a Signal System Coordination Project (Using PEMS)

	Ford Taurus				Oldsmobile Cutlass			
Time Period	Morning		Afternoon		Morning		Afternoon	
Direction	N	S	N	S	N	S	N	S
Trip duration (%)	−14	−24	−23	+0.3	−16	−17	−21	−0.9
Ave. speed (%)	+14	+32	+29	−1.8	+18	+20	+29	−1.8
Control delay (%)	−40	−63	−55	−4.6	−38	−50	−56	+8.5
Total stops (%)	−30	−60	−29	−2.3	−29	−46	−29	−11
HC emissions (%)	−12	−18	−11	+1	−12	−13	−12	−1
NO emissions (%)	−8	−12	−1	+1	−13	−14	−19	−1
CO emissions (%)	−12	−19	−5	+1	−4	−9	−1	−1

Note: From Ref. 25.

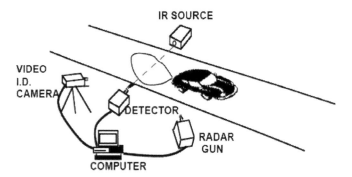

Figure 10 Schematic of operation of an emission RSD.

emissions. Unal et al. reported on the concurrent use of RSD and areawide vehicle detection for the simultaneous analysis of traffic performance and emissions.[26]

6.2 Mobile Emission Estimation Models

This section covers three models for estimating mobile emissions.

Mobile

By far, the most widely used family of mobile emission factor models is EPA's MOBILE series, of which MOBILE6.2 is the latest release.[27] These models represent the state of the practice for developing emission inventories and control strategies at the regional or state levels, as part of the required state implementation plans (SIP) under the 1990 Federal Clean Air Act. MOBILE emission factor values represent *national average* conditions, both in terms of vehicle fleet (classified into 28 vehicle categories), VMT distribution among roadway types, standardized driving cycles on these roadways, and speed distributions among roadway types. These averages can be modified to fit local conditions through a series of adjustments to the base emission factors on the basis of modified user input to better reflect local conditions. From a congestion management perspective, MOBILE inputs that are relevant include the types of roadways (freeways, ramps, arterials, and local streets), the VMT distribution across these facilities (national defaults are 34%, 3%, 50%, and 13% for the four facilities), and the average speed on those facilities (national default *daily* speeds are 36.5, 34.6, 31.2, and 12.9 mph, respectively). The pollutants of interest include HC, CO, NO_x, and CO_2. MOBILE can represent any vehicle fleet from 1952 through 2050, making assumptions on the distribution of vehicle age in a given calendar year (CY). Sample results of the sensitivity of MOBILE6 freeway and arterial emission factors to average speed are shown in Figure 11.

The data shown are for the projected fleet for CY2008, and include light-duty-vehicles *only*. Several patterns emerge from those data. First, MOBILE emission factors for freeways and arterials are not significantly different from each other when controlling for the pollutant type. This may be a result that the national default speed distributions between these two facilities are very similar as indicated earlier. Second, CO emissions factors exhibit a nonlinear relationship

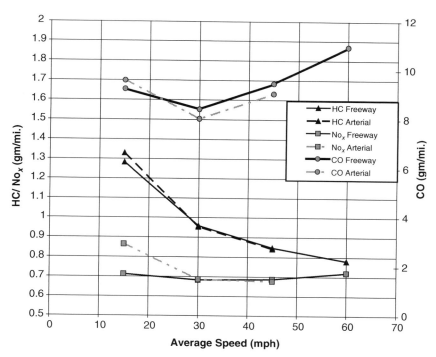

Figure 11 MOBILE6 emission factors' sensitivity to average speed on freeway and major arterials; in grams/mile of travel.

with speed. They tend to be lowest in a speed range of 25 to 35 mph, and increase below or above that range. However, HC and NO_x emissions factors decrease monotonically with speed for the low average speed range (15–30 mph) and then remain essentially unchanged for NO_x while HC emission rates continue to drop as speed increases. This indicates that stop-and-go traffic patterns are likely to be associated with high emissions of these two pollutants.

An important caution in the application of MOBILE6 factors for the assessment of traffic control measures is the need to acknowledge the sensitivity of travel demand (i.e., VMT) to level of service (LOS). The user must account for the possibility that improved traffic performance could generate additional (new) trips that could partially offset the benefits gained from reducing the emissions factors.

There are many more applications of MOBILE than can be covered in a comprehensive manner this chapter. The reader is strongly encouraged to consult the MOBILE6 user guide and associated documentation on the EPA Office of Transportation and Air Quality Web site (OTAQ).[28] It is noteworthy to mention that EPA is currently developing a new emission estimation framework called MOVES (**MO**tor **V**ehicles **E**mission **S**imulator) that will rely on more refined data, based partly on the VSP approach explained earlier. MOVES will eventually replace MOBILE6. The current release MOVES2004 provides estimates for fuel consumption and greenhouse gas emissions for on-road vehicles only.[29] Future releases will include the estimation of HC, CO, and NO_x, and for on-road vehicles.[6]

MODAL, VSP, and VSP-S Emission Factor Models

These models have been discussed earlier in the chapter (section 4) and are mostly derived from microscale speed profile data gathered in the lab using engine dynamometers, or in the field using PEMS.

COMMUTER2.0 Model

This model, developed for EPA, estimates the travel and emission impacts of many of the transportation demand management strategies that were described in Section 5.1.[30] It is designed to focus on small programs and transportation control measures that do not have large regional travel impacts, in which case traditional regional travel demand models should be used. In COMMUTER2.0 travel impacts are modeled using a logit model formulation that produces modal and temporal shifts based on the value of a utility function of modal travel times, transfer times, and costs. By specifying a baseline scenario of mode choice, trip lengths, and current VMT, COMMUTER2.0 will estimate *changes* in VMT's and associated emissions, the latter mostly derived from the MOBILE6 model. Figure 12 below shows a sample scenario output of this model, based on the input and traffic measures provided by the users. Additional details can be found in the model user guide.

6 Assessing Emission Impacts of Traffic Congestion Management

COMMUTER MODEL RESULTS

SCENARIO INFORMATION

Description	Example Scenario v2.0
Scenario Filename	Example Scenario v20.vme
Emission Factor File	
Performing Agency	Cambridge Systematics, Inc.
Analyst	C. Porter
Metropolitan Area	Boston
Area Size	1- Large (over 2 million)
Analysis Scope	1 -Area-Wide (e.g., MSA, county)
Analysis Area/Site	Cambridge
Total Employment	100,000

PROGRAMS EVALUATED

☐	Site Walk Access Improvements
☐	Transit Service Improvements
☒	Financial Incentives
☐	Employer Support Programs
☒	Alternative Work Schedules

☐ User-Supplied Final Mode Shares

MODE SHARE IMPACTS

Mode	Baseline	Final	%Change
Drive Alone	78.2%	75.4%	−2.9%
Carpool	12.1%	11.7%	−0.4%
Vanpool	0.5%	0.5%	−0.0%
Transit	4.9%	6.3%	+1.4%
Bicycle	0.4%	0.4%	−0.0%
Pedestrian	3.0%	3.0%	−0.0%
Other	0.8%	0.8%	−0.0%
No Trip	-	2.0%	+2.0%
Total	100.0%	100.0%	-

Shifted from Peak to Off-Peak	0.0%

TRAVEL IMPACTS (relative to affected employment)

Quantity	Peak	Off-Peak	Total
Baseline VMT	1,289,745	810,817	2,100,561
Final VMT	1,243,026	781,447	2,024,473
VMT Reduction	46,718	29,370	76,089
% VMT Reduction	3.6%	3.6%	3.6%
Baseline Trips	102,780	64,614	167,394
Final Trips	98,987	62,230	161,217
Trip Reduction	3,793	2,384	6,177
%Trip Reduction	3.7%	3.7%	3.7%

EMISSION REDUCTIONS (positive values are decreases)

lbs/day:

Pollutant	Peak	Off-Peak	Total
HC	21.60	12.94	34.53
CO	337.14	217.22	554.36
NO_x	22.20	13.92	36.12
PM2.5	1.16	0.73	1.90
Toxics			
Acetaldehyde	0.056	0.034	0.091
Acrolein	0.007	0.005	0.012
Benzene	0.605	0.369	0.974
1, 3-Butadiene	0.081	0.050	0.131
Formaldehyde	0.185	0.113	0.299
MTBE	1.195	0.713	1.908
CO_2	47,257	29,709	76,966

tons/day:

Pollutant	Peak	Off-Peak	Total
HC	0.011	0.006	0.017
CO	0.169	0.109	0.277
NO_x	0.011	0.007	0.018
CO_2 (metric tons)	21.4	13.5	34.9

GASOLINE CONSUMPTION AND COST SAVINGS

Reduction in gasoline consumption (gallons/day)	3,934
Gasoline cost savings ($/day)	$8,852

Figure 12 Example of travel and emission impacts derived from a COMMUTER2.0 model run.

6.3 Transportation Models

This section discusses the use of transportation models to assess the emission impacts of transportation control measures. For the most part, these models focus on estimating the travel or traffic impacts of regional and local transportation improvements, and, as such, the model selection depends on the nature and geographic impacts of the application. Models can be classified as travel demand models, macroscopic traffic operations models, and microscopic simulation models.

Travel Demand Models

This class of models is typically used for estimating long-term transportation system needs (15 to 20 years). Their focus is on predicting required link-level characteristics (type, capacity, etc.) that are needed to accommodate future traffic demands. Typically (with some exceptions), this class of models does have the capability of predicting emissions. Instead, a post-processing step is executed to use the travel model output (e.g., VMT by roadway class, vehicle type, and speed) as input into MOBILE6. Because MOBILE represents national average emission factors, it is important for the user to *not* rely on or use link-level emission estimates, since these are highly simplified. The intent is to obtain a fairly coarse estimate of the emission inventory on a regional scale. Commonly used travel demand models include TRANPLAN, EMME2, TransCad, and TRANUS.[31–34]

Macroscopic Traffic Operations Models

This class of models considers the effect of local or corridor-wide improvements (such as ramp metering, signalization, HOV lanes, etc.) on traffic performance with the implicit assumption that vehicle activities (i.e., VMT) remain fixed before and after the improvements. Many of these models have also incorporated emission prediction algorithms within the models, although it is often unclear what the assumed vehicle fleet is, and how the emission rates were calibrated. Some models, such as Synchro use a fuel-based approach that relates HC, CO, and NO emissions to fuel consumption (per hour or per mile).[35] Others, most notably aaSIDRA and SIGNAL200, incorporate four-mode (acceleration, deceleration, cruise, and idle) and three-mode (acceleration, idle, deceleration) modal emission models in their software, respectively.[36,37] The U.S. Highway Capacity Manual, the most widely used macroscopic traffic operational model in the United States (and possibly worldwide) has no provisions for emission estimation.[38] This has led users of the HCM to rely on other macroscopic or microscopic models to meet those needs.

Microscopic Traffic Simulation Models

This class of models traces individual vehicle movements over a traffic network, with vehicle position and speeds updated at one second or subsecond resolutions. The advance logic is based on driver behavioral models, such as car following, lane changing, and gap acceptance. Driver level of aggressiveness, which could have a significant impact on acceleration rates and emissions, are also considered, although rarely calibrated from empirical observations. Because these models can produce individual speed profiles their results can be easily integrated with microscale emission rate models such the MODE or VSP models described earlier to produce second-by-second emissions. Often, this integration is built-in the simulation, as in the CORSIM model which uses a speed-acceleration look-up table (by vehicle type) to calculate emissions based on unpublished dynamometer testing at the Oak Ridge National Laboratory.[39] A similar approach is adopted in the INTEGRATION model, the VISSIM model, and the PARAMICS model.[40–42]

It is often advisable to decouple the simulated speed profiles from the default emission estimation methods embedded in the simulators. This is because the emissions calibration data could be either out of date or not representative of the U.S. vehicle fleet. An example of representative speed profiles in the VISSIM model based on a study that compared traffic operations on an urban arterial

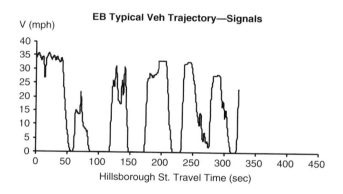

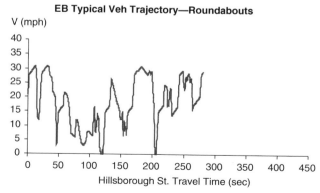

Figure 13 Representative VISSIM-generated speed profiles. (Source: From Ref. 18.)

Table 7 Unit Emissions for Traffic Signal and Roundabout Control

Direction and Type of Control		NO (gm/veh)	HC (gm/veh)	CO_2 (gm/veh)	CO (gm/veh)
EB	Signal	0.65	0.29	733.06	6.86
	Roundabout	0.59	0.23	693.38	4.71
		(−9.4%)	(−19.5%)	(−5.4%)	(−31.3%)
WB	Signal	0.63	0.24	630.22	7.65
	Roundabout	0.61	0.28	816.45	4.77
		(−2.3%)	(+15.4%)	(+29.6%)	(−37.7%)
Overall	Signal	1.28	0.53	1363.29	14.51
	Roundabout	1.20	0.51	1509.84	9.48
		(−.9%)	(−3.5%)	(10.7%)	(−34.7%)

Note: From Ref. 18.

using a series of signals (baseline) and roundabouts (proposed) is shown in Figure 13.[18] These vehicles were selected as their travel time coincided with the average travel time. While both profiles show considerable speed variations, there were more stops in the signalized arterial case. By applying the VSP approach described in section 4, VSP values, VSP bins, and associated emission rates were estimated. Table 7 summarizes the emission effects of replacing the signalized intersection system with a series of roundabouts. Multiplying these values by the corresponding VMT in each case gives the overall emission impacts.

7 SUMMARY

This chapter presents a survey of the impacts of traffic congestion management strategies on vehicular emissions and, by extension, on air quality. It begins with the premise that mobile source emissions are a significant contributor to the national emission inventory for some pollutants and that congestion mitigation strategies that focus on both transportation demand and supply sides can significantly reduce the harmful effects of airborne pollutants. The chapter also discusses tools that are available to the engineer and analyst to measure, estimate, and predict emissions in response to changes in supply or demand-oriented transportation strategies. Although the literature contains a wealth of information on such effects, there is no general consensus in the transportation and environmental community on which tools are appropriate for what types of analysis, especially those at the corridor and local intersection levels. The author hopes that the material presented herein, although by necessity much abbreviated, will be of assistance to the users in identifying and evaluating the available tools and in understanding the tools' strengths and limitations. Regardless of the method selected, the analyst must be cognizant of and account for the *long-term effects* of traffic control measures on overall emissions, particularly the impact of improved traffic flow conditions on latent demand and modal choice effects. Emerging EPA emission estimation paradigms exemplified by the MOVES model will in the future enable the analyst to carry out multi-scale level analyses that are sensitive to regional as well as local transportation improvements. Finally, whenever possible it is critical that local travel behavior and emission rates be used, especially if the analysis is in response to regulatory requirements such as state implementation plans (SIPs) mandated under the 1990 Clean Air Act.

ACKNOWLEDGMENTS

The author is grateful to many colleagues and graduate students whose work has been cited in this document. In particular, I am indebted to my colleague Chris Frey, professor of civil and environmental engineering at NC State University, who first raised my interest in the topic of transportation and air quality and whose continuous contributions in this field have been noted nationally and internationally. I would also like to thank Dr. Billy Williams, assistant professor of civil engineering for his insightful comments and suggestions on a previous draft, and Ms. Katie McDermott, communications manager at ITRE for being exceedingly thorough as my first-round editor. Many thanks to several former and current NC State students whose research was cited in this document: Alper Unal, Kosok Chae, James Colyar, and Haibo Zhai. Last but not least, I am grateful to my wife, Dr. Maria Rouphail, who persevered while this manuscript was being written, for her incisive editing and advice, without which this chapter could not have been completed.

REFERENCES

1. D. Schrank and T. Lomax, "The 2005 Urban Mobility Report", Texas Transportation Institute, Texas A&M University, May 2005.
2. U.S. Department of Transportation, Bureau of Transportation Statistics, *National Transportation Statistics 2004*, Washington, D.C., Government Printing Office, February 2005.
3. EPA, "8-hour Ozone Data Summary," http://www.epa.gov/oar/oaqps/greenbk/gnsum.html. Accessed December 5, 2006.
4. U.S. Department of Transportation, Bureau of Transportation Statistics, *National Transportation Statistics 2005*, Washington, D.C., Government Printing Office, 2005.
5. H. C. Frey, N. M Rouphail, A. Unal, and J. D Colyar, "Emission Reductions through Better Traffic Management," Report No. FHWA/NC/2002-001, December 2001, 368 p.
6. J. Koupal, E. Nam, B. Giannelli, and C. Bailey, "The MOVES Approach to Modal Emission Modeling," Presented at the CRC On-Road Vehicle Emissions Workshop, San Diego, CA, March 29, 2004.
7. H. C. Frey, A. Unal, J. Chen, S. Li and X. Xuan, "Methodology for Developing Modal Emission Rates for EPA's Multi-scale Motor Vehicle & Equipment Emission System," Report No. EPA420-R-02-027, August 2002.

8. H. C. Frey, A. Unal, N. Rouphail, and J. D. Colyar, "On-Road Measurements of Vehicle Tailpipe Emissions Using a Portable Instrument," *J. Air & Waste Management Association*, **53**, 992–1002 (2003).
9. H. Frey, N. Rouphail, and H. Zhai, "Speed and Facility-Specific Emission Estimates for On-Road Light-Duty Vehicles on the Basis of Real-World Speed Profiles," Transportation Research Record No. 1987, *Journal of the Transportation Research Board*, 128–137 (2006).
10. ICF International, "Multi-Pollutant Emissions Benefits of Transportation Strategies," Final Report Prepared for the USDOT-Federal Highway Administration, Report No. FHWA-HEP-07-004, November 14, 2006.
11. T. Litman, "London Congestion Pricing: Implications for other Cities", Victoria Transport Policy Institute, http://www.vtpi.org/london.pdf. Accessed February 19, 2007.
12. R. Johnston and R. Ceerla, "Land Use and Transportation Alternatives," in Sperling D. and S. Shaheen (eds.), *Transportation and Energy: Strategies for a Sustainable Transportation System*, ACEEE, Washington, D.C., 1995.
13. C. Rodier, R. Johnston, and, J. Abraham, "Heuristic Policy Analysis of Regional Land Use, Transit, and Travel Pricing Scenarios Using Two Urban Models, Trans. Res. Part D," *Transport and the Environment*, **7**(4), 243–254 (2002).
14. A. Unal, On-board Measurement and Analysis of On Road Vehicle Emissions, PhD Dissertation, North Carolina State University, Raleigh, NC, July 23, 2002.
15. U.S. Department of Transportation, ITS Benefits Page, http://www.itsbenefits.its.dot.gov/its/benecost.nsf/ByLink/BenefitsHome. Accessed on February 19, 2007.
16. R. Dowling, R. Ireson, A. Skabardonis, D. Gillen, and P. Stopher, Predicting Air Quality Effects of Traffic Flow Improvements, Final Report and User Guide, National Cooperative Highway Research Program Report 535, TRB Washington, D.C. 241p., 2005.
17. G. Piotrowicz and J. R. Robinson, "Ramp Metering Status in North America," Report No. DOT-T-95-17, FHWA, U.S. Department of Transportation, Washington, D.C., 1995.
18. N. Rouphail and K. Chae, Comparison of Traffic Operations with Traffic Signals vs. Roundabouts on Hillsborough Street, ITRE Technical Report submitted to The North Carolina Department of Transportation, April 2003, 18p.
19. B. M. Williams, Anthony P. Tagliaferri, Stephen S. Meinhold, Joseph E. Hummer, and Nagui M. Rouphail, "Simulation and Analysis of Freeway Lane Reversal for Coastal Hurricane Evacuation," *ASCE Journal of Urban Planning and Development*, **33**(1), 61–72 (2007).
20. M. Coelho, T. Farias, and N. Rouphail, "Measuring and Modeling Emission Effects for Toll Facilities," Transportation Research Record 1941, *Journal of the Transportation Research Board*, 136–144 (2005).
21. Institute for Transportation Engineers, "ITE Mega Issues, Signal Timing," http://www.ite.org/signal/index.asp. Accessed on February 19, 2007.
22. Kittelson and Associates Roundabout Database, http://roundabouts.kittelson.com/InvRoundabout.asp. Accessed February 19, 2007.
23. A. Varhelyi, "The Effects of Small Roundabouts on Emissions and Fuel Consumption: A case Study," *Transportation Research D: Transport and Environment*, **7**(1), 65–71 (2002).

24. A. Unal, N. Rouphail, and H. C. Frey, "Effect of Arterial Signalization and Level of Service on Vehicle Emissions," Transportation Research Record 1842, *Journal of the Transportation Research Board*, 47–56 (2003).
25. EPA Fact Sheet OMS-15, "Remote Sensing: A Supplemental Tool for Vehicle Emission Control," Report No. EPA-420-F-92-017, August 1993.
26. A. Unal, R. H. Dalton, H. C. Frey, and N. M. Rouphail, "Simultaneous Measurement of On-Road Vehicle Emissions and Traffic Flow Using Remote Sensing and an Area-Wide Detector," Paper No. 99–712, Proceedings of the 92nd Annual Meeting, St. Louis, June 20–24, 1999.
27. EPA User's Guide to MOBILE6.1 and MOBILE6.2, "Mobile Source Emission Factor Model," Report No. EPA-420-R-030-010, August 2003, 262p.
28. EPA Office of Transportation and Air Quality (OTAQ), "Mobile Model (on road vehicle)," http://www.epa.gov/otaq/mobile.htm. Accessed on February 20, 2007.
29. EPA, "A Roadmap to MOVES2004," Report No. EPA 420-S-05-002, Ann Arbor Michigan, March 2005.
30. EPA, "COMMUTER Model v2.0 User Manual," Report No. EPA420-B-05017, October 2005, 59p.
31. TranPlan homepage, http://www.citilabs.com/tranplan. Accessed February 19, 2007.
32. EMME2 Homepage, INRO Solutions, http://www.inro.ca/en/products/emme2/index.php.
33. TransCad homepage, Caliper Corporation, Newton, MA, http://www.caliper.com/tcovu.htm. Accessed July 13, 2005.
34. Modelistica, "Complete Mathematical Description of the TRANUS Model," http://www.modelistica.com. Accessed on February 20, 2007.
35. D. Husch, *Synchro 3.2 User Guide*, Trafficware, Berekely, CA, 1998.
36. A. Akcelik and M. Beskley, "Operating Costs, Fuel Consumption and Emissions in aaSIDRA and aaMotion." Presentation at the 25th Conference of Australian Institutes for Transportation Research (CAITR), U. Of South Australia, December 3–5, 2003. Available in pdf format, http://www.akcelik.com.au/documents/AKCELIK_COSTModels/CAITR%202003)v2.pdf.
37. D. Strong, *SIGNAL2000 Tutorial/Reference Manual*, Strong Concepts, 2007.
38. Transportation Research Board, *Highway Capacity Manual*, U.S. Government Printing Office, Washington, D.C., 2000.
39. Federal Highway Administration, *CORSIM User Manual*, U.S. DOT Office of Traffic Safety and Operations, Mc Lean, Virginia, 1997.
40. M. Van Aerde and Transportation Systems Group, *INTEGRATION User's Guide—Vol.1, Fundamental Model Features*, Queen's University, Ontario, Canada, 1995.
41. PTV, *VISSIM User's Manual,* Release 4.1.0, PTV-AG, Karlsruhe, Germany.
42. Paramics, www.paramics-online.com. Accessed February 19, 2007.

CHAPTER 6

ELECTRIC AND HYBRID VEHICLE DESIGN AND PERFORMANCE

Andrew Burke
Institute of Transportation Studies
University of California, Davis
Davis, California

1	INTRODUCTION	130
2	VEHICLE DESIGN OPTIONS	130
	2.1 Battery-powered Electric Vehicles (BEVs)	130
	2.2 Hybrid-Electric Vehicles	130
	2.3 Hydrogen Fuel Cell Vehicles	132
3	VEHICLE POWERTRAIN CONFIGURATIONS, COMPONENT SELECTION, AND CONTROL OPTIONS	134
	3.1 Battery-powered Electric Vehicles	134
	3.2 Hybrid-Electric Vehicles	136
	3.3 Hydrogen Fuel Cell Vehicles	139
4	DRIVELINE COMPONENT TECHNOLOGIES	140
	4.1 Motors	140
	4.2 Power Electronics	141
	4.3 Electrical Energy Storage—Batteries and Ultracapacitors	144
	4.4 Engines	150
	4.5 Auxiliary Power Units (APUs)	153
	4.6 Mechanical Components	154
5	FUEL CELLS AND HYDROGEN STORAGE	155
	5.1 Fuel Cell Technology	155
	5.2 Fuel Cell System Operation	157
	5.3 Hydrogen Storage in Vehicle Applications	160
6	POWERTRAIN CONTROL STRATEGIES	162
	6.1 Battery-powered Electric Vehicles (BEVs)	163
	6.2 Series Hybrid Vehicles	163
	6.3 Parallel Hybrid Vehicles	163
	6.4 Fuel Cell Electric Vehicles	165
7	SIMULATIONS OF ELECTRIC AND HYBRID VEHICLE OPERATION AND PERFORMANCE	166
	7.1 Battery-powered Electric Vehicles	166
	7.2 Parallel Hybrid Vehicles	167
	7.3 Series Hybrids	169
	7.4 Fuel Cell Vehicles	171
8	EMISSIONS CONSIDERATIONS	172
	8.1 Regulated Vehicle Exhaust Emissions	173
	8.2 Greenhouse Gas Emissions from the Vehicle	174
	8.3 Upstream Full Fuel Cycle Emissions	175
9	ECONOMIC CONSIDERATIONS AND MARKET PENETRATION	177
	9.1 Battery-powered Electric Vehicles	177
	9.2 Plug-in Hybrid Vehicles	179
	9.3 Charge-sustaining Hybrid Vehicles	181
	9.4 Series Hybrids	183
	9.5 Hydrogen Fuel Cell Vehicles	184
10	SUMMARY AND CONCLUSIONS—MARKET INTRODUCTION OF ADVANCED TECHNOLOGY VEHICLES	187

1 INTRODUCTION

This chapter discusses the design of electric and hybrid-electric vehicles as a pathway to the reduction of greenhouse gas emissions and the use of alternative energy sources for ground transportation. For each type of advanced vehicle there are multiple design options and driveline configurations and components that can be utilized. In addition, in the case of hybrid vehicles using either engines or fuel cells, there are different control strategies that can be employed that have significant effects on efficiency, emissions, and economics. The emphasis here will be on the vehicle design and performance, but some consideration will be given to fuel options and requirements as they relate to storage of the fuel on board the vehicle, especially for hydrogen for fuel cells, and the effect of the fuel/energy source on greenhouse gas emissions. Computer simulation results will be given for all the alternative vehicle designs, and available test data will be presented for comparison with the simulation results. Finally, component and vehicle costs will be discussed. For most of the vehicle technologies, mass marketing of the vehicles incorporating the new powertrain and fuel technologies will depend on their costs relative to conventional vehicles.

2 VEHICLE DESIGN OPTIONS

This chapter covers three options for vehicle design—battery-powered electric, hybrid-electric, and fuel cell.

2.1 Battery-powered Electric Vehicles (BEVs)

The simplest of the vehicles with an electric driveline is the battery-powered electric vehicle (BEV). The driveline of this vehicle consists of an electric motor, power electronics, and a large energy storage battery. All the energy to operate the vehicle is provided by the battery, which is recharged from the wall-plug. The range of the vehicle is determined by the energy storage capacity (kWh) of the battery and its acceleration and top speed characteristics are dependent on the power rating (kW) of the electric motor. As indicated in Table 1, electric vehicles have been designed and built in recent years by a number of auto companies. Consumer experiences with these vehicles have been favorable as far as performance and utility are concerned.

2.2 Hybrid-Electric Vehicles

There are several design approaches for hybrid-electric vehicles. In all cases, the vehicle has the capability to generate electricity onboard the vehicle from liquid or gaseous fuel and an electric motor provides at least part of the torque to propel the vehicle. The electricity can be generated either using an engine/generator or a fuel cell. The engine can be connected directly to the wheels or only to the

Table 1 Characteristics of Selected Battery-Electric Vehicles

Model/Manufacturer	Type	Curb Weight (kg)	Length/Width/Height (cm)	Battery Type (kWh)	Electric Motor (kW)	Range (Miles)	Max. Speed (kmph)
EV1/GM	full	1350	432/178/130	NiMthyd/29	104	140	>120
EV Plus/Honda	full	1634	405/174/162	NimtHyd/30	49	100	>120
RAV4/Toyota	full	1560	398/169/167	NiMtHyd/28	50	95	>120
Altra/Nissan	full	2080	487/177/169	Lithium-ion/32	62	120	>100
Smart/Mercedes	full	1380	357/172/160	NaAlCl/30	50	125	>120
Think/Ford	NEV	960	300/160/155	NiCad/12	12	50	40
E-com/Toyota	CEV	790	279/148/160	NiMtHyd/8	20	50	80
Hypermini/Nissan	CEV	840	266/148/155	Lithium-ion/10.5	24	60	80
Zenn/Feel good cars	NEV	510	258/138/139	VRLA lead-acid/7	8	35	40

Full—all roads and speeds, **NEV**—neighborhood EV, **CEV**—City EV

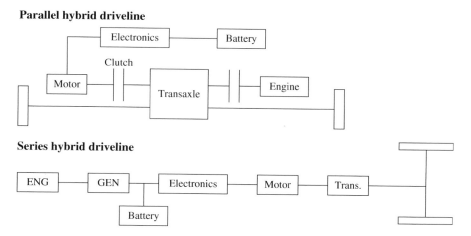

Figure 1 Driveline schematics for parallel and series hybrid vehicles.

generator. The later arrangement is referred to as a series hybrid and the former as a parallel hybrid. These powertrain arrangements are illustrated in Figure 1 and will be analyzed in some detail in a later section.

All the hybrid vehicles have an energy storage battery, but its function can be quite different for the various design approaches. If the battery is charged off the engine or fuel cell and maintained in a relatively narrow range of state-of-charge, the hybrid vehicle is referred to as a *charge-sustaining hybrid* (CSHEV) and all the energy to operate the vehicle is provided by the liquid or gaseous fuel.

Table 2 Fuel Economy and Emissions of the Toyota and Honda Hybrid Cars (2006)

Vehicle	Trans./Year	Electric Motor (kW)	0–60 mph accel. (sec.)	Emissions	Unadjusted mpg (City)	Unadjusted mpg (Hwy)
Honda Insight	M5	10	11.2	ULEV	67	87
	CVT	10	—	SULEV	63	72
Honda Civic (2002)	M5	10	—	ULEV	51	65
	CVT	10	12.0	SULEV	54 56%	61 22%
Honda Civic (2004)	CVT	15	–10	SULEV	55 62%	62 29%
Toyota Prius	Planetary/2000	33	12.6	SULEV	57 58%	58 18%
	Planetary/2004	50	10.1	SULEV	67 86%	64 26%
Toyota Camry	Planetary/2006	105	9.0	SULEV	47 81%	47 15%

*% improvement in fuel economy for the measured mpg for same car in HEV.

Refueling such a hybrid vehicle is not much different than refueling a conventional ICE vehicle. If the energy storage battery is depleted (net discharge) as the vehicle is driven and the battery can be recharged off the wall-plug, the hybrid vehicle is referred to as a *plug-in hybrid vehicle* (PHEV).

The intent of the CSHEV design is to improve the efficiency and fuel economy of the vehicle using gasoline. The intent of the PHEV design is to permit the substitution of electricity for gasoline for a significant fraction of the vehicle miles and in addition improve the efficiency of the vehicle when it is being operated primarily on gasoline.

All the hybrid-electric vehicles offered for sale by the auto companies as of 2007 are of the CSHEV type. The characteristics of selected hybrids are shown in Table 2. From the driver's point-of-view, these vehicles operate essentially the same as a conventional internal combustion engine (ICE) vehicle, but get much higher fuel economy. Markets for CSHEVs are growing as more models are becoming available from more auto companies.

As of 2007, there is considerable interest in PHEVs, but such vehicles have not been offered for sale by any auto company, and their development is in an early stage with possible products a number of years in the future. Nonetheless, it is likely that PHEVs will become important in future years.

2.3 Hydrogen Fuel Cell Vehicles

Electric vehicles powered by a fuel cell are also a design option that permits the use of other than petroleum-based fuels for ground transportation. The fuel cell is an electrochemical device that converts hydrogen directly to electricity with only water vapor as the exhaust product. The fuel cell used in vehicles is the proton-exchange membrane (PEM) type, which operates on hydrogen and air and has a high efficiency of 50 to 60 percent over a wide range of power. Hydrogen

Table 3 Characteristics of Selected Fuel Cell Vehicles

Manufact.	Vehicle/ Type	Year	Driveline Type	Fuel Cell Manuf.	Fuel cell kW	Fuel/ Storage	Range Miles
Diamler Chrysler	Necar 2/V-class	1996	PEM Fuel cell	Ballard	50	H$_2$/3,600 psi	155
	Necar 3/A-class	1997	PEM Fuel cell	Ballard	50	Methanol reformer	250
	Necar 4/A-class	2000	PEM Fuel cell	Ballard	85	H$_2$/ 5,000 psi	125
Ford	Focus/ sedan	2002	PEM Fuel cell/ nimth bt	Ballard	85	H$_2$/ 5,000 psi	180
General Motors	Hydrogen3 /Zafira van	2001	PEM Fuel cell	GM	94	H$_2$ liquid	250
	Hydrogen3 /Zafira van	2002	PEM Fuel cell	GM	94	H$_2$/ 10,000 psi	170
Honda	FCX-V3/ microvan	2000	PEM Fuel cell/ Ultracaps	Ballard	62	H$_2$/ 5,000 psi	108
	FCX-V4 microvan	2001	PEM Fuel cell/ Ultracaps	Ballard	85	H$_2$/ 5,000 psi	185
	FCX concept/sedan	2006	PEM Fuel cell/ Lith. Bat.	Honda	100	H$_2$/ compressed gas	350
Hyundai	Sante Fe/ SUV	2001	PEM Fuel cell/ Lith. Bat	UTC	75	H$_2$/ compressed gas	250
	Tucson SUV	2004	PEM Fuel cell/ Nimthy Bat	UTC	75	H$_2$/ compressed gas	185
Nissan	X-Trail/ SUV	2002	PEM Fuel cell/ Lith. Bat	UTC	75	H$_2$/ 5,000 psi	—
Toyota	Highlander/ SUV	2002	PEM Fuel cell/ Nimthy Bat	Toyota	90	H$_2$/ 5,000 psi	180

can be produced from many primary energy sources including natural gas, coal, biomass, and electricity from renewable solar and wind energy.

PEM fuel cells have been under development for about 20 years for vehicle applications and in recent years have been demonstrated in vehicles (passenger cars and transit buses) by a number of companies. The characteristics of a selected group of fuel cell vehicles are given in Table 3.

The fuel cell can be used alone to provide the electricity for the electric motor, but in most cases the fuel cell driveline is hybridized with a battery or ultracapacitor to provide the peak power pulses during accelerations and to recover energy during braking. In the hybrid fuel cell vehicle, the fuel cell can be downsized to reduce its weight, volume, and cost and to improve its time response to sudden changes in power demand. The fuel cell hybrid can be operated in the charge-sustaining mode or the plug-in mode similar to what is done for the engine-powered hybrid. With either control strategy, the vehicle operates as a series hybrid as all the torque to the wheels is provided by the electric motor and the fuel cell generates electricity as needed.

3 VEHICLE POWERTRAIN CONFIGURATIONS, COMPONENT SELECTION, AND CONTROL OPTIONS

This section focuses on key design features and components of BEVs, hybrid-electric vehicles, and hydrogen fuel cell vehicles.

3.1 Battery-powered Electric Vehicles

The driveline configuration of a battery-powered electric vehicle (BEV) is relatively simple in that it has only a few components (see Figure 2). The driveline consists of an electric motor, power electronics (DC/AC inverter), and a battery.

The battery pack for an EV is large (weighing at least 200 kg), since it is the primary energy storage unit and must provide all the energy needs of the vehicle. The electric motor(s) provide all of the wheel torque to accelerate the vehicle and for energy recovery during regenerative braking. The motor and electronics must be sized to meet the maximum torque/power required to accelerate/brake the vehicle and to maintain the maximum speed of the vehicle on a grade.

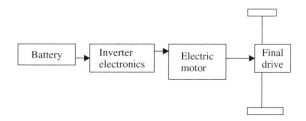

Figure 2 Battery-electric vehicle driveline schematic.

3 Vehicle Powertrain Configurations, Component Selection, and Control Options

The range of the electric vehicle is simply related to the energy use (kWh/mi.) of the vehicle and the energy storage capacity (kWh) of the battery. The energy use depends primarily on the weight of the vehicle and its road load characteristics (drag coefficient C_d, vehicle frontal area A, and tire-rolling resistance coefficient f_r). Considerable care should be taken in the design of electric vehicles to reduce their weight and road load compared to conventional ICE vehicles of the same size and type. Some of these trade-offs for light-duty vehicles based on simulation results are given in Table 4.

The results in Table 4 show the critical relationships between vehicle weight, road load parameters, energy consumption, and range.

For light-duty vehicles, usable energy storage of 15 to 40 kWh is needed to attain a range of 150 miles, depending on vehicle size and weight. This appears to be practical only for high-energy-density batteries such as lithium-ion or chemistries having a usable energy density greater than 100 Wh/kg. Ranges in excess of 200 miles will require energy densities greater than 150 Wh/kg. The United States Advanced Battery Consortium (USABC) has set a minimum goal of 150 Wh/kg for long-term commercialization of electric vehicles and a longer-term goal of 200 Wh/kg.

Even with high-energy-density batteries, one of the inherent limitations of electric vehicles is their limited range. In addition, the time needed to recharge the battery is usually several hours, or at least many minutes, due to the difficulty and expense of providing a high-power battery charger and the required electrical source for connecting it to the grid. For example, to charge a 30 kWh battery in 10 to 15 minutes would require an electrical source of at least 150 kW, which far

Table 4 Characteristics of Battery-powered Electric Vehicles of Different Design

Test Weight kg[1]	Drag Coefficent C_d [2]	Rolling resistance coefficient f_r	City Wh/mi. [3]	City Range miles	Highway Wh/mi	Highway Range Miles
2509	.3	.01	178	53	174	50
2509	.25	.01	172	55	158	56
2509	.2	.01	166	58	143	62.5
2809	.25	.01	184	51	166	52.5
2209	.25	.01	159	60	150	59
2509	.25	.008	161	59	147	60.6
2509	.25	.006	150	63.7	135	66.3
2509	.2	.006	144	66.8	120	75.9
2209	.2	.006	134	72.7	114	80

[1] All vehicles used a 75 kW AC induction motor, regenerative braking
[2] The frontal area is 21 ft^2
All vehicles used lithium-ion batteries, 100 Wh/kg, 320V, 30Ah, 95 kg, 9.5 kWh

exceeds that normally available for battery charging. Further, the battery thermal management system would have to be designed to handle the heat generated during the battery charging. This magnitude of heat generation is significantly higher than would be produced during normal use of the battery in driving the vehicle.

Another disadvantage of fast charging is that the battery cannot be fully charged at the high rate, which further limits the range of the BEV by 10 to 20 percent. Hence, it seems unlikely that the range limitations of electric vehicles can be effectively circumvented by providing fast recharge of the batteries.

3.2 Hybrid-Electric Vehicles

Series Hybrids

As shown in Figure 1, the series hybrid is essentially an electric, battery-powered vehicle with an engine/generator onboard to generate electricity as the vehicle is driven. When the power output capacity of the engine/generator is equal to or greater than the average power required for a particular use of the vehicle, the range of the vehicle for that use is set by the size of the fuel tank and not the battery. In most cases, the engine/generator rating is selected for range extension on some specified driving cycle. The advantage of the series hybrid is that the engine can be controlled to operate near its maximum efficiency independent of vehicle speed and power demand. The series approach is most suited for applications in which the peak power demand is relatively high compared to the average power and periods of high power demand are relatively short so that they can be met using energy from the battery. Otherwise, the power rating of the engine, generator, and electric motor have to be nearly the same, which results in the powertrain being heavy, large, and high cost.

The design of a series hybrid starts with the design of a battery-powered vehicle with a battery sized to a relatively short range. As for a BEV, the electric motor is sized (kW) so that the vehicle has a specified acceleration performance (0 to 60 mph time). The engine/generator is sized to meet a specified maximum electrical output (kW), which is often set by maintaining the vehicle at a specified constant speed on a grade. The battery energy storage capacity (kWh) is determined from the range (mi) requirement and the expected energy consumption (kWh/mi.) of the vehicle. The maximum power demand on the battery is less than would be the case in BEV of the same acceleration performance, because in the series hybrid, the engine/generator can provide part of the electrical demand. Hence, batteries in series hybrids can be designed to maximize energy density (Wh/kg), with less attention to power density (W/kg).

The series hybrid can be operated either as a charge depleting hybrid like a BEV with the battery being charged from the wall-plug or as a charge sustaining (CS) hybrid using only a liquid fuel. In the latter case, the primary objective of the hybridization is to achieve a large improvement in fuel economy. In the former case, the primary objective is range extension of a BEV and the substitution of electrical energy for the liquid fuel.

Parallel Hybrids

A schematic for a simple parallel hybrid-electric powertrain is shown in Figure 1. The key feature of the parallel configuration is that the engine torque can be connected directly to the wheels of the vehicle. Hence, the total torque and power to the wheels is the sum of the outputs of the electric motor and the engine. Figure 3 shows a more complex parallel arrangement in which there are two electric machines—one primarily used as a traction motor to power the vehicle and one as a generator to utilize part of the engine output to generate electricity onboard the vehicle. This is the arrangement is used by Toyota in the Prius. It has some of the characteristics of a series hybrid as in one mode of operation, it is possible to recharge the batteries even when the traction motor is being used to provide torque to the wheels. In this situation, the engine and traction motor are powering the vehicle at the same time that the batteries are being recharged. In the more complex arrangement (Figure 3), the combination of the engine, traction motor, and generator act as a continuous variable, electromechanical transmission whose effective gear ratio depends on the ratios of the powers of the three components. The traction motor is also used as a generator to recover energy during braking of the vehicle.

In the simple parallel arrangement (Figure 1), the engine output can be used to power the vehicle and/or to generate electricity using the traction motor as a generator to recharge the battery. A mechanical transmission is required for efficient operation of the engine even with the electric motor. The optimum choice of transmission for this parallel hybrid is a continuously variable, mechanical unit consisting of a variable geometry steel belt or linked chain operating between two adjustable pulleys (see Figure 4). The continuously variable transmission and utilization of part of the engine output to recharge the battery permits the engine to operate most of the time at high efficiency, resulting in a large improvement in the fuel economy of the vehicle. The continuous variable transmission also

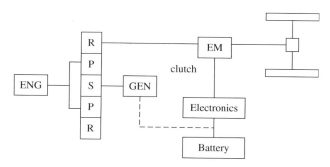

Figure 3 Planetary gear torque coupling for a parallel hybrid vehicle.

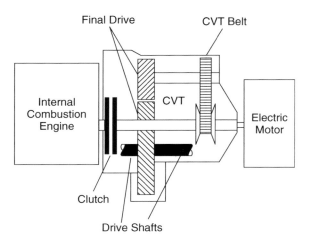

Figure 4 Schematic of a CVT in a hybrid driveline.

results in regenerative braking energy recovery down to near zero vehicle speed. Both parallel arrangements (Figure 3 and 4) utilize in essence a continuous variable transmission. It is likely that the system efficiency gains from the complex arrangement (Figure 3) will be slightly greater than with the simple arrangement (Figure 1), but it also seems likely that the cost of the complex arrangement will be higher.

One of the key issues in designing a parallel hybrid is the size (kW) of the engine and the resultant peak power requirement of the electric motor(s). One approach is to minimize the power of the electric motor and downsize the engine only slightly. With this approach, the power of the electric motor is selected such that the vehicle can be operated in the electric mode whenever operation of the engine would be inefficient. In this case, the peak power of the electric motor is only 15 to 20 percent of that of the engine and the vehicle is termed a *mild hybrid*. This approach, which is being pursued by Honda (see Table 2), also results in a relatively small, low-cost energy storage unit which could utilize either batteries or ultracapacitors. Both vehicle simulations and tests of the Honda hybrid passenger cars have shown that fuel economy improvements of 30 to 40 percent can be achieved using the mild-hybrid approach.[1,2]

Parallel hybrids can also be designed in which the engine is significantly downsized and a relatively large electric motor is used to augment the power of the engine. In this approach, which is termed a *full hybrid*, the electric motor and engine are of comparable size (kW) and the energy storage unit is significantly larger than for the *mild hybrid*, both in terms of kWh and power rating (kW). This approach is used by Toyota in the Prius (see Table 2 and Figure 3). The fuel economy improvement achieved with the full-hybrid approach will be somewhat larger than with the 'mild-hybrid' approach, but the incremental vehicle cost will also be significantly higher. These differences will be considered quantitatively in a later section of the article.

Full hybrids can be designed as either charge sustaining (CS) or plug-in hybrids (PHEVs). All hybrids currently (2007) marketed by the auto companies are CS hybrids, but some Toyota Priuses have been converted to PHEVs by several small vehicle engineering companies.[3] In order to convert a CS hybrid to a PHEV, it is necessary increase the onboard energy storage from 1–2 kWh to 5–10 kWh. In addition, the system control software must be altered to permit the batteries to be depleted as the vehicle is operated as an electric vehicle over some range of speed and power demand. Provision must also be made to recharge the battery from the wall-plug. Several hundred Prius have been converted to PHEVs with good success in that their fuel economy on gasoline have been increased to about 100 mpg for city driving. The cost of the conversions is high being at least $15,000 to $20,000.

3.3 Hydrogen Fuel Cell Vehicles

The fuel cell vehicle is essentially a series hybrid with the engine/generator replaced by a fuel cell, which generates electricity from the hydrogen stored on board the vehicle. As shown in Figure 5, the driveline configuration of a fuel cell vehicle is relatively simple in that it consists of an electric motor, power electronics (DC/AC inverter), and a fuel cell. In some cases, there is a relatively small battery pack to load level the fuel cell and permit energy recovery during braking of the vehicle. The battery pack in the fuel cell vehicle is not large (weighing less than 50 kg), since the primary energy source onboard the vehicle is hydrogen. The electric motor(s) provide all of the wheel torque to accelerate the vehicle and energy recovery during regenerative braking. The motor and electronics must be sized to meet the maximum torque/power required to accelerate/brake the vehicle and to maintain the maximum speed of the vehicle on a grade.

As discussed previously, the fuel cells being developed for vehicle applications are of the proton-exchange membrane (PEM) type. This type of fuel cell is

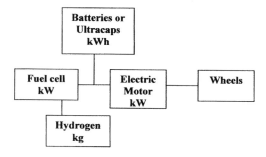

Figure 5 Fuel cell driveline schematic.

well suited for vehicle applications, because it has a high power density (in watts per liter), operates at relatively low temperature (80°C), and is tolerant of the impurities in air including CO_2. Fuel cells are inherently more efficient than engines, making reasonable the goal of increasing by a factor of three the equivalent fuel economy of fuel cell vehicles compared to gasoline ICE vehicles. Even with the increased efficiency of fuel cells, it is a large challenge to store enough hydrogen on board the vehicle to achieve a range of 300 to 400 miles.

4 DRIVELINE COMPONENT TECHNOLOGIES

This section covers different types of motors, power electronics, batteries and ultracapacitors, engines, auxillary power units (APU's), and mechanical components.

4.1 Motors

An electric motor is used to convert the electrical energy out of the battery to mechanical energy to power the vehicle. As shown in Figures 2 and 5, the torque from the electric motor is applied to the drive shaft of the vehicle, and the wheels in many cases, without a multigear ratio transmission. Electric motors have higher power density (power per unit weight and volume) than internal combustion engines and advantageous low-speed torque characteristics compared to IC engines. The result is the smooth, rapid acceleration common to electric vehicles.

There are a number of different types of electric motors (Figure 6) that have been used in electric vehicles. These include series and separately excited DC motors and induction, permanent magnet, and switch reluctance AC motors. Extensive information and data on electric motors for electric vehicles are given in References 4 to 6. The electric drive unit consists of the motor, power electronics, and microcomputer-based controller needed to meet the time-varying power demands of the vehicle. The power electronics take the DC output of the battery and convert it to the form needed by the various motor options over their complete range of torque and speed (RPM).

DC motors—both series and separately excited—utilize brushes for commutation. Power electronics are used to control the effective voltage applied to the armature and field windings of the motors. The lowest-cost electric drive units are those using series DC motors, but they are applicable only in low-speed vehicles. Separately excited DC motors can be used in higher-speed vehicles, but most EVs currently use one of the types of AC motors. The brushes in the DC motors limit their maximum RPM and to some extend the maximum system voltage and require periodic maintenance.

In general, the AC motor systems are smaller, lighter, more efficient, and lower cost than the DC systems, especially as the power requirements for the systems have increased.

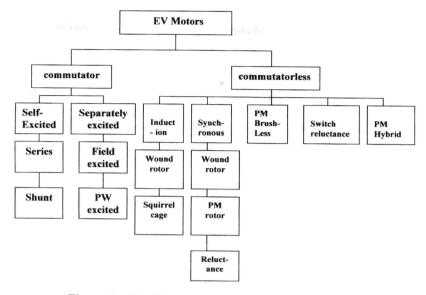

Figure 6 Classification of motors for electric vehicles.

High-performance, high-speed electric vehicles have been designed and built using both induction and permanent magnet types of AC electric motors. The permanent magnet type motor seems to be the choice for small- and moderate-power (25–150 kW) systems used in passenger cars, and the induction motor type is the choice for large vehicles such as heavy-duty trucks and transit buses. The permanent magnet (PM) motors tend to be smaller and easier to control than the induction motors at moderate powers, but the induction motors are more durable and lower cost when the power required is high (>200 kW).

Efficiency maps for induction and permanent magnet AC motors are shown in Figures 7 and 8. Note that the efficiencies vary significantly with torque and RPM and that no single value of efficiency is applicable for a motor in a vehicle operated over a driving cycle. Simulations of electric vehicles using both induction and PM motors have been performed.[4] The energy usage results (km/kWh) for various driving cycles are given in Table 5. For all the driving cycles, the energy usage of vehicles using the PM motors are lower than those using the induction motors. The differences vary with driving cycle, but are in the range of 10 to 20 percent, with the largest differences being on city cycles (stop/go driving). The improved efficiency with the PM motors would translate directly into longer driving range for the same size (kWh) battery.

4.2 Power Electronics

The peak power ratings of electric drivelines used in electric vehicles have increased greatly in the last 10 years. This is due primarily to the improved

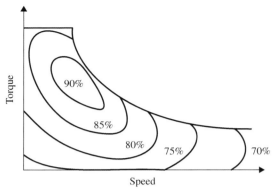

Typical power efficiency map of an EV induction motor.

Figure 7 Efficiency map for an induction motor.

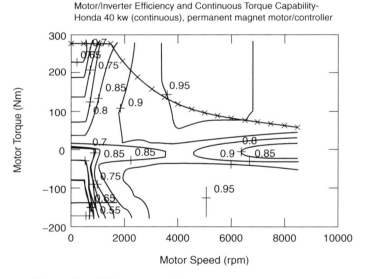

Figure 8 Efficiency map for a permanent magnet motor.

performance (current and voltage limits) of the semiconductor switching devices used in the power electronics. As shown in Figure 9, the DC/AC inverter in the driveline system includes at least six switching devices to control the time-varying voltage fed from the battery to the electric motor. The technology improvements in the switching devices has not only improved their performance, but lowered their cost and increased reliability significantly. The efficiency of the power electronics is in the range of 95 to 98 percent, meaning that most of the losses in the electric driveline are in the electric motor.

Table 5 Simulation Results for Vehicles Using Different Electric Motors

	Fuel economy of different EVs using different EV motors		
	Fuel economy (km/kWh)		
	Induction motor	PM brushless motor	Improvement (%)
Passenger car			
FUDS	5.19	6.18	18.30
FHDS	4.57	5.07	10.94
J227a-C	5.35	5.94	11.03
ECE Urban	5.85	6.95	18.80
Japan 10.15	6.03	7.07	17.25
Van			
FUDS	1.39	1.65	18.71
FHDS	1.04	1.16	11.54
J227a-C	1.56	1.74	11.54
ECE Urban	1.70	2.01	18.24
Japan 10.15	1.75	2.01	14.86
Bus			
FUDS	0.49	0.59	20.41
FHDS	0.50	0.55	10.00
1227a-C	0.47	0.53	12.77
ECE Urban	0.53	0.63	18.87
Japan 10.15	0.56	0.65	16.70

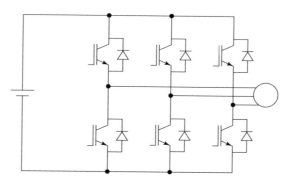

Three-phase full-brigde voltage-fed inverter.

Figure 9 DC/AC inverter schematic.

In addition, much progress has been made in developing and implementing new control algorithms for the various types of AC motors that permit motor operation at high efficiency over a large portion of the motor torque/RPM map.[4] Using current motor and power electronics technologies, average efficiencies of 85 to 90 percent over typical driving cycles are not uncommon.

4.3 Electrical Energy Storage—Batteries and Ultracapacitors

Since the development of batteries for electric and hybrid vehicles started over 30 years ago, many different battery types have been considered and development started. At the present time the battery types being developed are relatively few—namely, lead-acid, nickel metal hydride, lithium-ion and lithium polymer, and sodium nickel metal chloride. All of these battery types have advantages and disadvantages. Unfortunately, none of them are attractive in all respects for electric and hybrid vehicles, so there is no clear choice of the best battery for all applications. In addition, there are trade-offs between energy density, power density, cycle life, and cost, so that even for a particular type of battery, it is necessary to design the batteries for specific applications.

In more recent years (since 1990), *ultracapacitors* have been developed for vehicle applications. These devices function in the electric driveline like batteries, but have much higher power density, especially in pulsed modes of operation. Ultracapacitors are of particular interest for use in charge-sustaining mild hybrid vehicles and fuel cell vehicles. The characteristics and use of both batteries and ultracapacitors in energy storage units are discussed in this section.

The key requirements for the energy storage unit for a particular vehicle design are usable energy stored, peak power, and cycle and calendar life. These requirements must be met with a unit whose weight and volume are less than specified values, based on packaging the driveline in the vehicle.

Battery Technologies

There are a number of ways to express battery performance. The simplest approach is to state the energy density (Wh/kg) and peak power density (W/kg), as is done in Table 6. This approach is good for showing the relative performance of various types of batteries, but does not show the detailed performance of a particular battery, which requires knowledge of the Ragone curve (Wh/kg vs. W/kg for constant power discharges), battery open-circuit voltage and resistance versus state-of-charge, capacity (Ah) versus discharge current and temperature, and the charging characteristics of the battery at various rates and temperatures.

As indicated in Table 6, for a given battery type/chemistry, cells/modules can be designed with significantly different energy and power characteristics. There is a trade-off between energy density and power density, with the higher-power batteries having significantly lower energy density. This is true for all battery types. Since the battery pack in a fuel cell vehicle is sized by the power requirement

Table 6 Characteristics of Various Technologies/Types of Batteries for Use in Vehicle Applications

Battery Technology	Applic. Type	Ah	V	Wh/kg At C/3	Resist mOhm	W/kg Match. Imped.	W/kg 95% eff.	Max. Usable SOC
Lead-acid								
Panasonic	HEV	25	12	26.3	7.8	389	77	28%
Panasonic	EV	60	12	34.2	6.9	250	47	—
Nickel Metal Hydride								
Panasonic EV	EV	65	12	68	8.7	240	46	—
Panasonic EV	HEV	6.5	7.2	46	11.4	1093	207	40%
Ovonic	EV	85	13	68	10	200	40	—
Ovonic	HEV	12	12	45	10	1000	195	30%
Saft	HEV	14	1.2	47	1.1	900	172	30%
Lithium-ion								
Saft	HEV	12	4	77	7.0	1550	256	20%
Saft	EV	41	4	140	8.0	476	90	—
Saft	HEV	6.5	4	63	3.2	3571	645	20%
Shin-Kobe	EV	90	4	105	.93	1344	255	—
Shin-Kobe	HEV	4	4	56	3.4	3920	745	18%
A123	HEV	2.2	3.6	90	12			
Altairnano	EV	11	2.8	70	2.2	2620	521	60%
Altairnano	HEV	2.5	2.8	35	1.6	6125	830	60%

(kW), the lower energy density of the high-power batteries is not a significant disadvantage. If the fuel cell vehicle was to be designed as a plug-in vehicle, then it would be important to consider the energy density as well as the power density of the battery. The key performance characteristic of a high-power battery is its resistance as the peak power of the battery is given by the following expression:

$$P_{\text{peak}} = EF^*(1 - EF)^* V_0^2 / R$$

where EF is the efficiency of the power pulse.

For an efficiency of 95 percent, the pulse power of the battery is about one-fifth of $V_0^2/4R$, the matched impedance power of the battery. In practice, the matched impedance power of a battery is much greater than the usable power appropriate for use in an EV or HEV. For fuel cell vehicle applications, the batteries designated as HEV would be used.

In evaluating battery technologies, it is important to consider battery cycle life and safety, as well as battery performance. Cycle life is a key issue in evaluating the economic viability of a particular battery technology. Estimating cycle life for a particular vehicle design and application is not a simple matter. The cycle life depends critically on the rate of discharge, how the battery is recharged and the average depth-of-discharge before recharge, and the temperature of operation of the battery. All of these factors can affect cycle and calendar life of the battery. Of particular importance is the depth of discharge. As shown in Figure 10, the cycle life increases rapidly if the depth of discharge of the cycles is less than 50 percent. For batteries used in hybrid vehicles, the cycles are quite shallow (usually less than 10 percent) and the batteries are seldom if ever fully charged. Estimating cycle life under these conditions is very uncertain, and the battery life is often expressed as equivalent deep discharge cycles, or total energy throughput over the life of the battery. This is the meaning of the term '*percent swing*' in Figure 10. Note that for 'percent swings' less than 5 percent, cycle lives of 100,000 to 1,000,000 cycles are projected for nickel metal hydride and lithium-ion batteries.

Most battery packs have a battery management system (BMS) to monitor cell/module voltages and temperatures. In the case of lead-acid and nickel metal hydride batteries, the purpose of the BMS is to increase the life of the pack by assuring that the cells remain balanced and the temperatures do exceed a specified upper value. In the case of the lithium-ion batteries, the BMS is needed to assure

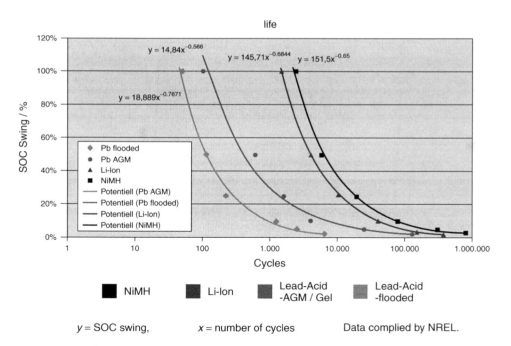

Figure 10 Cycle life correlations for various types of batteries.

that the pack is operated safely, as overcharging and/or overdischarging of the pack can result in a thermal runaway condition that can lead to an explosion or fire. Much of the current research on lithium-ion batteries stems from the desire to utilize electrode chemistries that do not have the inherent safety problems associated with graphite and NiCo electrode materials. Safety can be more of an issue with lithium batteries in battery-powered vehicles (BEV) than in hybrid vehicles like the Prius, because in the BEV the battery is deep discharged and fully charged after each cycle.

The cost of the battery is a major issue. Large batteries for electric vehicles are expensive, being about $700 to $800/kWh for nickel metal hydride and even somewhat higher for lithium-ion batteries. The cost of lead-acid batteries is about $100/kWh. It is expected that the cost of the advanced batteries will decrease markedly if they are manufactured in high volume. A key question is how low the cost/price of the advanced batteries, in particular the lithium-ion batteries, will fall in high volume. Most projections of the cost of lithium-ion batteries are in the range of $300 to $500/kWh in mass production (hundreds of thousands of packs per year).[7,8] There is little information on the cost of high-power batteries, but it can be expected they will be somewhat higher on a kWh basis than the lower-power BEV batteries.

Ultracapacitor Technologies

As noted previously, ultracapacitors could also be used for energy storage in charge-sustaining hybrid and fuel cell vehicles. *Ultracapacitors* are inherently high-power devices with very long cycle life for deep discharges and rapid charging. For many light-duty vehicles, an ultracapacitor unit could store only 100 Wh rather than the 1000 Wh needed in a battery-type energy storage unit.[9,10]

The characteristics of ultracapacitors are relatively straightforward to describe, as ultracapacitors are basically deep discharge devices whose power capacity for discharge and charge are essentially the same at all states of charge (voltages between the rated and minimum values). The energy density varies only slightly with discharge rate for power density values (W/kg), up to at least 1000 W/kg. Ultracapacitors of a number of different types are being developed. A summary of the physical and performance characteristics of various ultracapacitors is given in Table 7. The usable energy density of ultracapacitors varies between 4 to 8 Wh/kg, with a peak power capability (9 percent efficiency) of 800 to 1,400 W/kg.

Comparison of Ultracapacitors and Batteries for Energy Storage in Hybrid Vehicles

Lithium-ion batteries or ultracapacitors can be used for energy storage in mild hybrid and fuel cell vehicles. The characteristics of the energy storage unit for a mid-size passenger car requiring a usable energy storage of 100 Wh and a peak power of 27 kW are shown in Table 8 for both technologies. The weights

Table 7 Characteristics of Ultracapacitors for Vehicle Applications

Device	V rated	C (F)	R (mOhm)	RC (sec)	Wh/kg [1]	W/kg (95%) [2]	W/kg Match. Imped.	Wgt. (kg)	Density (gm/cm^3)
Maxwell**	2.7	2800	.48	1.4	4.45	900	8000	.475	1.48
ApowerCap	2.7	55	4	.22	5.5	5695	50625	.009	—
Ness	2.7	1800	.55	1.00	3.6	975	8674	.38	1.37
Ness	2.7	3640	.30	1.10	4.2	928	8010	.65	1.26
Ness	2.7	5085	.24	1.22	4.3	958	8532	.89	1.25
Asahi Glass (propylene carbonate)	2.7	1375	2.5	3.4	4.9	390	3471	.210 (estimated)	1.39
Panasonic (propylene carbonate)	2.5	1200	1.0	1.2	2.3	514	4596	.34	1.39
EPCOS	2.7	3400	.45	1.5	4.3	760	6750	.60	1.25
LS Cable	2.8	3200	.25	.80	3.7	1400	12400	.63	1.34
BatScap	2.7	2600	.3	.78	3.95	1366	12150	.50	—
Power Sys. (activated carbon, propylene carbonate)	2.7	1350	1.5	2.0	4.9	650	5785	.21	1.4
Power Sys. (graphitic carbon, propylene carbonate)	3.3	1800	3.0	5.4	8.0	825	4320	.21	1.4
Fuji Heavy Industry-hybrid (AC/C)	3.8	1800	1.5	2.6	9.2	1025	10375	.232	1.62

[1] Energy density at 400 W/kg constant power, Vrated-1/2 Vrated
[2] Power based on $P = 9/16 \times (1 - EF) \times V2/R$, EF = Efficiency of discharge
**Except where noted, all the devices use acetonitrile as the electrolyte

and volumes shown are for the cells alone and do not include packaging the cells into modules and the monitoring and management components needed for the various technologies. It is likely that the packaging factors for the batteries and ultracapacitors will not be too different and that the packaged weights and volumes should be proportional to those given in Table 8.

The results in Table 8 indicate that on the basis of performance, any of the technologies could be used to design the energy storage unit. In the final analysis, once a particular technology (ultracapacitors or batteries) is shown to be acceptable from a performance point of view, selection of the preferred technology most often is based on cost considerations. It is of interest to determine

Table 8 Energy Storage Unit Characteristics Using Various Technologies

Technology	No. of Cells	C or Ah	System Voltage	ΔSOC %	Total Energy Wh	Max. Power kW	EF %	Weight of Cells kg	Volume of Cells L
Ultracaps									
C/C	60	2400	156	75	100	40	90	22.5	15.2
C/graphiticC	50	2300	160	75	150	27	90	17.6	12.6
Li-ion batteries									
Graphite/NiCo	45	9	158	6.6	1500	31	91	24	11.4
Iron phosphate	50	11	160	5.5	2000	27	91	23	10.5
Li titanate	64	11	160	5.5	1800	27	90	26	14
Li titanate	64	4	160	16.6	600	27	90	16	8.5

the cost ($/Wh and cents/farad) of ultracapacitors that would result in the cost of an energy storage unit consisting of ultracapacitors being equal to that of one consisting of batteries when the unit cost of the batteries is specified as ($/kWh)$_{bat}$. If the energy stored in the ultracapacitor unit is 125 Wh and that in the battery unit is 1,500 Wh, the unit cost of the ultracapacitors is simply

$$(\$/Wh)_{cap} = .012(\$/kWh)_{bat}$$

The ultracapacitor cost in terms of cents/farad is given by

$$(cents/F)_{cap} = .125 \times 10^{-3} \times (\$/kWh)_{bat} \times V_{cap}^2$$

The relative costs of batteries and ultracapacitors that would result in equal initial costs of the energy storage units using the two technologies are shown in Table 9. For a battery cost of $500/kWh, the corresponding ultracapacitor cost is .4 to .5 cent/F, depending on the capacitor voltage. The price of ultracapacitors are 1 to 1.5 cents/F and that of batteries is $500 to 1000/kWh. It is difficult to project the future cost of either technology, but the results in Table 9 indicate that ultracapacitors are likely to be cost competitive with lithium-ion batteries as both technologies are further developed. This assumes that both units last the lifetime of the vehicle so that their life-cycle costs would also be equal. In order for ultracapacitors to compete costwise with batteries, they must be sized to store less energy than the batteries and advantage must be taken of their higher-power capability.

Table 9 Relative Costs of Batteries and Ultracapacitors

Battery Cost ($/kWh)	Ultracap Cost (cents/F)	Ultracap Cost (cents/F)
300	.25	.34
400	.34	.45
500	.42	.56
700	.59	.78
900	.76	1.0
1000	.84	1.12

4.4 Engines

The engine characteristics of most interest for the design of hybrid-electric vehicles are torque/speed maps that show the engine efficiency or specific fuel consumption and emissions as a function of torque and RPM. The specific fuel consumption and emissions are expressed as gm fuel or emissions/kWh engine output. The brake specific fuel consumption (bsfc) is related to engine efficiency as

$$\text{Efficiency} = \text{bsfc}/(\text{kWh/gmfuel})$$

where kWh/gmfuel is the heat content of the fuel. For gasoline, the expression becomes

$$\text{Efficiency} = 81.8/\text{bsfc}$$

For example, if the engine specific fuel consumption is 230 gm/kwh, the efficiency is .355. A typical efficiency map for a port fuel injected gasoline engine is shown in Figure 11.

Note that there is a large variation in engine operating efficiency over the map and that the maximum efficiency occurs only in a small region of engine torque and RPM. At low torque (power), the efficiency is much less than the maximum value. The objective of hybridizing the powertrain of the vehicle is to have the engine spend a much greater fraction of its operating time at high efficiency than is the case in a conventional ICE vehicle. Needless to say, it is highly desirable to develop engines with a high maximum efficiency and a less variation in efficiency over the map. As is discussed later in this section, such engines are being developed.

Most light-duty vehicles sold in the United States currently utilize stoichiometric, port-fuel-injected engines having two or four valves per cylinder. This type of engine has been significantly improved in recent years by the manufacturers in terms of both efficiency and emissions, but the engine maps (brake-specific

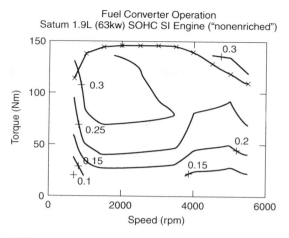

Figure 11 Efficiency map for a typical port fuel-injected gasoline engine.

values—gm/kWh vs. torque and RPM) of the improved engines are not generally available in the open literature. Hence, for studies of hybrid vehicles outside the auto industry, it has not been possible to use maps for the latest production engines, and the maps often used are somewhat dated. Nevertheless, it is possible using simulations to assess the relative fuel economy of conventional ICE and hybrid vehicles using different engines for various powertrain configurations and control strategies.

Engine characteristics can be found in the literature and as input files that are part of vehicle simulation computer programs such as Advisor and PSAT.[11-14] The engine maps shown in Figures 12 to 16 were taken from Advisor.[13] Maps are

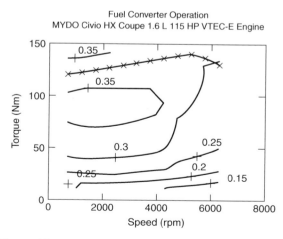

Figure 12 A map for a Honda VTEC gasoline engine.

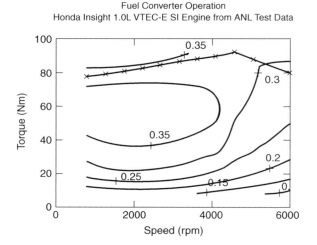

Figure 13 A map for the Honda Insight VTEC engine.

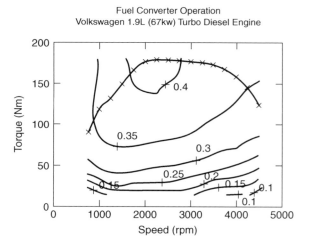

Figure 14 A map for the VW turbo diesel engine.

shown for the following types of engines: port-injected gasoline, turbo-charged diesel, Atkinson cycle (Prius) gasoline, lean burn (Insight) and VTEC gasoline, and direct-injection gasoline (GDI). These maps show significant differences in both the maximum efficiency and the pattern of the changes in efficiency over the map. The maximum efficiency is the highest for the turbo-charged diesel and direct-injection gasoline engines, and those engines have high efficiency over relatively large portions of the map. Of the gasoline engines, the Atkinson cycle Prius engine exhibits the most efficient operation, approaching that of the

4 Driveline Component Technologies   153

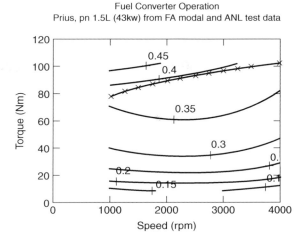

Figure 15 An engine map for the Prius engine.

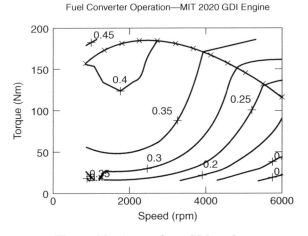

Figure 16 A map for a GDI engine.

diesel and GDI engines. The standard port-injected gasoline engines show the lowest efficiency and largest variation in efficiency, with the lean-burn, VTEC engine from the Honda Insight having the most efficient operation over its map. Development of all these engine types is continuing.

4.5 Auxiliary Power Units (APUs)

In a series hybrid, an auxiliary power unit (APU) or engine/generator is needed for on board generation of electricity. In principle, any of the engine types can be used with a generator attached. The engine in the APU operates over a narrower range of torque and RPM than would be the case in a conventional ICE or parallel hybrid vehicle. In addition, the engine is not turned off and on frequently, as would be the case in a parallel hybrid. Two engine types that are especially suited for use in series hybrids are the Wankel and gas turbine engines. These engines operate purely in a rotary motion at high RPM just as the generator does, so they are well suited for this application. This is especially true of the gas turbine engine, which has been utilized in several series hybrid transit buses.[15] A key issue concerning the gas turbine engine is the presence and material of the recuperator, which permits the recovery of exhaust energy for the turbine (see Figure 17). With a ceramic recuperator, maximum efficiencies approaching 40 percent can be achieved. Gas turbine engines are external combustion engines, so they can utilize a wide range of liquid fuels. Since the efficiency of the generator can be over 95 percent, the engine/generator can be an efficient means of producing electricity onboard the vehicle. Inefficiency of electricity generation is not the reason for the limited applicability of the series hybrid approach for transportation applications.

4.6 Mechanical Components

The mechanical components include the transmission and gearing that are used to combine the outputs of the engine and electric motor and to transfer them to the wheels to power the vehicle. The gear ratios are selected to permit the engine and motor to operate in the efficient portions of their maps. All parallel hybrids require a transmission or planetary gear set for efficient operation of the engine. The battery-powered electric vehicle (BEV) and the fuel cell vehicle do not require a transmission because the power electronics/controller acts as

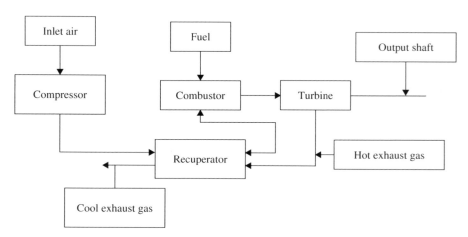

Figure 17 Schematic of a gas turbine engine with recuperator to improve efficiency.

an electronic transmission to maintain efficient motor operation at all vehicle speeds. The final drive ratio in the differential is selected to maintain the engine and motor in a specified range and to achieve a desired acceleration performance of the vehicle.

One type of transmission that is particularly well suited for hybrid vehicles, especially single-shaft parallel designs, is the continuously variable transmission (CVT) utilizing adjustable pulleys and a variable geometry steel belt or chain (see Figure 4). The use of this transmission in hybrids is advantageous because it permits the effective recovery of energy during regenerative braking down to very low vehicle speeds and the smooth operation of the electric motor at all speeds without the sudden, nonuniform speed changes inherent in a shifting, geared transmission. Honda has incorporated the CVT in its Insight and Civic hybrids (see Table 2).

Toyota has incorporated the planetary gear set in all its vehicles (see Figure 3). As indicated in the Figure 3, the electric motor is attached to the ring gear, the generator to the sun gear, and the engine to the pinion gear. This arrangement is often referred to as a continuously variable transmission, but its function is much different than the mechanical CVT shown in Figure 4. The effective gear ratios of the Toyota power split unit (Figure 3) result from controlling the torques and speeds of the motor, generator, and engine, and not from altering the mechanical configuration of the unit. Hence the selection of the maximum powers of the three components attached to the planetary set cannot be arbitrary selected, as is the case for the engine and electric motor in the mechanical CVT in the single-shaft arrangement.

5 FUEL CELLS AND HYDROGEN STORAGE

5.1 Fuel Cell Technology

As with batteries, a number of types of fuel cells can be developed. The characteristics of these fuel cells are shown in Table 10. Although R&D continues on each of these technologies, only the proton exchange membrane (PEMFC) technology appears to be potentially commercially viable for vehicle applications. Hence, this article will focus only on developments related to the PEMFC technology.

The basic operation of a hydrogen PEM fuel cell is illustrated in Figure 18. The hydrogen molecule (H_2) is split into two protons (H^+) and two electrons (e^-) at the negative electrode of the cell upon reaction with the membrane catalyst (platinum). The electrons are conducted through a metal conductor (or load) to the positive electrode of the cell. The protons (H^+) are conducted as ions through the membrane to the positive electrode, where they react with the oxygen in the

Table 10 Characteristics of Various Types of Fuel Cells

Fuel cell Type	Electrolyte	Anode Gas	Cathode Gas	Temperature	Efficiency
Proton exchange membrane (PEM)	solid polymer membrane	hydrogen	pure or atmospheric oxygen	75°C (180°F)	35–60%
Alkaline (AFC)	potassium hydroxide	hydrogen	pure oxygen	below 80°C	50–70%
Direct methanol (DMFC)	solid polymer membrane	methanol solution in water	atmospheric oxygen	75°C (180°F)	35–40%
Phosphoric acid (PAFC)	phosphorous	hydrogen	atmospheric oxygen	210°C (400°F)	35–50%
Molten carbonate (MCFC)	alkali-carbonates	hydrogen, methane	atmospheric oxygen	650°C (1,200°F)	40–55%
Solid oxide (SOFC)	ceramic oxide	hydrogen, methane	atmospheric oxygen	800–1,000°C (1500–1,800°F)	45–60%

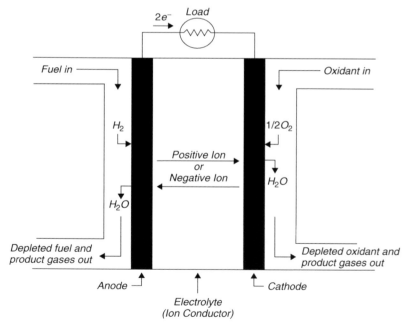

Figure 18 A schematic of the processes occurring in a PEMFC.

air and the electrons, with the aid of the catalyst layer, form water. The reactions are shown below:

$$\text{Negative electrode}: H_2 \rightarrow 2H^+ + 2\ e^-$$

$$\text{Positive electrode}: \tfrac{1}{2}O_2 + 2H^+ + 2\ e^- \rightarrow H_2O$$

This describes cell operation, but all practical systems are made up of a number of cells in series. The cells are combined in a bipolar arrangement with the electrons being passed from cell to cell through a conducting plate having a negative electrode on one side and a positive electrode on the other side.

Stacking cells in series provides an increase in system voltage. The maximum current capacity of the fuel cell stack is governed by the geometric area of the electrodes in the cell. Hence, the maximum power of the fuel cell stack depends on the number of cells in series and the cross-sectional area of electrodes. Both the current and voltage of the fuel cell stack vary with the power demand. Since higher current results in lower voltage (increased losses), the efficiency of the fuel cell is higher at relatively low power.

Using hydrogen in a fuel cell is more efficient than converting hydrogen to useful mechanical energy through combustion in an ICE. Typical fuel cell efficiencies are in the range of 40 to 60 percent, which is significantly higher than the 25 percent to 35 percent efficiency of IC engines.[16,17]

5.2 Fuel Cell System Operation

Although the stack is the heart of the fuel cell system, it requires additional auxiliary equipment to operate efficiently. The additional components include a compressor or blower to provide air, a hydrogen storage and metering system, humidifiers for the hydrogen and air, pumps for the cooling system, and power electronics to control and filter the power output. The complete system and its layout are shown in Figure 19.

Several subsystems that make up the *balance of plant* (BOP):

- *Hydrogen storage and fuel supply.* Hydrogen is stored on board the vehicle in any number of ways (high pressure gas, liquefied, or in a metal hydride). Fuel is delivered to the fuel cell stack by the control of valves and differential pressure. Pressure regulation outside of the storage tank is important to ensure safety and system integrity.
- *Air supply.* The oxygen required at the cathode catalyst is extracted from ambient air supplied to the fuel cell stack by an air compressor (or blower). Nitrogen gas is not required for the electrochemical reaction, but aids in the removal of water from the cathode channels. Air preparation is required, including particle filtration and humidification.
- *Water management.* Water is generated in the cathode reaction and must be removed from the fuel cell to prevent blockage of reaction sites. Proper

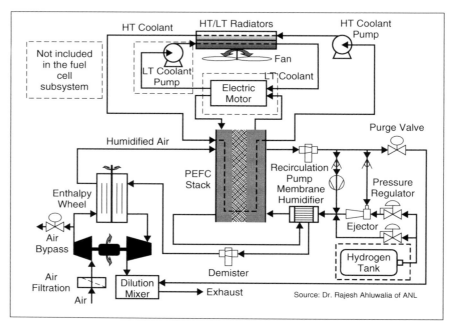

Figure 19 System diagram for a typical PEM fuel cell and its auxiliaries. (Adapted from Ref. 27.)

humidification of the membrane is necessary to improve proton transfer. This is done by recycling water from the exhaust gases to the air and hydrogen inlet gas streams that carry the water to the membrane.

- *Thermal management.* Even though the fuel cell reaction is more efficient than the combustion engine, heat is generated at the fuel cell reactions. This thermal energy is removed from the fuel cell (in most designs) with the use of a coolant fluid that flows between solid plates sandwiched between some of the cells. This coolant picks up the thermal energy at the fuel cell and exhausts it with a radiator at the front of the vehicle. This subsystem requires pumps, a radiator unit, and control valves.
- *Power management.* The electrical current generated by the fuel cell stack is delivered to the drive motor through power conditioning devices, including DC/DC converters, that control the AC voltage to the motor. A portion of the fuel cell stack current is used to power fuel cell auxiliary devices such as blowers, pumps, and fans.

Fuel cell systems require constant monitoring and control in order to maintain efficient operation. Each of the reactant streams must be extremely clean and free of impurities, which would degrade fuel cell performance. A *well-balanced* and properly controlled fuel cell will maintain adequate fuel flow, relative humidity, temperature, and pressure for any desired power request within the fuel cell's operating range. There is generally a short lag (fraction of a second) in the response of the fuel cell to sudden changes in power demand. This is due primarily to the response of the air supply, which is dependent on the spooling rate of the compressor or flow rate of the blower. The effect of these lags in power output is mitigated by the batteries or ultracapacitors, which have a very short response time. In combination, the fuel cell and the electrical energy storage unit can respond to changes in power demand even more rapidly than the IC engine system.

The operating characteristics of the fuel cell are similar to those of a battery in that they are expressed in terms of voltage, current, and resistance. The fuel cell has an open circuit voltage of about 1.2 V/cell and the cell voltage decreases as current is drawn from the cell. The voltage-current characteristic (V vs. A/cm^2) of a proton exchange membrane (PEM) fuel cell operating on hydrogen and air is shown in Figure 20.[16,17]

The data shown in Figure 20 are for illustration only and do not indicate the state of the art in cell performance. Note that the cell efficiency is much higher than for IC engines and that it is highest at low power (small currents) rather than at relatively high power as in the case of the engine. It is the high values of cell efficiency that has lead to the expectation that fuel cell vehicles will have significantly higher equivalent fuel economy than gasoline-fueled vehicles of the same size and performance. There has been continuing improvement in the cell performance in terms of W/cm^2 at a specified voltage. For example, the maximum power shown in Figure 20 is .6 W/cm^2 at .6 V/cell, but it is projected

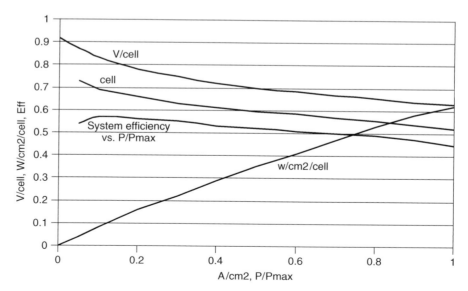

Figure 20 Cell voltage, efficiency, and power vs. A/cm^2.

that future fuel cells will attain that power at .8V/cell, which corresponds to an efficiency of 65 percent.[18]

5.3 Hydrogen Storage in Vehicle Applications

Hydrogen storage is needed on board the vehicle and at distribution and refueling stations. Storage of hydrogen on board the vehicle is difficult, because hydrogen is a low molecular weight gas and has a tendency to diffuse through metals. Both of these characteristics contrast markedly with gasoline, which is a liquid fuel and is easy to contain at ambient conditions in a simple metal tank.

The Department of Energy (DOE) has set goals for the development of hydrogen storage systems[19]. Primary consideration is given to the weight and volume of the system to store sufficient hydrogen for a vehicle range approaching that of conventional ICE vehicles using gasoline. The characteristics of the state of the art for various hydrogen storage technologies are shown in Table 11, compared with the DOE goals for 2005 to 2015.

The basis for the hydrogen storage goals set by DOE is the desire to have fuel cell vehicles with the same range as ICE vehicles. The weight and volume of the hydrogen storage system should be such that it does not adversely affect the performance and utility of the vehicle. There is, of course, considerable judgment involved in setting the limits for acceptable weight and volume for the hydrogen storage system. In any analysis of hydrogen storage on board vehicles, it is convenient to reference the weight and volume limits in terms of the weight

Table 11 Comparisons of the System Metrics for Various Technologies and DOE Goals (2005 to 2015)

Hydrogen Storage Approaches	Wt%H$_2$/ Tank	kg System DOE Goal: 6–9 wt.%/kg	gm H$_2$/ Tank	Liter sys. System DOE goal: 45–80 gm H$_2$/L
Compressed gas				
5,000 psi	6	4–5	20	15
10,000 psi	5	3–4	32	25
Liquid (LH$_2$)	20	15	63	52
Activated carbon (77 deg K)	6	5	30	25
Hydrides				
Low temperature (<100° C)	2	1.8	105	70
High temperature (>300° C)	7	5.5	90	55

Table 12 Baseline Vehicle Characteristics

Vehicle Class	Curb Wgt. (kg)	W_B (inches)	Width (inches)	F. E. (mpg)	Fuel Tank (gal.)	Range (miles)
Cars						
Compact	1136	102	67	34	13	442
Mid-size	1409	108	72	28	17	476
Full-size	1590	113	75	25	18.5	462
SUVs						
Small	1455	103	70	22	15	330
Mid-size	1682	107	72	21	20	420
Large	2364	118	78	16	28	448
Pickup Trucks						
Compact	1455	118	69	20	18	360
standard	1910	138	80	18	26	468

fraction of a conventional ICE vehicle and the space available under the vehicle to store hydrogen.[20] The baseline characteristics of the ICE vehicles are given in Table 12.

For these vehicles, the maximum volume available for the hydrogen storage can be estimated using the following relationship:

$$\text{Maximum volume available for hydrogen storage} = 9 \text{ inches} \times \text{Wheelbase} \times \text{Width of the vehicle} \times 0.25$$

It is assumed that a storage unit of this volume could be packaged under the vehicle without significantly affecting overall design of the vehicle. The maximum weight allowed for the hydrogen storage unit can be taken to be 10 percent of the curb weight of a comparable ICE vehicle. This increase in weight for fuel storage can be accommodated without a significant effect on the power of the electric driveline components. The amount (kg) of hydrogen to be stored can be estimated based on a doubling of the energy efficiency of the fuel cell vehicles compared to comparable ICE vehicles and the fact that 1 kg of hydrogen is approximately equal to 1 gallon of gasoline. Hence, the number of kilograms of hydrogen required is set equal to half the gallon capacity of the gasoline tank. The range of the fuel cell and ICE vehicles would then be equal for each vehicle type.

Estimates for the hydrogen storage requirements for eight classes of vehicles—compact, mid-size and full-size cars, small, mid-size, and large SUVs, and compact and standard-size pickup trucks—are given in Table 13.

Table 13 Vehicle Hydrogen Storage Characteristics and Attribute Requirements

Vehicle Class	kg H_2 Stored[1]	Wgt. kg[2]	Volume Liter[3]	Volume Liter[4]	% wt. H_2 sys.	gm H_2/L sys.[3]	gm H_2/L sys.[4]	Range Miles
Cars								
Compact	6.5	114	255	99	5.7	26	66	442
Mid-size	8.5	141	285	130	6.0	30	65	476
Full-size	9.85	159	314	141	5.8	30	65	462
SUVs								
Small	7.5	145	265	115	5.2	28	65	330
Mid-size	10	168	283	153	6.0	35	65	420
Large	14	236	340	214	5.9	41	65	448
Pickup trucks								
Compact	9	145	300	137	6.2	30	66	360
Standard	13	191	407	198	6.8	32	66	468

[1] kg H_2 needed in fuel cell vehicle to get the same range as a comparable ICE vehicle, assuming the fuel cell vehicle is two times as efficient as the ICE vehicle.
[2] Weight of the H_2 storage system, assuming it weighs 10% that of the vehicle. The system includes the tank and all balance of plant components.
[3] Volume of the H_2 storage system if its volume is given by $W_B \times$ width $\times$ 9" $\times$ .25
[4] Volume of the H_2 storage system if it is twice the volume of the gasoline tank in the ICE vehicle.

Comparing the DOE hydrogen storage system goals in Table 11 with the vehicle system requirements in Table 13 indicates that, in terms of weight percentage, the DOE goals are adequate to achieve a hydrogen system weight of 10 percent of the vehicle weight or less. In comparing the volume goals (gm H_2/L), it appears that the goals are adequate for under-the-vehicle storage, but require meeting the 2015 goal of 80 gmH_2/L to have the volume of hydrogen storage only twice that of the current gasoline tanks in the vehicle. The data in Table 13 indicate that of the various H_2 storage technologies being developed, only liquid H_2 storage seems to offer good prospects of meeting both the near-term and long-term goals and requirements for system weight and volume.

6 POWERTRAIN CONTROL STRATEGIES

Control strategies are discussed for BEVs, series hybrid vehicles, parallel hybrid vehicles, and fuel cell electric vehicles.

6.1 Battery-powered Electric Vehicles (BEVs)

The control strategy for a BEV is very simple in that all the electrical energy to operate the vehicle comes from the battery. The primary control issues are the maximum current and minimum voltage that are permitted for the battery. These values are set so as not to damage the battery. When the battery voltage falls to the specified value, the controller limits the current and thus the performance of the vehicle. When the reduction in performance is noted by the driver, the batteries either need recharged or replaced.

6.2 Series Hybrid Vehicles

There are several control strategies that can be utilized for a series hybrid. The first consideration is whether the intent is to deplete the charge in the battery by operating the vehicle on the battery alone until the battery state-of-charge (SOC) decreases to a specified value (e.g., 30 percent) and then use the engine/generator to maintain the SOC near that value. In that case, the hybrid would be operated in the plug-in mode and the engine/generator would be used to extend the range of the vehicle on primarily battery stored energy. This design of the series hybrid would require a relatively large battery storing 10 to 20 kWh.

A second approach for control is to use the engine/generator to maintain the battery SOC in a specified relative narrow range (e.g., 40 to 60 percent) with the engine operated in an on/off mode close to the maximum efficiency/power condition. This approach will result in long battery life and high energy efficiency, but will not result in the substitution of electricity for petroleum in the transportation sector. The battery size (kWh) in a vehicle operated this way can also be smaller, as there is not an all-electric range consideration as there would be for the plug-in hybrid vehicles. The primary consideration is to utilize the control strategy that maximizes the fuel economy of the vehicle.

6.3 Parallel Hybrid Vehicles

As noted earlier in the article, a parallel hybrid can be designed to be either a charge-sustaining (CS) or a plug-in hybrid (PHEV). In the case of the CS hybrid, the control strategy is selected to maximize the fuel economy of the vehicle. This is done by operating the engine only when it can be operated efficiently (in the region of the map with low specific fuel consumption). When the power demanded is too low for efficient engine operation, it is turned off and the vehicle is operated as an electric vehicle using the electric motor to power the vehicle. The electric motor is also used as a generator to charge the battery and to recovery energy during regenerative braking.[21,22] A simulation

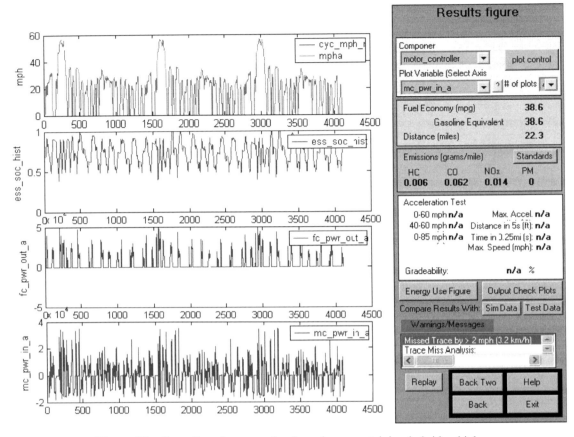

Figure 21 On–off engine operation in a charge sustaining hybrid vehicle.

result showing the on–off operation of the engine on the Federal Urban Driving Schedule (FUDS) is given in Figure 21. The energy storage unit (either a battery or ultracapacitor) is designed such that it can provide the power needed by the electric motor when the engine is off. This approach can be used in either mild or full hybrid vehicles, but it is best suited for mild hybrids in which it can be utilize to minimize the size and thus the cost of the electric driveline.

In the case of plug-in hybrids, the electric motor and energy storage unit (batteries) are sized to provide all-electric operation of the vehicle for most driving conditions with the intent of discharging the battery as the vehicle is driven until the battery reaches some specified low SOC. In designing the PHEV and its control strategy, the key issues are the all-electric range and under what power demand conditions the engine is used when the battery SOC is higher than the minimum value. In a parallel hybrid, the engine and electric motor can provide power to accelerate the vehicle. A maximum effort acceleration would require the engine to operate regardless of the battery SOC.

Hence, if all-electric range means that the engine is never used under any circumstances when the battery SOC is greater than the minimum value, the control strategy would limit the vehicle acceleration to electric motor only, even though greater acceleration performance would be possible by turning on the engine. It seems more likely that a control strategy would be implemented that minimizes engine use when the battery SOC is high, but does not forbid it, and only recharges the battery from regenerative braking until its SOC has reached the minimum specified value. This approach would result in a near maximum substitution of electricity for petroleum and at the same time offer the driver the maximum vehicle acceleration performance. It would also lead to a lower-cost hybrid powertrain because the power of the electric driveline (motor and batteries) could be smaller than would be the case if all-electric range meant no engine operation under any circumstances. After the batteries are discharged to the minimum SOC, the control strategy would be similar to that outlined for the CS hybrid with the intent of maximizing the engine operating efficiency and vehicle fuel economy. This should be more easily done with the larger battery and higher-power electric motor in the PHEV than in the CS hybrid.

6.4 Fuel Cell Electric Vehicles

The control strategy in the fuel cell vehicle is not much different in principle than that used in the engine/generator series hybrid. The fuel cell vehicle could be designed as either a CS hybrid or a plug-in hybrid (PHEV). In either design, the power of the electric motor must be selected to provide the vehicle with the required acceleration performance, and the combination of the fuel cell and battery must provide the electrical power required by the electric motor. If the power of the fuel cell is downsized, the battery will have to provide a greater fraction of the peak power. In that case, a PHEV control strategy would likely be used with the electric motor/battery providing most of the power/energy until the battery SOC reaches the minimum value, after which the fuel cell would try to keep the battery at an acceptable SOC.

Most of the fuel cell vehicles built to date by the auto industry have been CS hybrids in which the battery is used to provide peak power during acceleration and recovery energy during braking. The primary function of the control strategy is to control the power split between the fuel cell and battery during periods in which the power demand is high and changing rapidly. During periods of low power demand, the fuel cell will recharge the battery to maintain it in the specified relatively narrow range of SOC. Maximum fuel cell power would be demanded only when the vehicle power demand is high and the battery SOC is low.

7 SIMULATIONS OF ELECTRIC AND HYBRID VEHICLE OPERATION AND PERFORMANCE

Various aspects of the design of electric and hybrid vehicles have been discussed in the previous sections. In this section, the performance of the various vehicle designs is considered based on computer simulation results. Several simulation programs have been developed that can be applied to vehicles with electric drivelines. These include Advisor, PSAT, and SIMPLEV.[13,14,23] The simulation results presented in this section have been obtained using Advisor and SIMPLEV. A large volume of computer simulations have been performed in recent years by a number of groups.[24,25] The simulations presented in this article for the most part have been done at the University of California-Davis, Institute of Transportation Studies.[26,27] They are representative of results obtained by other groups. Comparisons of simulation results from different groups are difficult because the results are highly dependent on the vehicle road load parameters and acceleration performance assumed, input component characteristics used, control strategy, and driving cycle. For the simulation results given in this section, all of these inputs are well documented.[26,27] The Advisor program was developed by the National Renewable Energy Laboratory (NREL) and is readily available. Hence, it is possible for others to duplicate the simulations presented. The characteristics of the baseline ICE vehicles used in the present simulations are given in Table 14.

7.1 Battery-powered Electric Vehicles

The simulation of battery-powered electric vehicles is relatively simple because there is only one energy source, the battery, and one source of torque to the wheels, the motor. Simulation results for electric vehicles of various types using lithium-ion batteries are shown in Table 15. The simulation results in the table were obtained using the SIMPLEV program.[23] The key results to note are the vehicle energy use (Wh/mi.) and the acceleration times.

Table 14 Characteristics of Baseline ICE Vehicles of Various Types

Type	Curb Weight (kg)	C_D	A_f Ft^2	Rolling resist. coeff.	P_{max} kW	0–60 mph per Second	EPA mpg City/hw*	Price 2003$
Compact car	1160	.3	21.4	.007	95	10	25/31	16200
Mid-size car	1500	.3	23.1	.007	135	8.5	20/28	20250
Full-size car	1727	.32	23.7	.007	180	8.0	17/25	26700
Small SUV	1590	.38	26.4	.008	135	10	19/25	21925
Mid-size SUV	1910	.42	28.0	.008	165	9.5	15/19	28510
Large SUV	2500	.45	34	.008	200	9.5	14/16	35155

7 Simulations of Electric and Hybrid Vehicle Operation and Performance

Table 15 Characteristics of Electric Vehicles of Various Types

Vehicle Type	Vehicle Test Weight (kg)	Battery Wgt. (kg)[1]	Battery kWh Stored[2]	Electric Motor kW[3]	Required Battery Pulse Power W/kg[4]	Wh/mi from Battery[5]	0–60 mph per Second
Cars							
Compact	1373	168	20.2	65	387	202	11.3
Mid-size	1695	208	24.9	102	490	249	8.9
Full	1949	238	28.5	122	513	285	8.6
SUV							
Small	2103	266	31.9	128	481	319	9.6
Mid-size	2243	278	33.3	143	514	333	9.3
Full	2701	317	38.0	160	501	380	9.6

[1] Lithium-ion battery with an energy density of 120 Wh/kg
[2] All vehicles have a range of 100 miles
[3] Peak motor power
[4] Peak pulsed power required from the battery at 90% efficiency
[5] Average energy consumption on the FUDS and FHWAY drive cycles

7.2 Parallel Hybrid Vehicles

Next, consider simulation results for a plug-in hybrid obtained using Advisor. The vehicle can operate as an electric vehicle on both the Federal Urban and Highway driving cycles for about 20 miles, or until the battery is depleted to 20 percent state of charge, and then as a charge-sustaining parallel hybrid. The lithium-ion battery pack in the plug-in hybrid is about a third the weight of the pack in the electric vehicle. The simulation results for a compact-size passenger car are summarized in Figure 22 and Tables 16. The simulation results for the plug-in hybrid vehicle indicate that their effective fuel economy can be very high, even for long daily driving ranges, and that as a result they can use grid electricity for a large fraction of the miles traveled in place of gasoline. This can be accomplished with a relatively small battery pack (98 kg in the case of the compact car simulated).

Simulations have also been performed for hybrid vehicles that have essentially zero all-electric range. The electric driveline is used to permit more efficient operation of the engine and to recover energy during braking. In these vehicle designs, the engine operates in an on–off mode, but it is not off for long periods of time, as would be the case in a plug-in hybrid. The electrical energy storage unit is not recharged from the grid, but is maintained in a specified range of state-of-charge by the motor/generator using power from the engine. Hence, these hybrid vehicles use only gasoline (or other fuel). Simulations have been performed for mild

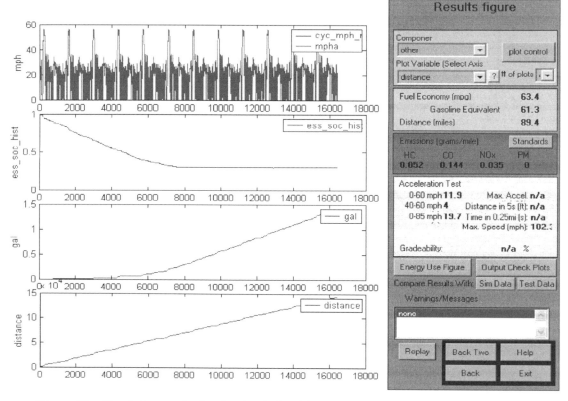

Figure 22 Simulation results for the plug-in hybrid compact car on the federal highway cycle.

Table 16 Simulation Results for the Plug-in Hybrid on the FUDS Driving Cycle

Distance Traveled Miles—FUDS	Fuel Economy mpg	kWh Electricity	Gallons Gasoline
22.4	ZEV	4.05	0
37.3	230	6.15	.16
52.2	105	6.3	.50
67.1	78	6.3	.86
74.5	71	6.3	1.05
90	63	6.3	1.43

Compact PHEV, test weight 3,025 lb., 9 kWh energy storage, 98 kg lithium-ion batteries, 55 kW engine, 50 kW electric motor 0 to 60 mph acceleration 11 sec
Conventional ICE compact car: 29 mpg FUDS, 43 mpg federal highway cycle, 0 to 60 mph 10 sec

7 Simulations of Electric and Hybrid Vehicle Operation and Performance

Table 17 Characteristics of the Hybrid Vehicles of Different Types

Vehicle Class	Test Weight (kg)	Full Hybrid			Mild Hybrid		
		Engine (kW)	Motor (kW)	Batteries (V/Ah)	Engine (kW)	Motor (kW)	Batteries (V/Ah)
Compact car	1350	60	40	335/12	85	10	150/8
Mid-size car	1660	75	65	335/20	120	15	150/13
Full-size car	1865	100	85	335/27	160	20	150/18
Small-SUV	1726	75	65	335/20	120	15	150/13
Mid-SUV	2170	90	75	335/24	150	20	150/18
Large-SUV	2636	110	95	335/30	180	25	150/22

All vehicles have CVT transmissions and nickel metal hydride batteries.

and full hybrid vehicles of various types. The characteristics of the vehicles and drivelines are given in Table 17.

The detailed simulation results for the mid-size car are shown in Table 18 for all the engines, hybrid drivelines, and driving cycles.[25] It is clear from the table that large improvements in fuel economy can be achieved with hybridization, but that the magnitude of the improvement is highly dependent on both the driveline components and the driving cycle. In addition, when expressing the improvement in percentage or fractional times, the baseline from which the improvement is given must be specified, otherwise the results are difficult to interpret.

Improvement factors for the combined Federal Urban and Highway cycles are given in Table 19 based on the conventional ICE baseline vehicles with the PFI gasoline.[25] The results in the table indicate that significant improvements in fuel economy can be achieved by using diesel and lean-burn engines in conventional ICE vehicles, and large improvements can be achieved by hybridizing the drivelines using all the different engines. The fuel economy improvements for the full hybrids are somewhat greater than those for the mild hybrids, but the differences may not be large enough to justify the higher cost of the full hybrid driveline components. Note also that the gain in fuel economy using the diesel engine in the ICE vehicles is comparable to that gained by hybridizing the PFI gasoline-powered vehicles.

7.3 Series Hybrids

Simulations were performed for series hybrid vehicles for comparison with the battery-powered vehicle and parallel hybrid results. The electric drivelines of the series hybrid vehicles were the same as that of the battery-powered vehicles, except that the batteries stored only 40 percent of the energy in the BEVs. The

Table 18 Simulation Results from Advisor for Mid-size Cars—Full and Mild Hybrids and Conventional ICE—with Various Engines

Type of Driveline	Engine Type	FUDS	Highway	Mpg US06	Japan 10/15	ECE-EUDC
Full hybrid	Gasoline PFI	35.8	44.2	30.0	33.2	35.0
	Lean burn	44.3	55.8	37.5	40.1	44.4
	TC Diesel	40.1(45.1)	53.7(60.4)	38.3(43)	36(41)	40(45)
Mild hybrid	Gasoline PFI Bat.	33.8	37.3	25.1	31.8	30.7
	Ultracap	37.2	42.9	29.4	34.2	34.7
	Lean burn Bat.	42.1	48.7	35.1	39.3	41.7
	Ultracap	45.4	54.8	38.7	43.0	45.0
	TC Diesel Bat.	37.3(42)	45(50.6)	33.2(37)	34(38)	36(41)
	Ultracap	41.2(46.3)	51.9(58.4)	36.8(41)	36(41)	40(45)
Convent. ICE—CVT	Gasoline PFI	20.4	32.3	23.3	16.5	20.2
	Lean burn	29.7	44.4	29.4	25.0	29.5
	TC Diesel	24.5(27.7)	35.1(39.5)	24.2(27)	20(23)	24(26)

All vehicles use CVTs and nickel metal hydride batteries and have 0–60 mph acceleration times of about 9 sec.
For diesel engine powered vehicles, the first mpg given is the gasoline equivalent mpg and the second number in parentheses.

Table 19 Fuel Economy (mpg) Improvement Ratios Relative to PFI Engine-powered Vehicles

Vehicle Class	ICE Diesel	ICE Lean-burn	PFI Mild	PFI Full	Lean-burn Mild	Lean-burn Full	Diesel Mild	Diesel Full
Compact Car	1.46	1.42	1.28	1.44	1.64	1.77	1.66	1.96
Mid-size car	1.53	1.42	1.42	1.60	1.81	1.98	1.81	2.04
Full-size Car	1.40	1.42	1.42	1.60	1.81	1.98	1.81	2.04
Small-SUV	1.50	1.42	1.35	1.52	1.72	1.87	1.74	2.0
Mid-SUV	1.40	1.42	1.42	1.60	1.73	1.85	1.8	2.0
Large-SUV	1.40	1.42	1.42	1.60	1.81	1.98	1.81	2.04

Table 20 Characteristics of Series Hybrid Vehicles

Vehicle	Test Weight (Kg)	Engine/generator (kW)	Battery (kWh)	FUDS (mpg)	Highway (mpg)
Mid-size car					
Series HEV	1830	40	10	40	47
CS HEV	1640			36	44
Conventional ICE	1640			20	32
Mid-size SUV					
Series HEV	2150	55	14.7	29	31
CS HEV	1910			28	32
Conventional ICE	1910			16	25

engine/generator power was selected such that the vehicles had acceptable steady gradeability on generator electricity alone. The characteristics of the series hybrid configurations are given in Table 20 for a mid-size passenger car and SUV.

Also shown in Table 20 is the fuel economy of the series hybrid vehicles compared with that of a charge-sustaining hybrid and a conventional ICE vehicle. All the vehicles use a PFI gasoline engine with a maximum efficiency of 34 percent, and both hybrids used a charge-sustaining control strategy. The series hybrids could have been operated in a plug-in mode with a range of about 30 miles if the batteries were discharged to 80 percent of their rated capacity. The simulation results indicate that the fuel economy of the series hybrids is slightly higher than that of the parallel charge-sustaining hybrids when both are operated in the charge sustaining mode. The engine/generator was sized such that the series hybrids were full-function vehicles.

7.4 Fuel Cell Vehicles

A key question concerning fuel cell vehicles is estimation of the factor by which their equivalent fuel economy will be higher than that of a baseline gasoline-fueled vehicle. Most studies of fuel cell vehicles and related hydrogen demand using those vehicles assume an improvement factor of two to three. It is of interest to consider whether available test data and simulation results support this assumed improvement.

The most relevant test data currently available are for the Honda FCX vehicle, which is shown in Table 21.[28] Based on a baseline vehicle fuel economy of 25/34 mpg for the urban/highway cycles, the improvement factors for the Honda FCX in 2005 are 2.5 on the city cycle and 1.5 on the highway cycle, resulting in an average improvement of 2.0.

Table 21 EPA Fuel Economy Ratings for the Honda FCV

	mpg 2005	mpg 2004
Federal city cycle	62	51
Federal highway cycle	51	48

Baseline vehicle: Saturn Vue (A–4, 2.2L/4 engine).

Table 22 Fuel Economy Improvement Factors Based on Simulation Results for Hydrogen Fuel Cell Vehicles (Mid-size Passenger Cars)

Source	Fuel Economy Improvement Factor*	Reference
UC Davis	2.44	28
General Motors	2.38	29
MIT	3.12	30
Directed Technologies	2.73	29
AD Little	2.52	29

*Average on the federal urban and highway cycles.

The fuel economy (efficiency) of hydrogen fuel cell vehicles have been simulated in a number of studies.[29–31] A summary of the results are given in Table 22.

The simulation results for the fuel cell vehicles are reasonably consistent and are only slightly more optimistic than the test data for the Honda FCX vehicle. Hence, it seems likely that fuel economy improvement factors in the range of two to three are achievable with hydrogen (fueled) fuel cell vehicles in the relatively near-term future. It is likely that those vehicles will incorporate energy storage (batteries or ultracapacitors) to permit sizing the fuel cell to lower power than needed to meet the peak power of the electric drive system and to recover energy during braking. Energy storage in fuel cell systems is not utilized as in the hybrid engine-electric drivelines to improve the average efficiency of primary energy converter (fuel cell or engine), because the fuel cell efficiency is a maximum at a relative small power fraction (see Figure 20). In the case of fuel cell systems, energy storage is used to recover braking energy and to reduce the size and cost of the fuel cell unit needed.

8 EMISSIONS CONSIDERATIONS

Thus far in the chapter, little has been said about vehicle emission or full cycle fuel-related (upstream) emissions, but emissions considerations are important in the assessment of advanced vehicle and alternative fuel technologies. Of particular interest are greenhouse emissions, particularly CO_2, which are closely related to fuel economy.

8.1 Regulated Vehicle Exhaust Emissions

Consider first the regulated vehicle exhaust emissions, which are hydrocarbons (HC), carbon monoxide (CO), and nitric oxides (NO_x). It seems reasonable to expect that future vehicles (ICE and HEV) with engines will have to meet the stringent California ultra low-emission vehicle (ULEV) and super ultra low-emission vehicle (SULEV) standards shown in Table 23.

At the present time (2007), the emission standards in California are more stringent than in the United States as a whole and those in Europe. The emission standards in Europe, the United States, and California are compared in Table 23. The driving cycle for the United States and California emissions tests is the FUDS (Federal Urban Driving Schedule) and the driving cycle for the tests in Europe is the ECE-EUDC schedule. This new European cycle is comparable to the FUDS cycle in that fuel economy values obtained for the same vehicle on the two cycles differ by a small percentage (less than 10 percent).

The large differences in the standards occur for NO_x and to a lesser extent for particulates. The NO_x standards are .07 and .02 gm/mi. for Tier 2 and SULEV, respectively, while the Euro 4 standard for NO_x is .4 gm/mi. and .13 gm/mi. for the proposed Euro 5 standard. These differences in the NO_x standards will result in a large challenge for companies developing diesel-powered light-duty vehicles desiring to enter the U.S./California market.

In both the United States and Europe, the future standards are being set independent of engine type and light-duty vehicle class. This compounds the difficulty of marketing diesel engines in the larger light-duty vehicle classes. The particulate standards in the United States will be more stringent than in Europe unless the Euro 5 standards (.004 gm/mi. compared to .01 gm/mi.) are adopted. Meeting the particulate standards with diesel engines appears to be less difficult than meeting the most stringent NO_x standards in the United States.

Table 23   Federal, California, and European Emissions Standards

Standard	Year	CO	HC	NO_x	PM	HC + NO_x
Fed. Tier 1 Gasoline	—	4.2	0.32	0.6	0.1	0.92
Fed. Tier 1 Diesel	—	4.2	0.32	1.25	0.1	1.6
Euro 3	2001	1.0	0.09	0.81	0.08	0.9
NLEV	—	4.2	0.09	0.3	0.08	0.39
Euro 4	2005	0.81	0.08	0.4	0.04	0.48
Fed. Tier 2 (Bin 5)	2007	4.2	0.09	0.07	0.01	0.16
Euro 5 (proposed)	2008	1.6	0.08	0.13	0.004	0.21
California ULEV	2004	1.0	0.04	0.05	0.01	0.09
California SULEV	—	1.0	0.01	0.02	0.01	0.03

As indicated in Table 2, the charge-sustaining hybrid vehicles being marketed in California by Toyota and Honda meet the SULEV standard. This indicates that the engine control and exhaust emission after-treatment technologies for reducing emissions from gasoline engines to very low levels are available and being used in production vehicles. This is not the case for diesel or lean-burn engines using gasoline. Those engines are very attractive for increasing fuel economy, but further R&D is needed to reduce vehicle emissions from vehicles using those engines to SULEV levels.

8.2 Greenhouse Gas Emissions from the Vehicle

The CO_2 emissions from a vehicle depend on the fuel converter (engine or fuel cell) on board the vehicle, as well as the fuel being used. Battery-powered and hydrogen fuel cell vehicles, of course, will exhaust no CO_2 because there is no carbon associated with the energy stored on board the vehicle. Hence, greenhouse gas exhaust emissions are an issue only for hybrid-electric vehicles using liquid hydrocarbon fuels such as gasoline, diesel, and ethanol. The properties of these fuels are given in Table 24.

Table 24 indicates the inverse relationship between the vehicle exhaust CO_2 emissions and fuel economy. The mpg value to be used is not the gasoline equivalent, but the mpg for the fuel of interest. In the case of diesel fuel relative to gasoline, the fuel economy of the diesel-fueled vehicle must have a fuel economy of about 18 percent higher than the gasoline vehicle before the diesel-fueled vehicle has the lower CO_2 emissions. If the fuel economy of diesel engine vehicles is 30 to 40 percent greater than the comparable gasoline engine vehicles, the resultant reduction in the CO_2 emissions is 11 to 19 percent.

Next, consider the case of the use of ethanol compared with gasoline. If the efficiency of the two vehicles is the same (that is, equal MJ/mi, where MJ = mega joules or 10^6 joules), the mpg will be the ratio of the MJ/liter of the two fuels, which is .64. Hence, even though the ethanol has a much lower carbon content than gasoline, its use results in about 2 percent higher exhaust CO_2 emissions.

Table 24 Chemical and Physical Properties of Gasoline and Diesel Fuel

Property	Gasoline	Diesel Fuel	Ethanol
Density gm/cm^3	.75	.86	.79
Lower HV			
MJ/kg	44	43.2	26.7
MJ/liter	33	37.15	21.1
Composition for calculating CO_2	C_8H_{18}	$C_{13}H_{24}$	C_2H_5OH
MW	114	180	46
gmCO$_2$/mi.	8,820/mpg	10,400/mpg	5,760/mpg

Calculation of the effect of implementation of the different hybrid vehicle technologies on greenhouse gas emissions on new vehicles sold is a complex matter, as it depends on how many of the various types of vehicles are sold as well as their fuel economy improvement.[25] Of particular interest is the CAFÉ (Corporate Average Fuel Economy) that could result from using hybrid electric technologies for all the vehicles sold (see Table 14 for a list of various types of vehicles). The CAFÉ and CO_2 emissions that would be achieved is shown in Table 25. Note the large improvement in fleet average fuel economy and greenhouse gas emission reductions that could be achieved compared to business as usual with conventional ICE powertrains in the vehicles. The price factors indicate the fractional change in the retail prices of the hybrids compared to the baseline PFI ICE vehicles. The result shown in Table 25 indicate that a 37.5 percent increase in average fuel economy and a comparable reduction in CO_2 emissions could be achieved with only a 7 percent increase in average vehicle price.

8.3 Upstream Full Fuel Cycle Emissions

The emissions discussed in the previous sections were from the vehicle exhaust. These emissions are often referred to as the *tank-to-wheels* emissions. Next, consider the emissions due to the production and distribution of the fuel or energy used by the vehicles. These are often referred to as the *well-to-tank* emissions. There have been a number of studies of *well-to-wheels* emissions.[32,34] Those references are used to compile a summary (Table 26) of the well-to-tank (upstream) emissions for gasoline, diesel, ethanol, hydrogen, and electricity, as they are the fuels/energy utilized in the previous discussions of battery-powered electric, hybrid, and fuel cell vehicles. The upstream emissions are usually expressed as gm CO_2/MJ energy tank of the vehicle. The efficiency of the fuel/energy production is given as MJ primary energy/MJ energy tank. Note from Table 26 that

Table 25 CAFÉ, CO_2 Emission, and Price Values for Various Technologies

Technology	Mild	Hybrid		Full	Hybrid	
Engine	CAFÉ (mpg)*	CO_2 (gm/mi.)	Price Factor	CAFÉ (mpg)	CO_2 (gm/mi.)	Price Factor
PFI	37.7	234	1.07	42.4	208	1.16
Lean-burn	47.8	185	1.09	52.0	170	1.18
TC diesel	48.1	216	1.17	55.0	189	1.23
Baseline PFI ICE Vehicles	27.4	322	1.0			

*The fuel economy values are based on the unadjusted EPA test data and simulation results for the various vehicle classes.

Table 26 CO$_2$ Emission Factors for Fuels from Various Pathways

Pathway	(MJ)in/(MJ)out	(gmCO$_2$)/(MJ)out
Gasoline	1.23	20
Crude oil		
Diesel	1.19	13
Crude oil		
CNG	1.15	16
Methanol NG	1.59	24
GH2/ Central RF/NG	1.7	104
GH2/ Station RF/NG	1.85	109
GH2/ Station electrolysis/ NG/CC	3.13	190
Electricity/ US mix	2.44	190

both the efficiency and the CO$_2$ emissions are strongly dependent on the pathway for the production of the fuel and the primary energy source.

The baseline fuels are gasoline and diesel from crude oil and natural gas. The efficiency and emissions for these fuels are favorable, being in the range of 80 to 85 percent efficient and 15 to 25 gm CO$_2$/MJ tank. The total full-cycle emissions for a vehicle are the sum of the vehicle exhaust emissions and the upstream emissions from the production of the fuel, which can be written as

$$(\text{gm CO}_2/\text{mi.})_{total} = (\text{gm CO}_2/\text{mi.})_{exhaust} + \{(MJ_{tank}/\text{mi.})^*(\text{gm CO}_2/MJ_{tank})\}$$

where $MJ_{tank}/\text{mi.} = \text{gal/mi.}^* 3.82^* (MJ/L)_{fuel}$

For example, in the case of gasoline,

$$(\text{gm CO}_2/\text{mi.})_{total} = 1/\text{mpg}\{8820 + 3.82 \times 33 \times 20\} = 1/\text{mpg} \times 11,341$$

and the upstream emissions increase the CO$_2$ emission by 28.5 percent. The effect on total CO$_2$ emissions of using the various hybrid vehicle technologies can be calculated using the equation above and the CAFÉ fuel economy values given in Table 25.

Consider a BEV having an energy consumption of 200 Wh/mi. from the battery with a battery charging efficiency of 85 percent. The total CO_2 emissions can be written as an equation:

$$(\text{gm } CO_2/\text{mi.}) = (\text{Wh/mi.})_{bat}/\text{EFF}_{bat\ chg} \times (\text{gm } CO_2/MJ) \times 1MJ/278\ Wh$$

$$(\text{gm } CO_2/\text{mi.}) = 200/.85 \times 190 \times 1/278 = 161$$

This compares to 324 gm CO_2/mi. for a gasoline vehicle getting 35 mpg.

Next consider a hydrogen fuel cell vehicle that has a gasoline equivalent fuel economy of 87.5 mpg (2.5 times that of the gasoline vehicle). All the CO_2 emissions for this vehicle are the result of the production of the hydrogen from another energy source such as natural gas (steam reforming) or electricity (electrolysis). The CO_2 emissions can be calculated from the following equation:

$$(\text{gm } CO_2/\text{mi.})_{H_2} = 1/\text{mpg}_{gasoline,equiv.} \times 3.82$$
$$\times (MJ/L)_{gasoline} \times (\text{gm } CO_2/\text{mi.})_{H_2}$$

Using this equation and the CO_2 emission factors for hydrogen production from Table 26, the following CO_2 emissons are calculated.

H_2 from reforming natural gas: 150 gm CO_2/mi.

H_2 from electrolysis: 274 gm CO_2/mi.

These results show that the CO_2 emissions vary markedly, depending on the powertrain technology and the fuel used in the vehicle and how the fuel is produced from a primary energy source.

9 ECONOMIC CONSIDERATIONS AND MARKET PENETRATION

In the final analysis, whether a new technology has an important effect on the environment depends on the extent to which the new technology is economically competitive with the current incumbent technology and its resultant market penetration. In this section, the various advanced vehicle technologies are compared with the baseline technologies in the different vehicle types in terms of their economic attractiveness. The characteristics of the baseline ICE vehicles were given in Table 14. Each of the vehicle technologies is discussed separately.

9.1 Battery-powered Electric Vehicles

In order to determine the economic attractiveness of a technology, it is necessary calculate the effect on both the initial price of the vehicle and the operating costs, primarily the cost of energy (gasoline and electricity). This has been done for a family of vehicle types having a range of 100 miles using lithium-ion batteries. The characteristics of the electric vehicles were given previously in Table 15. All these vehicles are full-function BEVs that can be used on all types of roads. The

Table 27 Battery-powered Vehicle Cost Characteristics for Vehicles of Various Types

Vehicle types	Retail cost of the battery $[3]	Retail cost of the motor $[4]	Retail Price differential $	Cost of Electricity for 100,000 miles at 6 cents/kWh $[1]	Gallons of gasoline used by the baseline ICE	Break-even gasoline price ($/gal)[2]
Cars						
Compact	7070	1532	6280	1424	2941	2.62
Mid-size	8627	1764	6543	1763	3448	2.41
Full–size	9915	1879	6664	2010	4000	2.17
SUVs						
Small	11089	1914	9164	2256	3846	2.97
Mid-size	11585	1994	8734	2348	5000	2.22
Large	13220	2085	9462	2679	5555	2.19

[1] Cost of electricity = 100 × Wh/mi. × Cents/kWh × .01/ battery charging eff.; battery charging efficiency = .85, 6 cents/kWh electricity

[2] gallons gasoline by ICE × ($/gal)$_{breakeven\ gasoline}$ = Delta vehicle price + Electricity cost for 100,000 miles

[3] OEM battery cost $250/kWh: all vehicles have a range of 100 miles

[4] Mark-up factor from OEM cost to retail price is 1.4

economics of the vehicles are given in Table 27. The retail cost of the battery and electric motor, as well as the net retail price difference (after the cost of the ICE driveline components is subtracted), are shown in the table. The cost of the battery is dominant even when the OEM cost of the battery is only $250/kWh.

The economic attractive of the BEV can be expressed in terms of the breakeven gasoline, which is the price of gasoline at which the difference in cost of gasoline for the baseline ICE vehicle over 100,000 miles and electricity for the same miles in the BEV exactly offsets the initial retail price differential for the vehicles. The breakeven gasoline prices calculated are relatively low, being in the $2 to $3/gallon range for all the vehicle types. Note that it was assumed in the calculations that the battery OEM cost was $250/kWh and the life of the batteries was at least 100,000 miles, electricity was 6 cents/kWh, and the initial price differential of the BEV was recovered over 100,000 miles. All of these are relatively optimistic assumptions. In addition, the range of the vehicles was 100 miles, which may limit the market for a full-function BEV. Longer range would increase the cost of the battery and increase the breakeven gasoline price. The assumed battery cost of $250/kWh is above the USABC goal of $150/kWh in large volume production, but well below the present cost of large lithium-ion batteries, which is currently $700 to $800/kWh.

9.2 Plug-in Hybrid Vehicles

The economics of the plug-in hybrid can be studied using the same approach as followed for the battery-powered electric vehicle in the previous section. That is, add the cost of the driveline components for the nonconventional vehicle and subtract off the cost of the conventional engine/transmission driveline. The battery in the plug-in hybrid was sized to give the hybrid vehicle a 20-mile range as an electric vehicle. It was assumed that only 50 percent of the kWh capacity of the battery could be used in order that the life of the battery would be at least 100,000 miles. The electric motor and the engine were sized to be the same as discussed previously (Table 17) for the full charge-sustaining hybrid. In other words, the driveline of the plug-in hybrid is essentially the same as the full CS hybrid except that the battery energy-storage capacity is much greater. The component cost values used in the calculations are for high volume production.[35,36] The mark-up factor of 1.4 used in the BEV analysis to relate OEM cost and retail price is used in this calculation.

As in the battery-powered vehicle analysis, the cost calculations are done for 100,000 miles. Further, it is assumed that the use pattern of the plug-in hybrid is such that 70 percent of the miles will be driven on electricity and 30 percent will be driven with the vehicle in the HEV charge-sustaining mode. The fuel economy in this mode was assumed to be the same as that shown in Tables 18 and 19 for the full hybrid vehicle. The electrical energy consumption (kWh/mi) was assumed to be the same as that given in Table 15 for the BEVs of the same vehicle type.

The driveline and cost characteristics of the plug-in hybrid vehicles of various types are given in Tables 28 and 29. Note that even though the battery in the plug-in hybrid is much smaller than in the BEV, it is still the highest cost component in the driveline. That is the reason that the all-electric range of the vehicle was limited 20 miles. A battery unit cost of $350/kWh was assumed, which was higher than the $250/kWh assumed for the BEV because the plug-in hybrid requires a high-power battery with a P/E (power to energy) ratio of 8 to 10.

The breakeven gasoline price was calculated for the plug-in hybrid, much as was done for the BEV by equating the life cycle operating cost, including the differential initial cost, of the plug-in hybrid and the comparable conventional ICE vehicle. The results of the calculation are given in Table 30. The values of the breakeven gasoline cost are relatively low, being in $1.50 to $2/gallon range. These values are lower than often quoted, for a couple of reasons. First, the mark-up factor used in these calculations was only 1.4, compared to values of 20 to 2.5 often used. The lower value was used based on the detailed analysis of vehicle costs given in References 35 and 36. For example, if a mark-up factor of 1.75 had been used,[37] the breakeven gasoline prices calculated would have been in the range of $3 to $3.5/gallon. The breakeven gasoline prices are also sensitive to all the other assumptions made to perform the calculations. Of

Table 28 Driveline Characteristics of the Plug-in Hybrid Vehicles

Vehicle	Electric Motor (kW)	Engine (kW)	Battery (kWh)[1]	Wh/mi Electric Alone	HEV (mpg)	ICE (mpg)
Car						
Compact	40	60	8.2	202	44	28
Mid-size	65	75	10	249	38	24
Full-size	85	100	11.4	285	33	21
SUV						
Small	65	75	12.8	329	33	22
Mid-size	75	90	13.3	333	27	17
Large	95	110	15.2	380	24	15

[1] SOC range is 50%.

Table 29 Cost Characteristics of the Plug-in Hybrid Vehicles

Vehicle	Battery Cost ($)[1]	Motor Cost ($)	Engine Cost ($)[3]	Total OEM Cost ($)	Total Retail Cost[2]	ICE Engine Cost ($)	Differential Retail Cost ($)
Cars							
Compact	2870	960	1200	5030	7042	2280	4762
Mid-size	3500	1099	1500	6099	8539	3848	4691
Full-size	3990	1196	2000	7186	10060	5130	4930
SUV							
Small	4480	1099	1500	7079	9911	3848	6063
Mid-size	4655	1154	1800	7609	10653	4845	5808
Large	5320	1240	2200	8760	12264	5843	6421

[1] High volume battery cost $350/kWh, P/E = 8–10
[2] Mark-up factor from OEM to retail 1.4
[3] OEM engine cost $20/kW

particular importance is the fraction of battery capacity that can be used on a regular basis as that affects strongly the battery capacity required for a specified range and thus the battery cost. In the current calculation, the usable fraction was taken to be 50 percent. If that fraction had been 80 percent, the breakeven gasoline price would have been close to $1/gallon.

Table 30 Breakeven Gasoline Prices for Plug-in Hybrids of Various Vehicle Types

Vehicle [1]	Differential Retail Cost ($)	kWh/mi. Battery Alone	Electricity Cost($)[2]	Gallons of Gasoline/ yr.[3]	Gallons of Gasoline ICE[4]	Breakeven Gasoline/price ($/gal)[5]
Cars						
Compact	4,762	.202	998	682	3,571	1.99
Mid-size	4,691	.249	1,230	789	4,166	1.75
Full-size	4,930	.285	1,408	909	4,761	1.65
SUV						
Small	6,063	.319	1,576	909	4,545	2.10
Mid-size	5,808	.333	1,645	1,111	5,882	1.56
Large	6,421	.382	1,878	1,250	6,666	1.53

[1] Vehicle mileage 100,000 miles; 70% all-electric and 30% as an HEV
[2] Battery charging efficiency 85%; electricity cost is 6 cents/kWh
[3] Gallons of gasoline operating in the HEV mode
[4] Gallons of gasoline for the conventional ICE vehicle in 100,000 miles
[5] Breakeven gasoline price = (Delta retail price + Cost of electricity)/(ICE gal. − HEV gal.)

9.3 Charge-sustaining Hybrid Vehicles

The approach to evaluate the economics of charge-sustaining hybrids is not much different than that for plug-in hybrids, except that all the energy/fuel used by the vehicle is gasoline. In the analysis it is of particular interest to compare the economics of mild and full hybrid designs. The characteristics of the vehicle and driveline of the hybrids are given in Table 17. Note that in these vehicles the battery technology used is nickel metal hydride, which is the preferred technology in all the hybrids being marketed. The assumed retail price of the batteries was $350/kWh, which is the same as assumed for the lithium-ion batteries in the plug-in hybrid analysis. The fuel economy values used in this analysis are those given in Table 18 and 19.

The costs and fuel savings results for the mild and full hybrids are given in Table 31 for compact and mid-size cars and a mid-size SUV. The breakeven gasoline price is also shown in the table. This price is the value for which the increased price of the hybrid will be recovered in 100,000 miles due to its higher fuel economy. The breakeven gasoline prices are in the range of $2 to $2.60/gallon for the full hybrids and $1.20 to $2 /gallon for the mild hybrids. This indicates that mild hybrids are more cost effective than full hybrids. It is also of interest to compare the cost characteristics of the charge sustaining full hybrids

Table 31 Cost Results for Various Engines and Vehicle Classes for Full and Mild Hybrids

Engine/Vehicle	Retail Price diff. ($)[2]	Full HEV Fract. Fuel Saved[1]	Fuel Saved (gal)	$/gal Breakeven	Retail price diff.$	Mild HEV Fract. fuel Saved	Fuel Saved gal	$/gal Breakeven[3]
PFI								
Compact	2461	.31	1104	2.65	1063	.23	800	1.58
Mid-car	3371	.37	1699	2.36	1441	.30	1370	1.25
Mid-SUV	4273	.36	2125	2.39	2139	.28	1638	1.55
Lean-burn								
Compact	2701	.45	1574	2.04	1403	.40	1395	1.20
Mid-car	3631	.49	2248	1.92	1921	.45	2060	1.11
Mid-SUV	4613	.47	2770	1.98	2739	.43	2536	1.28
TC-Diesel								
Compact	3541	.47	1660	2.53	2593	.41	1430	2.16
Mid-car	4541	.52	2373	2.27	3601	.46	2099	2.04
Mid-SUV	5803	.51	3021	2.28	4839	.45	2696	2.13

[1] All fuel use is based on the FUDS/Highway composite driving cycle and 100,000 miles
[2] The baseline vehicle in all cases is the conventional vehicle using a gasoline PFI engine
[3] The breakeven gasoline price is calculated for a use period of 8 years and mileage of 12,000 miles/yr. and a discount rate of 4%

with those of the plug-in hybrids. As expected, the retail price differences are significantly higher for the plug-in hybrids due to the larger batteries required. Also as expected, the fuel savings of the plug-in hybrids are much higher than for the charge-sustaining hybrids because they used electricity for 70 percent of the miles traveled. What might be unexpected is that the breakeven gasoline prices for the plug-in hybrids are lower than for the charge-sustaining hybrids for an electricity price of 6 cents/kWh. These results indicate that the economics of the plug-in hybrids can be favorable when lower-cost off-peak electricity is used to recharge the batteries at night.

It is also of interest to consider the total CO_2 emissions of the charge-sustaining and plug-in hybrids for the current assumptions. Using the relationships cited previously for calculating the total CO_2 emissions, one finds the following emissions for a mid-size car and electrical powerplant emissions of 190 gm CO_2/MJ:

Charge-sustaining hybrid: 298 gm CO_2/mi.

Plug-in hybrid: 229 gm CO_2/mi.

Battery-powered electric: 200 gm CO_2/mi.

ICE vehicle: 472 gm CO_2/mi.

These emission comparisons indicate that the plug-in hybrids are a cost-effective approach to reducing CO_2 emissions, because their emission reductions are close to those of battery-powered electric vehicles and their breakeven gasoline prices are significantly lower.

9.4 Series Hybrids

The cost characteristics of series hybrids have been calculated using the same approach as used for the other hybrid vehicles. The vehicle characteristics used are those shown previously in Table 17. The unit costs of the components assumed are the same as those used in the previous analyses. The results for a mid-size passenger car and a mid-size SUV are shown in Tables 32 and 33. The retail price increase of the series hybrid is about the same as the plug-in hybrid and significantly higher than that of the charge-sustaining hybrid. This is the case for three reasons:

1. The electric motor in the series hybrid must be the same size in a battery-powered vehicle.
2. There is the cost of the generator attached to the engine in the series hybrid.
3. The battery in the series hybrid is much larger than in the charge-sustaining hybrid.

The breakeven gasoline price for the series hybrids is in the range of $3 to $4/gallon, even though it has a fuel economy improvement slightly higher than

Table 32   Cost Characteristics of the Series Hybrid Vehicles

Vehicle	Battery Cost ($)[1]	Motor Cost ($)	Engine/ Generator Cost ($)[3]	Total OEM Cost ($)	Total Retail Cost[2]	ICE Engine/trn Cost ($)	Differential Retail Price ($)
Mid-size car	3500	1260	1400	6160	8624	3848	4776
Mid-size SUV	5145	1424	1925	8494	11892	4845	7047

[1] High-volume battery cost $350/kWh, P/E = 8–10
[2] Mark-up factor from OEM to retail 1.4
[3] OEM engine/generator cost $35/kW

Table 33 Breakeven Gasoline Prices for Series Hybrid Vehicles

Vehicle [1]	Differential Retail Cost $	gallons of Gasoline/yr. [2]	gallons of Gasoline ICE [3]	Breakeven Gasoline Price ($/gal) [4]
Mid-size car	4,776	2,299	3,846	3.09
Mid-size SUV	7,047	3,333	4,878	4.56

[1] Vehicle mileage 100,000 miles
[2] Gallons of gasoline operating in the HEV mode
[3] Gallons of gasoline for the conventional ICE vehicle in 100,000 miles
[4] Breakeven gasoline price = (delta retail price)/(ICE gal. − HEV gal.)

the full charge-sustaining hybrid (see Table 20). This breakeven gasoline price is much higher than that of either the plug-in hybrid or the charge-sustaining hybrid. In the case of the plug-in hybrid, the reason is that it has substituted lower-cost electrical energy for gasoline used by the series hybrid. In the case of the charge-sustaining hybrid, its retail price differential is lower due to its smaller battery and no need for a separate generator. Hence, there appears to be no good reason to use the series hybrid configuration rather than the parallel configuration when an engine is used as the primary energy converter on the vehicle.

9.5 Hydrogen Fuel Cell Vehicles

Next consider the economics of hydrogen fuel cell vehicles. The cost of the fuel cell driveline is the sum of the costs of fuel cell, the hydrogen storage system (tanks and regulators), the electric motor and electronics, and the electrical energy storage unit (batteries or ultracapacitors). The costs of the electric driveline components can be calculated as follows:[35,36]

Electric motors: OEM motor cost($) $= -111.3 + (127.7 \times \ln(kW_{peak}))$

Power electronics/controller $= 480 + (2.95 \times (kW_{peak}))$

These relationships are valid for high production rates of 200,000 units/yr.
The cost of the electrical energy storage units are

Batteries: OEM battery cost($) $= (kWh)_{bat} \times (\$/kWh)_{bat}$

Ultracapacitors: OEM ultracap cost($) $= (Wh)_{cap} \times (96^*cents/F)/V_0^2$

The unit cost values used for the batteries and ultracapacitors are

Batteries: $250 − 500/kWh   Ultracapacitors: 0.5 − 1.0 cents/F

The OEM fuel cell system cost is assumed to be $45/kW, based on the analysis of fuel cell costs.[38] Estimates of the cost of the hydrogen storage system (tanks and regulators) span a wide range, from the current cost of $400 to 500/kg H_2

9 Economic Considerations and Market Penetration

Table 34 Characteristics of the Baseline and Fuel Cell Vehicles

Vehicle type	Fuel Cell (kW)	Batteries					Ultracapacitors	
		H_2 Stored (kg)	Electric Motor (kW)	Energy Storage (kW)	Energy Storage (kWh)	kg	Energy Storage (Wh)	kg
Cars								
Compact	30	3.5	65	55	3.3	55	185	31
Mid-size	50	4.2	102	72	4.3	72	240	40
Full	65	4.8	122	85	5.1	85	285	47
SUV								
Small	65	4.6	128	90	5.4	90	300	50
Mid-size	75	6.0	143	100	6.0	100	340	56
Full	85	6.7	160	115	6.9	115	385	64

to the 2015 DOE goal of $67/kgH$_2$. For purposes of this analysis, a value of $125/kgH$_2$ is assumed. The vehicle range is taken to be 300 miles.

The characteristics of the fuel cell vehicles analyzed are given in Table 34. Those of the baseline IC vehicles have been given previously in Table 14. The road load characteristics and acceleration performance of the fuel cell and baseline vehicles are the same.

The results of the cost calculations are shown in Table 35. The highest-cost components are the fuel cell and the batteries. The batteries are relatively large because they have to provide the power to supplement the output of the fuel cell. The electric motors have high power in order to meet the acceleration performance of 0 to 60 mph in 9 seconds. The costs of the fuel cell and hydrogen storage are much higher than shown in Table 35, but with large-volume production it is projected that the costs will be reduced to those shown in Table 35.

The next step in the cost analysis is to estimate the breakeven fuel cost that would repay the fuel cell vehicle purchaser over 100,000 miles for the higher price of the fuel cell vehicles. Gasoline and hydrogen are clearly two very different fuels whose price on an unit energy basis could be different. However, in this analysis it is assumed that this is not the case, and therefore the price ($/kg) of the hydrogen is taken to be the same as the $/gallon price of gasoline. The breakeven fuel price is strongly dependent on the number of miles over which the additional cost of the vehicles is to be recovered. The fuel savings over 100,000 miles and the corresponding breakeven fuel price are shown in Table 36. Note that the breakeven fuel price varies from about $1.80/gallon equivalent to $2.50/gallon equivalent with the higher prices for the larger vehicles. Also shown in the table is the fractional increase in the cost of fuel cell vehicle for the various types of

Table 35 Component and Driveline Costs for Various Types of Fuel Cell Vehicles

Vehicle Type	Electric Drive ($)	Fuel Cell ($)	H$_2$ Storage ($)	Battery ($)	Ultracaps ($)	Manuf. Cost ($)	Retail Price ($)	ICE ($)	Retail Price Difference ($)
Cars									
Compact	1,094	1,350	438	1,320	1,203	4,132	5,785	2,280	3,505
Mid-size	1,261	2,250	525	1,720	1,560	5,676	7,946	3,848	4,098
Full	1,343	2,925	600	2,040	1,710	6,728	9,419	5,130	4,289
SUV									
Small	1,367	2,925	575	2,160	1,950	6,917	9,684	3,848	5,836
Mid-size	1,425	3,375	750	2,400	2,210	7,850	10,990	4,845	6,145
Full	1,489	3,825	838	2,760	2,502	8,782	12,295	5,843	6,451

Table 36 Fuel Savings and Breakeven Fuel Price for Fuel Cell Vehicles

Vehicle Type	Retail Cost Diff. ($)	Fraction of the Vehicle Retail Price	ICE (mpg)	Fuel Cell Vehicle (mpg Equiv.)	Fuel Gallon Equivalent Savings (100,000 miles)	Breakeven Fuel Price ($/gal equiv.)
Cars						
Compact	3,505	.22	34	85	1,765	1.99
Mid-size	4,098	.20	29	72	2,059	1.99
Full	4,289	.16	25	62	2,387	1.80
SUV						
Small	5,836	.26	26	65	2,308	2.52
Mid-size	6,145	.22	20	50	3,000	2.05
Full	6,451	.18	18	45	3,333	1.94

vehicles. This varies from 16 to 26 percent with, in general, the fractional costs being slightly lower for the passenger cars.

The breakeven fuel prices are in the range of $1.8 to $2.5/gal equivalent even though the percent vehicle price differential is high. This is the case because the fuel cell vehicles had an equivalent fuel economy 2.5 times that of the baseline gasoline vehicles. The price differential of fuel cell vehicles is higher than that of hybrid vehicles, but because the fuel economy improvement factor for the fuel cell vehicle is higher than for hybrids (Tables 18 and 19), the breakeven fuel prices are in the same range. If the equivalent price of hydrogen is higher than that of gasoline, then the breakeven price of gasoline for use in the baseline vehicles would also have to be higher.

10 SUMMARY AND CONCLUSIONS—MARKET INTRODUCTION OF ADVANCED TECHNOLOGY VEHICLES

The process by which a new technology is ultimately introduced into the market and its sales increase to the point that it can be said to be mass produced is very complex and takes many years.[39,40] This is a complex process for a number of reasons. First, the new technology must be developed to the point that it is shown to be technically feasible and attractive at an affordable cost. Second, there must be a reasonable pathway to reduce the cost of the advanced vehicles in high production volumes to be competitive with comparable vehicles that use conventional technology. Third, tests and demonstrations of the advanced vehicles must be done over some period of time to show their utility and durability to the potential market.

Advanced vehicles, such as hybrids and fuel cell vehicles, will be offered for sale even in relatively small volume by the auto companies only after they have concluded that eventually a mass market will develop for the new technologies. Currently, this is happening in the case of charge-sustaining hybrid vehicles, with sales of several hundred thousand vehicles likely in 2007.[41] This is not likely to happen in the case of fuel cell vehicle for at least 10 to 15 years because of the high cost of fuel cells and the need to develop a hydrogen fuel infrastructure. The increasing sales of charge-sustaining hybrids could lead to the introduction of plug-in hybrid vehicles in the next 5 to 10 years and eventually to the marketing of city battery-powered electric vehicles with limited range and performance.

The materials presented in the previous sections have shown that most of the driveline and energy storage technologies to design and manufacture advanced vehicles from small passenger cars to large SUVs are now available. The cost of the battery and fuel cell technologies is still high, but it is projected that their costs can be reduced to the point that vehicles using them can be cost competitive over the lifetime of the vehicles. It seems likely that the initial prices of vehicles using the advanced technologies will be higher than those using conventional ICE engines/transmissions, but calculations of fuel use indicate that those cost differences can be recovered from reduced fuel use in the advanced vehicles in less than 100,000 miles at relatively low gasoline prices ($1.50 to $2.50/gallon).

REFERENCES

1. A. F. Burke, "The Future of Hybrid-Electric ICE Vehicles and Fuels Implications," ITS-Davis Report UCD-ITS-RR-02-09, October 2002.

2. A. F. Burke, "Saving Petroleum with Cost-Effective Hybrids," SAE Paper 2003-01-3279, paper presented at the Powertrain and Fluids Conference, Pittsburgh, Pa. October 2003.
3. M. Duoba, H. Ng, and R. Larsen, "In-situ Mapping and Analysis of the Toyota Prius HEV Engine," SAE paper 2000-01-3096, March 2000.
4. C. C. Chan and K. T. Chau, *Modern Electric Vehicle Technology,* Oxford University Press, 2001.
5. S. A. Nasar, *Electric Machines and Power Systems, Volume 1—Electric Machines*, McGraw-Hill, New York 1995.
6. P. T. Krein, *Elements of Power Electronics,* Oxford University Press, 1998.
7. T. E. Lipman, "The Cost of Manufacturing Electric Vehicle Batteries," Institute of Transportation Studies, University of California-Davis, Report UCD-ITS-RR-99-5, May 1999.
8. M. Anderman, Proceedings of the Seventh Advanced Automotive Battery and Ultracapacitor Conference, Long Beach, California, May 2007.
9. A. Burke, M. Miller, and Z. McCaffrey, "The Worldwide Status and Application of Ultracapacitors in Vehicles: Cell and Module Performance and Cost and System Considerations," Proceedings of EVS-22, Yokohama, Japan, October 2006.
10. A. F. Burke and M. Miller, "Supercapacitor Technology-Present and Future," proceedings of the Advanced Capacitor World Summit 2006, San Diego, California, July 2006.
11. R. Van Basshuysen and F. Schafer, *Internal Combustion Engine Handbook,* published by SAE International, 2002.
12. J. B. Heywood, *International Combustion Engine Fundamentals*, McGraw-Hill, New York, 1988.
13. T. Markel et al., "ADVISOR: A Systems Analysis Tool for Advanced Vehicle Modeling," *Journal of Power Sources,* **110**(2), 255–266 (August 2002).
14. PSAT documentation, www.transportation.anl.gov/software/PSAT/. Accessed on December 14, 2007.
15. P. Craig, "The Capstone Turbogenerator as an Alternative Power Source," SAE paper 970292, February 1997.
16. J. Larminie and A. Dicks, *Fuel Cell Systems Expanded,* John Wiley & Sons, Hoboken, NJ, 2002.
17. S. Srinivasan, *Fuel Cells—From Fundamentals to Applications,* Springer Science, 2006.
18. E. J. Carlson, P. Kopf, J. Sinha, S. Sriramulu, and Y. Yang, "Cost Analysis of PEM Fuel Cell Systems for Transportation," NREL Report SR-560-39104, December 2005.
19. A. Bouza, C. J. Read, S. Satyapal, and J. Milliken, *2004 Annual DOE Hydrogen Program Review, Hydrogen Storage*, Office of Hydrogen, Fuel Cells, and Infrastructure Technologies, 2004.
20. A. F. Burke and M. Gardiner, "Hydrogen Storage Options: Technologies and Comparisons for Light-duty Applications," Report No. UCD-ITS-RR-05-01, January 2005.
21. A.F. Burke, "Application of Ultracapacitors in Mild/Moderate Hybrid-Electric Vehicles," paper presented at EssCap06, Lusanne, Switzerland, November 2006.
22. A. F. Burke, "Engine and Fuel Cell Powered Vehicles using Batteries and Ultracapacitors," Proceedings of the IEEE, **95**(4), April 2007.

23. G. H. Cole, "SIMPLEV: A Simple Electric Vehicle Simulation Program-Version2," EG&G Report No. DOE/ID-1,0293-2, April 1993
24. S. Plotkin, D. Santini, et al. "Hybrid Electric Vehicle Technology Assessment: Methodology, Analytical Issues, and Interim Results," Argonne National Laboratory Report ANL/ESD/02-2, October 2001.
25. M. A. Kromer and J. B. Heywood, "Electric Powertrains: Opportunities and Challenges in the U.S. Light-duty Vehicle Fleet," MIT publication LFEE 2007-02 RP, May 2007.
26. A. F. Burke, "Engine and Fuel Cell Powered Vehicles using Batteries and Ultracapacitors," paper to be published in the *IEEE Journal*, 2007.
27. A. F. Burke and A. Abeles, "Feasible CAFÉ Standard Increases using Emerging Diesel Hybrid-electric Technologies for Light-duty Vehicles in the United States," *World Resource Review*, **16**(3) (2004).
28. EPA, Honda FCV fuel economy data, 2005.www.corporate.honda.com/environment/fuelcell/2006.
29. D. J. Friedman, "Maximizing Direct Hydrogen PEM Fuel Cell Vehicle Efficiency–Is Hybridization Necessary?" SAE Paper 1999-01-0530, February 1999.
30. *On the Road to 2020*, MIT Energy Laboratory, October 2000.
31. S. R. Ramaswamy, *Understanding Fuel Cell Vehicles*, Handbook for FCV Workshops, ITS-Davis Report UCD-ITS-RR-01-08, January 2002.
32. M. Wang, GREET Spreadsheet Model, Argonne National Laboratory, January 2006, www.transportation.anl.gov/software/GREET. Accessed on December 14, 2007.
33. *Well-to-Wheels Analysis of Advanced Vehicle Systems—North American Study of Energy Use, Greenhouse Gas Emissions, and Criteria Pollutant Emissions*, prepared by General Motors, Argonne Laboratory, BP, Exxon Mobil, and Shell, May 2005.
34. *Well-to-Wheels Analysis of Advanced Vehicle Systems—EUCAR, EU*, prepared by General Motors, EU Joint Research Center, Conawe, December 2003.
35. M. A. DeLucchi et al. "Electric and Gasoline Vehicle Lifecycle Cost and Energy-Use Model," ITS-Davis report UCD-ITS-RR-99-4, April 2000.
36. T. E. Lipman, "Hybrid-electric Vehicle Design—Retail and Life-cycle Cost Analysis," Report No. UCD-ITS-RR-03-01, April 2003.
37. T. Markel et al., "Plug-in Hybrid Vehicle Analysis," NREL/MP-540-40609, November 2006.
38. E. J. Carlson, P. Kopf, J. Sinha, S. Sriramulu, and Y. Yang, "Cost Analysis of PEM Fuel Cell Systems for Transportation," Report NREL/SR-560-39104, September 2005.
39. A. Teotia and P. Raju, "Forecasting the Market Penetration of New Technologies using a Combination of Economic Cost and Diffusion Models," *Journal of Product Innovation Management* **4**, 225–237, (1986).
40. D. Santini, R.Vyas, and M. Singh, "The Scenario for the Rate of Replacement of Oil by H2 FCV in Historical Context," private, available from Argonne National Laboratory, April 2003.
41. Updated sales information on hybrid vehicles, www.hybridcars.com/market. Accessed in December 2006.

CHAPTER 7

HYDRAULIC HYBRID VEHICLES

Amin Mohaghegh Motlagh, Mohammad Abuhaiba, Mohammad H. Elahinia, and Walter W. Olson
Department of Mechanical, Industrial and Manufacturing Engineering
The University of Toledo
Toledo, Ohio

1	**INTRODUCTION**	191
2	**HYDRAULIC HYBRID SYSTEM**	192
	2.1 Parallel Hydraulic Hybrids	193
	2.2 Series Hydraulic Hybrids	195
	2.3 Benefits of a Hydraulic Hybrid System	197
3	**COMPONENTS OF A HYDRAULIC HYBRID**	198
	3.1 Hydraulic Pump/Motors	198
	3.2 Accumulators	200
	3.3 Valves	201
	3.4 Hydraulic Vehicle Control Systems	204
4	**HYDRAULIC HYBRID VEHICLE RESEARCH AREAS**	**206**
	4.1 Increasing Storage Capacity	206
	4.2 Reliability	207
	4.3 Optimizing Control for Fuel Economy	207
	4.4 Noise Reduction	209
5	**SUMMARY**	**210**

1 INTRODUCTION

Hydraulics (often called fluid power) offers the best solution for hybridizing heavier vehicles such as SUVs, trucks, and buses to improve fuel economy. Using conventional gasoline engines under a parallel hybrid, US EPA/NVFEL testing and modeling programs project a 34 percent fuel economy improvement for a large four-wheel-drive (4WD) SUV. For a 2WD mid-size automobile the same technology provides a 50 percent improvement in fuel economy.[1] Furthermore, this technology exists as a result of years of experience with hydraulics and fluid power in the aircraft industry, agriculture, mining, and construction industries. This technology has the interest of the major automotive manufacturers for use in larger vehicles, particularly Ford Motor Company and General Motors Corporation. SUV, bus, and delivery truck prototypes already exist.

Today's designs are largely based on off-the-shelf components. The selection of sizes and capacities of the hydraulic components and the conventional automotive components are made from an experiential rationale rather than a true design-optimization process. The control systems are largely *bang-bang* (e.g., two parameter on/off states,) based on decision rules regarding state of charge, vehicle velocity, accelerator position, and so on, while some advanced designs do make use of proportional integrative derivative (PID) techniques. Existing automotive units for powertrain, braking, power steering, and suspension control have been left in place. However, as noted by a number of different researchers in the area, there is a need for optimizing the design selections and the controls systems of these vehicles. Such a system includes innovations in the braking system and the power train by *right-sizing* components such as the internal combustion engine and the hydraulics to meet vehicle performance standards.

Hydraulic hybrids have not received the visibility of electric hybrids for two reasons: (1) There is very little funding available for development of this type of hybrid, and (2) there is very little awareness of the research opportunities in the field of fluid power, as most universities today do not teach the required subject matter in a mechanical engineering curriculum. Investment in hydraulic hybrid research and development is required to provide true optimization to accelerate improvements in current *proof-of-concept* prototypes. Optimized control systems based on a rigorous design process will take the proof-of-concept vehicles to new levels of efficiency. This investment provides the structure, the motivation, and the pathway to fully realizing this technology in both personal and commercial transportation.

2 HYDRAULIC HYBRID SYSTEM

A *hybrid vehicle* is defined as one with two sources of power. In the past, a traditional vehicle featured an internal combustion engine fueled by a petroleum product, with the power transmitted to the driving wheels using a mechanical drive train. Thus, when an additional power source was added to a traditional vehicle, it was termed *hybrid*. The additional power source can be electrical, chemical, hydraulic, fly wheel, or any other form of power storage and production.

There are many reasons for hybridization. In some cases, the exhausts of the conventional system cannot be permitted in certain spaces, thus requiring an alternate source. In other applications, the noise of the conventional engine may be too great. However, today, hybridization almost always refers to the attempt to extract better overall fuel economy of the vehicle, and that is the application here.

It is well recognized that the gasoline- or diesel-fueled internal combustion engine is very inefficient and that the inefficiency becomes worse when operated in certain regimes. A typical engine map for engine efficiency is presented in Figure 1. In normal city operations, it is rare to achieve engine efficiencies greater than 30 percent, although the engine is capable of an efficiency of almost 42 percent. The objective of hybridization is to realize better efficiency by operating the engine nearer to most efficient operating point.

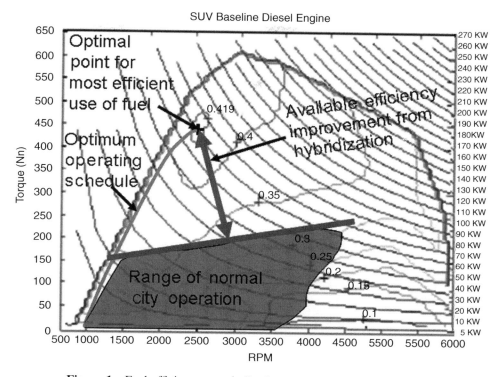

Figure 1 Fuel efficiency map indicating ranges of vehicle operation.

A second source of energy is to recapture the power that is consumed in braking. The term used to describe such systems is *regenerative braking*. It is not uncommon to expend more power during a hard braking cycle than the engine is capable of producing due to the short duration of the event. For example, stopping a modern SUV from 60 mph in a panic stop requires more than 300 HP. Most of this energy is expended as heat to the external environment. Thus, almost all hybrids today have regenerative braking as an integral part of their systems.

2.1 Parallel Hydraulic Hybrids

The most basic hydraulic hybrid system places a hydraulic pump/motor into the driveline of the vehicle, as shown in Figure 2. In pumping mode, hydraulic fluid (similar to the transmission fluids of automatic transmissions) is pumped into an accumulator under pressure. This requires power from the draft shaft of the vehicle provided by either the internal combustion engine or by braking. The accumulator holds the pressurized fluid until it is needed. The accumulator is a storage device that stores power in the form of hydraulic fluid under pressure. In motor mode, the hydraulic pump/motor uses the pressurized fluid to drive the

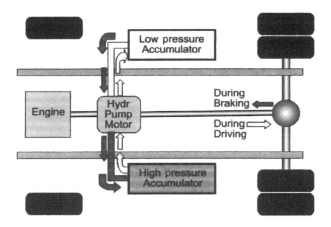

Figure 2 Parallel hybrid architecture.

vehicle. In this way, the hydraulic pump/motor returns power to the vehicle. This is shown in Figure 2.

A common method for this is the *parallel hydraulic hybrid*. In a parallel hybrid system, the original drive system of the vehicle is left largely intact, with the hybrid system added on. Thus, the systems operate in parallel. Parallel systems have the capability with conventional engine technology to realize an

Figure 3 The U.S. Army hydraulic hybrid shuttle bus.

Figure 4 The Hydraulic Launch Assist (HLA) module.

improvement of fuel usage of between 20 percent and 50 percent, depending on vehicle usage. Because the engine must follow the speed of the road geared up through the transmission, higher efficiencies cannot be achieved.

One such parallel system is the Eaton Corporation Hydraulic Launch Assist (HLA) system. A prototype shuttle bus was developed for the U.S. Army by a consortium of Impact Engineering, Eaton Corporation, General Dynamics Corporation, MKP Company, and the University of Toledo. The prototype is shown in Figure 3. A model of the HLA is shown in Figure 4.

This vehicle, delivered to the U.S. Army in May 2006 was able to achieve over 25 percent savings in fuel economy on the EPA city driving cycles as tested by EPA. It also improved the acceleration performance of the vehicle, reduced emissions and extended brake life service.

2.2 Series Hydraulic Hybrids

A series hydraulic system capable of more than 70 percent improvement in fuel savings. Its design envisions the removal of a direct link between the internal combustion engine (ICE) and the driveline components of the vehicle. Since the internal combustion engine is now separated from the road, higher efficiencies can be gained by operating the engine only at the torque and speed needed to

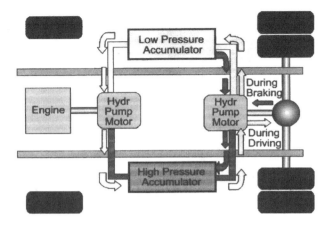

Figure 5 Series hydraulic hybrid architecture.

achieve maximum efficiency (known colloquially as the *sweet spot*), as shown in Figure 1. This system is shown schematically in Figure 5.

The engine is operated to pump hydraulic fluid at pressure to the high-pressure accumulator. When operating conditions are such that there is sufficient pressure in the high-pressure accumulator, the ICE is placed in an off condition. When

Figure 6 UPS package delivery parallel hydraulic truck.

it is necessary to again operate the engine, a small amount of fluid is directed to the pump/motor to restart the engine, and this unit returns to pumping mode. The rear pump/motor operates the drive train of the vehicle. In motoring mode, it takes high-pressure hydraulic fluid from the accumulator to drive the axle. When braking, the axle drives the unit in pumping mode to repressurize the high-pressure accumulator.

A prototype UPS package delivery truck developed by the National Vehicle Fuels and Emissions Laboratory of the U.S. Environmental Protection Agency and delivered to UPS on June 21, 2006, is in proof-of-concept testing mode by UPS shown in Figure 6. This vehicle is expected to improve fuel economy improvement by over 70 percent and reduce CO_2 emissions by 40 percent.

2.3 Benefits of a Hydraulic Hybrid System

The major benefits of a hydraulic hybrid system to hybridize vehicles are as follows:

- *Greatly improved fuel economy.* Due the very high efficiencies achieved with hydraulic units in conversion between storage and usage, hydraulic hybrids are able to achieve far greater fuel economies than any other known technology, including electric hybrids. Whereas electric hybrids have a wheel-to-wheel efficiency of less than 40 percent, hydraulic hybrids are able to produce efficiencies greater than 70 percent. This immediately translates into less oil needed for transportation.
- *Less pollution.* This applies generally to all hybrids, as the internal combustion engine is used less. However, hydraulic hybrids offer greater savings because of the lesser need to operate the engine. Because of the ultralow emissions, there are far less greenhouse gases generated.
- *Improved vehicle performance.* Hydraulic hybridization has the highest power density per unit mass of any hybridization technology. Nothing competes with the ability of fluid power to deliver high torque where it is needed. This translates directly into better acceleration of vehicle using this technique.
- *Lower incremental costs.* The components of a hydraulic are readily available, as the technologies have been in use in various industrial and transportation applications for several decades. As a result, the acquisition costs are low. In addition, the unit costs of a hydraulic pump/motor with accumulators are considerably less expensive than corresponding electric motors and batteries.
- *Less vehicle maintenance.* Brake life in a hydraulic hybrid is extended by more than two times. In addition there is less wear in the internal combustion engine and in the drive trains of hydraulic hybrids.

3 COMPONENTS OF A HYDRAULIC HYBRID

The components of hydraulic hybrid consist of hydraulic pump/motors, accumulators, power management systems, and piping/hosing. All of these components suitable for installation of a vehicle can be purchased commercially from several different vendors today at reasonable prices, allowing system paybacks within three years in saved fuel costs. However, research is being performed to achieve even higher efficiencies and to improve the units for transportation applications.

3.1 Hydraulic Pump/Motors

The key to achieving high efficiencies between the conversions of torque to fluid pressure are the hydraulic pump/motors (see Figure 7). These units can be operated reversibly either as a pump to generate fluid pressure given an input torque or as a motor to produce torque from fluid pressure. Typical maximum pressures used in this application are 5,000 pounds per square inch (350 atmospheres), with flows up to 100 gallons per minute. Under most operating conditions, these maximum conditions are never realized.

Two pump/motor technologies are currently used for hybridization: swash plate pump/motors and bent axis pump/motors (Figures 7 and 8). Both have high conversion efficiencies of greater than 85 percent. The bent axis design has greater potential for higher efficiencies, often about 90 percent. Both pump/motor designs are axial units based on pistons inside of a rotating housing, with fluid pressure controlled by a fixed kidney valve at the top of the rotating housing.

The swash plate pump/motor is considered to be the most durable design and is used extensively in construction, mining, and agriculture. The swash plate may be at a fixed angle to the drive shaft or variable driven by an actuator. For hybrid use, that angle is varied to change pump displacement to match speed and torque requirements. The ends of the pistons ride on the swash plate surface.

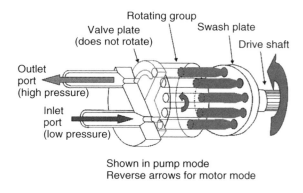

Swash Plate Pump/Motor

Figure 7 Swash plate hydraulic pump/motor.

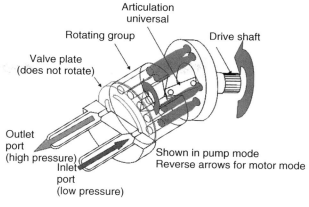

Bent Axis Pump/Motor

Figure 8 Bent axis hydraulic pump/motor.

Because of the nutation of the swash plate, the piston positions reciprocate in the rotating group, which is also fixed to the drive shaft. As the cylinder holes pass under the ports of the valve plate, fluid flows in or out of the cylinder, depending on which port is encountered. In pump mode, low-pressure fluid flows in the cylinders as the pistons retract within the cylinder. Near bottom dead center, the cylinder passes from the low-pressure port to the high-pressure port. As the shaft continues to rotate fluid is then forced out into outlet port as the pistons are force upward in the cylinder by the swash plate. Near top dead center, the cylinder now passes from the high-pressure port to the low-pressure port and the cycle begins anew.

The flow rate for these two types of pump/motors is

$$Q = \left[\frac{\pi}{4}d^2 n_p d_p \tan(\theta)\right]\omega$$

where Q is the flow rate, d the diameter of the pistons, n_p the number of pistons, d_p the piston pitch circle diameter, θ the swash plate or bent axis angle, and ω the rotation speed of the pump/motor.

The volumetric displacement (V_d) of the pump/motor is

$$V_d = \frac{\pi d^2 n_p d_p \tan(\theta)}{4} = \frac{Q}{\omega}$$

The torque for these motors is

$$T = \frac{D\Delta p}{2\pi}$$

where T is the torque, D the pump displacement, and Δp the pressure difference across the pump.

Other pump/motor technologies being considered for hydraulic hybrids are radial hydraulic/pump motors and variable displacement gear motors. However, very little information is available on these units. For example, a firm in Scotland, Artemis Intelligent Power, Ltd., is advertising a radial pump/motor with individual solenoid-driven cylinder valves that can be sequenced according to power needs. It is expected that more information will become available in the near future as work continues in their development.

3.2 Accumulators

The accumulator in hydraulic hybrid vehicle is the power storage unit (see Figure 9). There are three types of accumulators that have been used for hydraulic hybrids: the gas bladder accumulator, the gas piston accumulator and the spring piston accumulator. In addition to storing power, accumulators act to damp fluid-borne noise and vibration in the system.

The preferred accumulator for hybrid vehicles is the *gas bladder design*. The reason for this is that the accumulator almost always is in a horizontal position due to vehicle packaging requirements. (Other types of accumulators can be mounted horizontally but experience undesirable wear in this position.) In this accumulator, a rubberlike elastomeric bladder that is impervious to hydraulic fluid contains nitrogen gas. As hydraulic fluid is pumped into the accumulator, the gas is compressed. The compressed gas maintains pressure on the hydraulic fluid until the hydraulic is released from the accumulator. The containment shell may either be a carbon fiber composite or steel or a combination of these two materials.

Three problems occur with the gas bladder accumulator. Over time (measured in years), the nitrogen permeates through the elastomeric bladder. Therefore, the

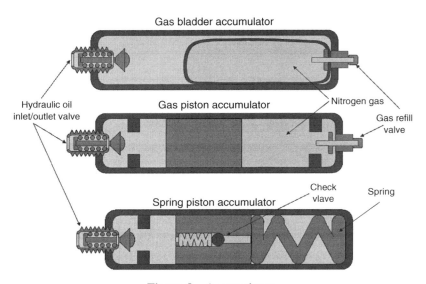

Figure 9 Accumulators.

accumulator needs to be recharged. Although not costly or difficult, the permeated gas reduces the efficiency of the hybrid system. Second, the bladder may fold inside the accumulator if it is not designed correctly. This leads to sharp angle fatigue cracking and will lead to failure of the accumulator. The third problem, while relatively minor results form the heat released as a gas compresses. This heat, if not recaptured, reduces the overall efficiency of the system. To trap this heat within the system, polymer foams are used inside the bladder so that the heat energy is retained and not lost.

Generally, the accumulator has to be large enough to ensure that internal pressure does not drop below the minimum operating pressure and that the accumulator never runs out of hydraulic fluid. The volume of the accumulator is based on a derivation of the ideal gas law, assuming adiabatic conditions:

$$V = \Delta V \left(\frac{\left[\frac{P_{high}}{P_{low}}\right]^{1.4}}{\left[\frac{P_{high}}{P_{low}}\right]^{1.4} - 1} \right)$$

where ΔV in gallons is the maximum change permitted in the gas times a safety factor, typically 1.4, P_{low} in pounds per square inch is the least pressure, and P_{high} in pounds per square inch is the maximum pressure of the accumulator. Note that ΔV is also the storage capacity of useable hydraulic fluid stored in the accumulator.

In the piston accumulators, a piston separates the fluid form either a gas or a spring. As the fluid is pumped into the accumulator, the pressure is maintained by the spring or gas. In piston accumulators, the internal containment surface is steel, although in some designs, bronze is used to improve lubricity. The inner steel container can be wrapped with carbon fiber composite to add strength. Piston accumulators can be used in the horizontal position. However, contamination in the hydraulic fluid will settle to the bottom of the accumulator. This debris then interferes with the operation of the piston and causes abrasive wear. Eventually, the wear reduces the effectiveness and the storage duration of the piston design.

3.3 Valves

The control system of a hydraulic hybrid system uses active components valves. Four different valve functions exist within the hydraulic circuits of the hybrid vehicle. The first application is to change the flow in response to a control input. In a hydraulic hybrid system, flows alternate between low pressure and high pressure, depending on whether the system is collecting and storing energy or using energy. For these purposes, control is imposed using a sequence of spool valves or poppet valves. The second application is to ensure that flow reversal does not occur in certain sections of the system. For example, back flushing of filters would return collected debris to the system. Typically, check valves (a form of the poppet valve) are used to ensure that the fluid does not reverse direction. The third is to prevent overpressuring certain components of the system. As a safety function, the components of the system are protected by pressure-relief valves that shunt express pressure by releasing hydraulic fluid into the reservoir. The fourth application is to cut off flow for long periods of time. These cutoff valves, typically ball valves, are needed so that parts of the system can be taken out of the pressure loops and maintained without loss of hydraulic fluid or pressure in the system.

Control valves may be either spool valves or poppet valves. Generally, spool valves are preferred for their smoothness of operations, the ability to proportionally control flow, and ease of control. However, at high pressures, spool valves may allow bypass leakage and may become more difficult to control. The spool element in a spool valve may be designed such that the spools are wider than the ports (closed center), are equal to the port width (critical center), or are somewhat less than the port width (open center).

Both the closed-center and the critically center valves will completely cut flow. The open-center valves allow for a small amount of fluid to flow to balance the axial forces on the spools. However, this convenience comes at a loss of overall system efficiency; therefore, open-center valves are generally not preferred for hydraulic hybrid applications. Although critical-center valves are ideally the best choice from both efficiency and controllability for hydraulic hybrids, critical-center valves wear over time and become open-center valves.

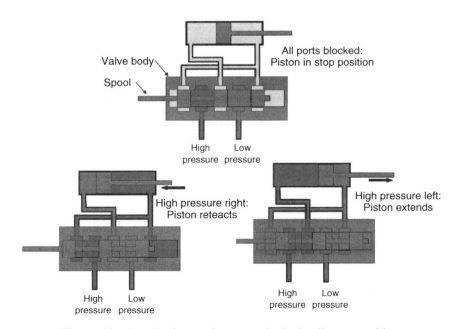

Figure 10 Spool valve used to control a hydraulic ram position.

Thus, closed-center spool valves are typically selected for hydraulic vehicle applications. However, because there is a dead band in flow as the spool passes the port opening during which there is no flow, the control is inherently nonlinear, posing a complexity to obtaining proportional flows.

In the application shown in Figure 10, a closed-center spool valve is used to control a hydraulic ram position. This valve is technically known as a *three-position four-port valve* and is a very common application for controlling flow directions. If the spool is in its leftmost position, high-pressure hydraulic fluid flows into the piston to cause it to retract. At the same time, the outlet ports are open to release the fluid into a low-pressure area. If the spool is in its center position, the flow is cut off and the ram is stationary. If the spool is moved to its right position, high-pressure hydraulic fluid enters the rear of the piston and drives the ram out, while low fluid from the piston is allowed to discharge.

The singlemost common type of valve is the poppet valve (see Figure 11). This valve consists of a body with a flow-through orifice with a movable center element (called the *poppet*) that in one position completely blocks the orifice but when moved away from the orifice allows flow through the orifice. It has advantages of positive flow interruption that improves with higher pressures, quick operation, and very little or no leakage. Several researchers and developers are considering use of poppet valves for hydraulic hybrid vehicle control because poppet valves are less affected by high pressure than spool valves. It is also easy to design and manufacture than a spool valve. However, because of its quick and positive throttling action on the fluid, it is more likely to suffer from cavitation,

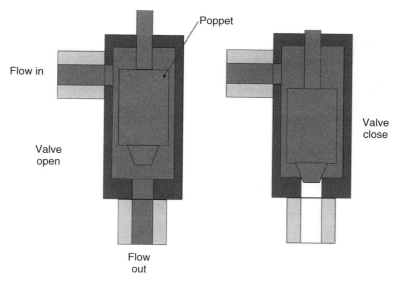

Figure 11 Poppet valve.

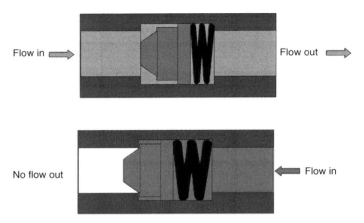

Figure 12 Check valve design.

and it also requires more force to operate than a spool valve. Common applications include check valves (Figure 12) and relief valves. For control purposes, poppet valves may be sequenced and serve the same functions as spool valves.

The steady state flow rate across a valve is defined by the following equation:

$$Q = AC\sqrt{\frac{2\Delta P}{\rho}}$$

where A is the valve internal cross section area, C is the discharge coefficient of the valve, typically 0.7, ΔP is the pressure drop across the valves, and ρ is the density of the fluid.

3.4 Hydraulic Vehicle Control Systems

The function of the control system in a hybrid vehicle is to regulate the power split between the two main sources of energy. The main control objective for hydraulic hybrid vehicles is to maximize fuel economy, to use regenerative braking to store energy as much as possible, and to use the stored energy for acceleration or cruising as soon as possible.[4] The control strategy should maximize the available energy storage capacity for capturing from regenerative braking. In the acceleration phase, a torque request from the driver is sent to the control system. The stored energy is released to the hydraulic pump/motor. If additional torque is needed, the control system brings the engine torque in.

Effective control algorithms are essential to maintain high efficiency, improved fuel economy, and reduced pollution through hydraulic hybridization. Monitoring the state of the charge of the secondary source of energy (battery or accumulator) is another main aspect of hybrid vehicle control systems.

City buses are good candidates for hydraulic hybridization. There are several characteristics for a city bus:

- The city bus traffic includes very transient driving (i.e., lots of accelerations and decelerations).
- The route is known in advance and often repeated many times each day.
- The bus stops are known in position and, in some extension, also in time (the time table).
- The weight of the vehicle changes during the ride due to changes in the number of passengers.

These characteristic are some of the reasons for developing hybrid buses. In an effort to develop whole vehicle control strategies Andersson et al. studied control of a hybrid bus.[5] To this end, the driver as well the environment was included in a simulation model. Using this tool, the Volvo Environmental Concept Bus was studied in city traffic. The main task was to improve the control performance by using additional information. The additional information includes the location of the bus in the route and time. The controller also includes a filter for driver commands. In the simulation, this combined control strategy showed superior results in terms of better fuel economy.

The hydraulic pressure of wheel cylinders is controlled individually and smoothly. This brake system also operates ABS, VSC, and TRC functions. By using this new brake system, it is possible to make maximum use of the regenerative brake, which reduces exhaust gas and improves fuel consumption with the same brake pedal operation as in conventional vehicles.

Kepner in his pioneering publication on hydraulic hybrid vehicles introduced the importance and functionality of the control systems in interacting and arbitrating among the hydraulic, powertrain and braking systems.[4] "As the research findings from hybridizing a Ford SUV demonstrated, the control system must be able to achieve smooth transitions even under the occurrence of a torque reversals as might be generated from driveline backlash."

The hydraulic pump/motor provides a feedback on the displacement through a potentiometer. The other control feedback variables include pressures, vehicle, driver, and powertrain system controller inputs. The control system regulates the solenoid valves to achieve the various operational states and transitions.[4] Actuation of the regenerative braking in this vehicle was achieved through the use of a dead band parallel regenerative braking system. A standard brake pedal and master cylinder were modified so that the first portion of the pedal travel is a dead band that does not actuate the master cylinder. This first portion of travel is sensed electronically with a potentiometer, and the pedal position is converted directly to a desired braking torque by the controller. The controller converts the braking torque to the appropriate solenoid commands to achieve the desired pumping torque. If the desired braking torque cannot be fully supplied by the hydraulic system, the braking system relies on the driver to further depress the brake pedal to actuate the friction brakes.

At zero speed the regenerative braking is reduced smoothly to zero. At zero speed, the regenerative braking function is unsatisfactory, however, because it allows the vehicle to creep forward, and also because any control flow is a stored energy loss.[4] Both of these situations of regenerative braking back-out require the driver to further depress the pedal to maintain braking force.[4] A possible solution is to have the transition between friction and regenerative braking with an electro-hydraulic braking system, such as the one investigated by researchers at Toyota.[6]

4 HYDRAULIC HYBRID VEHICLE RESEARCH AREAS

Research in hydraulic hybrid vehicles focused on the following areas:
- Increasing the storage capacity
- Increasing the reliability of the components of the vehicle
- Optimizing the control systems for the vehicle to improve fuel economy
- Reducing noise

4.1 Increasing Storage Capacity

The storage capacity of the hydraulic vehicle is limited by the maximum pressure and the volume of the accumulators. In today's SUV-class vehicle, packaging space limitations permit a maximum two accumulators of approximately 22 gallons each. This allows the transfer of a maximum of 18 gallons of hydraulic fluid from 5,000 pounds per square inch to 1,250 pounds per square inch. At 1,250 pounds per square inch, the accumulator must be closed or there is a danger of feeding the bladder through the outlet port. It is unlikely that more space would be available; therefore, research is largely directed at increasing the upper end of the pressure and at alternative forms of energy storage.

The power storage increases linearly with pressure. Current systems operate at 5,000 pounds per square inch (350 atmospheres). Several researchers are considering increasing the high-pressure systems in hydraulic vehicles to as high as 20,000 pounds per square inch (1,400 atmospheres). This pressure is considerably higher than the design pressures of most common components available for hydraulics. In addition, conventional design practices for hydraulic components would greatly increase the amount of material and therefore the size and weight of the components used in hydraulic hybrid vehicles. Research is needed to address the materials used and the design of hydraulic components able to safely withstand both the force and the capacity needed for hydraulic hybrid vehicles at these pressures.

An additional area of intense interest is finding another reversible mechanism that efficiently stores large amounts of power with little loss in conversions. Currently, considerations have been given to materials that undergo phase changes, such as metal hydrides. The use of flywheels and other mechanical means has also been investigated. Unfortunately, there has been little success to date.

4.2 Reliability

Hydraulics systems are already considered to be highly reliable and have been in use in a large number of critical systems such as aircraft, construction equipment, and manufacturing equipment. Therefore, the application to hydraulic hybrids does not immediately pose increased reliability concerns. Work in this area addresses improving hydraulic fluid cleanliness, reducing wear in the moving components of the system, reducing erosion due to moving fluids, reducing the potential for cavitation, and reducing the permeability of the accumulator bladders.

The close tolerances in the hydraulic pump/motors and the control valves require that the hydraulic fluid be maintained in high state of cleanliness. However, there will be wear and erosion in the moving parts of the system as well as aging and deterioration of seals and other components. This debris must be filtered from the fluid. Additionally, water and excessive heat will cause the fluid to deteriorate. Therefore, while filtering is effective in removing most contaminants, further research is needed for developing in-line sensors to continually monitor fluid performance, as well as to extend the life of the fluids. In addition, improving materials to reduce wear and erosion is a continuing field of research and development.

Cavitation is one of the most damaging effects that can occur with hydraulic systems. Common sites for cavitation are in the valve plates of the hydraulic pump/motors and in the poppet valves of the systems. To reduce cavitation, most hydraulic hybrid systems on the low-pressure side are operated at approximately 250 pounds per square inch (17 atmospheres). Further research is needed to identify the causes of cavitation in these components and to provide redesign of these components to limit cavitation while maintaining the highest possible efficiencies.

4.3 Optimizing Control for Fuel Economy

The overall purpose of hybridizing a vehicle is to improve fuel economy. Considerable success has been shown with relatively simple vehicle control systems based on feedback control, including two state controllers (bang-bang or on/off control), proportional controllers, and proportional-integrating-derivative controllers. However, there is considerable room for improvement. Using Stochastic Dynamical Programming, the best possible fuel economy for a parallel hydraulic hybrid on a closed cycle was 47 percent improvement in fuel economy.[7] The best operational parallel hybrid vehicles are currently able to only obtain approximately 35 percent. However, the method used by Wu et. al. requires precise detail of the driving cycle before the vehicle is operated and therefore is not practical in the sense of application.[7] But their method did establish that optimization of controls is a serious subject that needs research to obtain the greatest efficiencies for real vehicles. Current methods in research include use of neural net controllers, fuzzy logic controllers, and advanced predictor/estimator controllers.

In 1997, Kusaka and Nii investigated three different control methods for hybrid electric vehicles.[8] They had three main objectives in developing the control system: (1) to develop a motor control without a speed sensor; (2) to have a robust output torque control for the motor despite fluctuation in the temperature and in the power source; and (3) to maintain high efficiency. These goals were achieved by estimation techniques and by using available sensory data from the engine.

In order to prevent damaging the battery in its low state of charge, Kimura et al. developed a control design methodology for parallel hybrid vehicles.[9] A backward design procedure is used by this group from Toyota to design each component of the hybrid vehicle based on the required drive force. Similar control functionality is needed for the high-pressure accumulator in a hydraulic hybrid vehicle.

Smoothing the regenerative braking through proper control strategies is another area of control research, which was recently studied by a research group from Toyota.[6] Nakamura and his colleagues developed a brake by wire system by incorporating hydraulically controlled friction brakes to achieve two goals:

1. To compensate the changes of the regenerative brake force of front and rear motors, the friction brake force is controlled by adjusting the wheel cylinder hydraulic pressures
2. The pressure of each wheel cylinder is controlled by linear solenoid valves.

The availability of feedback signals is important for achieving better control performance in HHVs. Hahn et al. developed an observer to estimate the pressure information of a hydraulic actuator. The proposed algorithm is based on the slip velocity and the models of a hydraulic actuator and a mechanical subsystem. The resulting robust observer is guaranteed to be stable against possible parametric variations and torque estimation errors. The hardware-in-the-loop studies demonstrate the viability of the proposed algorithm in the field of advanced vehicle power transmission control and fault diagnosis.[10]

Matheson and Stecki at Monash University in Australia developed simulation models for hydraulic hybrid vehicles.[11] The control system uses inputs from the vehicle (driver pedal positions, engine, and gearbox states) and environmental inputs (road grade) to calculate the optimal power distribution between the hydraulic pump/motor and the engine to achieve the required road speed, while maximizing fuel economy. The control strategy, which is implemented in the simulation and validated in experiment, is of a basic generic format. It uses the backward-facing driveline request for torque and speed to calculate the swash plate angular position and the final power contribution from the hydraulic system. This control strategy uses simple Boolean logic to determine when the pump/motor can and cannot be used, depending on the state of the system and the overall driveline.

Another feature of the control system is the smoke control, which reduces the amount of torque available from the engine in the first and second gear. The authors consequently implemented a fuzzy logic controller through simulations.[12] The goals of the controller are to increase the fuel economy and to increase the maximum acceleration of the vehicle.

The proposed fuzzy controller, in order to provide the optimal power split between the diesel engine and the hydraulic pump/motor, uses a power-limiting strategy. In this method, the output power of the engine is limited based on the capacity of the hydraulic system at each instant of driving. Three variables define the power output of the pump/motor: shaft speed, the accumulator pressure, and the efficiency of the system. The controller also limits the power output of the hydraulic system where the output power becomes small.

4.4 Noise Reduction

Hydraulic systems when used in vehicles produce a tonal noise pitch that is particularly noticeable during braking when the other major noise producing elements of a vehicle (mainly the engine and the exhaust system) are at their most quiescent states. During motoring, the noise from the engine and the exhaust are usually greater than that of the hydraulic system. Thus, it is the noise of pumping that tends to be most offensive.

The pump/motor is the most significant source of noise in hydraulic systems.[13] It is coupled with motor as a pump/motor assembly in the hydraulic hybrid vehicle and is driven by the main engine. Pump speed has a strong effect on noise, whereas pressure and pump size have about equal but smaller effects. The noise created can be separated according to the path of transmission: audible airborne noise, structurally transmitted noise in the form of vibration, and fluid transmitted noise in the form of pressure pulses. Pumps commonly generate as much as 1,000 times more energy in the form of structurally and fluid transmitted noise as they do airborne noise. These forms act on other components and frequently end up generating more noise than that coming directly from the pump.

The current methods in noise reduction depend on identifying one or two of the strongest noises and finding a way of reducing them.[14] Figure 13 is a typical pump/motor airborne noise pressure spectrum. While moving components within the pump vary in frequency from 30 to 300 hertz during operation, the frequencies with the greatest power usually exist in the fourth through the eighth harmonics of the fundamental frequency, or in the range 500 to 2000 Hz. This phenomena is not understood fully but has been confirmed over a wide range of sound tests with bent axis and swash plate hydraulic pumps.

Pump noise is created as the internal rotating components abruptly increase the fluid pressure from inlet to outlet. The abruptness of the pressure increases

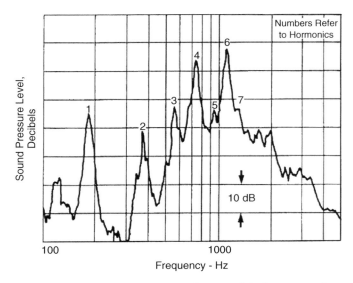

Figure 13 Noise pressure spectrum for an axial piston pump/motor.

plays a major role in the intensity of the pump noise. It is clear that pump/motors with odd numbers of cylinders generate less airborne noise and less fluid ripple than do pump/motors with an even number of cylinders. Thus, pump/motors for hybrid use are designed with an odd number of cylinders. However, little work has been performed to precisely identify the pump/motor components leading to noise.

5 SUMMARY

In summary, hydraulic hybridization is a solution to improving the fuel economy of heavier vehicles. It is cost effective, applicable today with off-the-shelf components, and has demonstrated greater savings than any other form of hybridization in use. The components are reliable and have been proven in years of experience in various industries. The results have been dramatic: The initial systems are proven at improving fuel economy by more than 25 percent in basic parallel hybrid systems and considerably more in series hybrid configurations. In addition, these systems produce much less pollution while improving vehicle performance and reducing operational and maintenance costs.

Research and development is continuing in hydraulic hybrids exploring pump/motor technologies, better control methodologies, reducing noise, and improving components to store higher power densities. Although little external support exists for continued research in this area, advancements are being made by small working groups of interested parties. There is still work to do, including educating the public that hydraulic hybrids offer the best solution to reducing transportation fuel costs for large vehicles.

REFERENCES

1. Progress Report on Clean and Efficient Automotive Technologies, Interim Report, EPA 420-R-02-004, January 2004.
2. "What is the Eaton HLA System?" http://www.etopiamedia.net/mtw/pdfs/EatonHLA1.ppt#608,2. Accessed on. December 14, 2007.
3. http://www.epa.gov/otaq/technology/recentdevelopments.htm. Accessed on December 14, 2007.
4. R. Kepner, Hydraulic Power Assist: A Demonstration of Hydraulic Hybrid Vehicle Regenerative Braking in a Road Vehicle Application, Proceedings of the SAE International Truck and Bus Meeting and Exhibition, Detroit, Michigan, November 18–20, 2002.
5. J. Andersson, R. Axelsson and B. Jacobson. "Route Adaptation of Control Strategies for a Hybrid City Bus," *JSAE Review*, **20**: 531–536 (1999).
6. M. Nakamura, E. Soga, A. Sakai, A. Otomo and T. Kobayashi, Development of Electronically Controlled Brake System for Hybrid Vehicle, Proceedings of the SAE World Congress, Detroit, Michigan, March 4–7, 2002.
7. Bin Wu, Chan-Chiao Lin, Zoran Filipi, Huei Peng and Dennis Assanis, "Optimal Power Management for a Hydraulic Hybrid Delivery Truck," *Vehicle System Dynamics*, **42** (1–2), 23–40 (2004).
8. Y. Kusaka and Y. Nii, "Control Methods for Induction Machine Directly Coupled to Engine," *JSAE Review*, **18**: 289–293 (1997).
9. A. Kimura, T. Abe and S. Sasaki, "Drive Force Control of a Parallel-series Hybrid System," *JSAE Review*, **20**: 337–341 (1999).
10. J. Hahn, J. Hur, Y. Cho and K. Lee, "Robust Observer-based Monitoring of a Hydraulic Actuator in a Vehicle Power Transmission Control System," *Control Engineering Practice*, **10**: 327–335 (2002).
11. P. Matheson and J. Stecki, Development and Simulation of a Hydraulic Hybrid Powertrain for use in Commercial Heavy Vehicles, Proceedings of the SAE International Truck and Bus Meeting and Exhibition, Fort Worth, Texas, November 10–12, 2003.
12. P. Matheson and J. Stecki, Modeling and Simulation of a Fuzzy Logic Controller for a Hydraulic-Hybrid Powertrain for Use in Heavy Commercial Vehicles, Proceedings of the SAE International Powertrain & Fluid Systems Conference & Exhibition, Pittsburgh, October, 2003.
13. M. H. Elahinia, W. Olson, T. M. Nguyen and P. Fontaine, Chassis Vibration Control for Hydraulic Hybrid Vehicles, SAE 2006 Automotive Dynamics Stability and Controls Conference and Exhibition, Novi, Michigan, February 14–16, 2006,
14. S. Skaistis, *Noise Control of Hydraulic Machinery*, Marcel Dekker, Inc., New York, 1988.

CHAPTER 8

BIOFUELS FOR TRANSPORTATION

Aaron Smith, Cesar Granda, and Mark Holtzapple
Texas A&M University
College Station, Texas

1	**INTRODUCTION**	213
2	**ETHANOL**	**215**
	2.1 Introduction	215
	2.2 Biomass Sources	215
	2.3 Manufacturing Methods	216
	2.4 Manufacturing Research and Development	217
	2.5 Quality Standards	217
	2.6 Vehicle Modifications and Use	217
	2.7 Performance	219
	2.8 Emissions	221
	2.9 Fuel Transportation and Distribution	221
	2.10 Fuel Storage	222
	2.11 Safety	222
	2.12 Subsidy	222
	2.13 Current Availability	223
	2.14 Ethanol Links	224
3	**BIODIESEL AND VEGETABLE OIL**	**224**
	3.1 Introduction	224
	3.2 Biomass Source	225
	3.3 Manufacturing Methods	225
	3.4 Quality Standards	227
	3.5 Vehicle Modifications	228
	3.6 Performance	229
	3.7 Emissions	229
	3.8 Fuel Transportation Issues	231
	3.9 Storage Issues	231
	3.10 Current Availability	231
	3.11 Production Capacity	232
	3.12 Subsidy	233
	3.13 Safety	233
	3.14 Current Research and Future of Biodiesel	233
4	**HYDROGEN**	**234**
	4.1 Introduction	234
	4.2 Manufacturing Methods	234
	4.3 Quality Standards	237
	4.4 Hydrogen Vehicles	237
	4.5 Emissions	242
	4.6 Fuel Transmission and Distribution	243
	4.7 Storage Issues	245
	4.8 Current Availability	248
	4.9 Safety	248
	4.10 Current Research	249
	4.11 Hydrogen Links	250
5	**OTHER BIOFUELS**	**250**
	5.1 Methanol	250
	5.2 Butanol	250
	5.3 Mixed Alcohols	251
	5.4 Petroleumlike Biofuels	251
	5.5 Synthesis Hydrocarbons	251
	5.6 Methane	251
6	**CLOSING REMARKS**	**251**

1 INTRODUCTION

Why biofuels? Biofuels offer a responsible solution to potential crises in the near future. Fossil fuels have been the dominate energy source since the Industrial Revolution because they are inexpensive, energy dense, and easy to use. Biofuels are renewable and provide a means to become independent from fossil fuels. Oil exploration is becoming more costly as more remote locations must be explored to find oil. As oil prices increase, biofuels will become economically competitive. The United States depends on foreign oil, so its oil supplies are not secure. There is a growing consensus that carbon emissions from fossil fuels are a large contributor to global warming. Biofuels are carbon neutral and, therefore, address global warming. Other benefits of biofuels are improved emissions, healthier rural economies, and the potential to reduce waste.

The purpose of this chapter is to educate practicing engineers and students about biofuels for transportation. Biofuels are made from renewable materials such as plants and organic waste. Conventional transportation fuels such as gasoline or diesel are made from petroleum. Although biofuels are alternative fuels, not all alternative fuels are biofuels. For example, natural gas and propane are alternative fuels, but not biofuels.

In this chapter, ethanol, biodiesel, and hydrogen will be discussed in detail. A fourth section is included to enumerate other biofuels that are less well known, yet potentially valuable. Each of the three central biofuels is reviewed objectively. Advantages and disadvantages are discussed, as well as engineering and research issues. Although some detailed information is provided, our goal is for the reader to see the big picture.

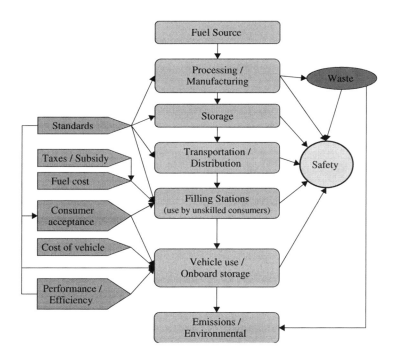

Figure 1 Considerations when selecting a biofuel.

Figure 1 illustrates the major considerations when selecting a biofuel. After reading this chapter, you should have a firm understanding of the ins and outs of each biofuel described.

It is the view of the authors that no single biofuel is a silver bullet; however, alcohols and vegetable oils have clear advantages. We believe the future of biofuels will be diverse, and each fuel will be selected for specific applications.

2 ETHANOL

2.1 Introduction

In 1826, Samuel Morey of Oxford, New Hampshire, built the first prototype internal combustion engine in the United States. He used alcohol biofuel in his experiments.[1,2] In 1876, when Nikolaus Otto invented the four-stroke internal combustion engine, he designed it to run on ethanol.[2] Why? Because gasoline did not exist. Henry Ford promoted ethanol and built the famous model-T to run on ethanol, gasoline, or any ratio of the two with manual adjustment to the carburetor. Ford felt so strongly about ethanol fuel that he built an ethanol plant in the Midwest.[1,2] Eventually, the plant was closed due to competition with petroleum. In modern times, gasoline has been cheaper than alcohol, so it dominated; however, alcohols such as ethanol are excellent motor fuels.[3]

Ethanol is typically blended with gasoline rather than used *neat*. Throughout this section, blends such as E10 are mentioned. This nomenclature indicates that the blend is 10 percent percent ethanol by volume. Likewise, E20 is 20 percent ethanol by volume, E85 is 85 percent ethanol, and so on.

2.2 Biomass Sources

Ethanol can be produced from any biodegradable source, if the appropriate method for conversion is used. At present, sucrose (e.g., from sugarcane) and starch (e.g., from corn) are the main feedstock for fuel ethanol. Sucrose is squeezed or extracted from sucrose-bearing crops and fermented into ethanol. Before starches can be fermented, they must be enzymatically *saccharified*, or converted into simple sugars. Both of these two sources directly compete with food. Further, they have a relatively low productivity because only a portion of the crop is converted to ethanol.

Sugar and corn alone cannot significantly impact the world energy economy. A more prolific feedstock is needed, such as lignocellulose. The most abundant biological material, it is composed of lignin, cellulose, and hemicellulose. The conversion of lignocellulose to ethanol is the future of ethanol manufacturing, and many methods to achieve this conversion are being investigated.

There are obvious advantages of being able to convert the entire plant, and not just the easily digestible biomass (e.g., sucrose or starch) into liquid fuels. For example, on a dry basis, sugarcane is roughly 40 percent sugars and 60 percent fiber (i.e., lignocellulose), including tops and leaves. Using only sugars, the average ethanol yield is about 6,370 L/(ha·yr) (680 gal/(acre·yr)) in Brazil.[4] Using the lignocellulose adds about 5,930 L/(ha·yr) (630 gal/(acre·yr)), assuming 85 percent of theoretical yield for bagasse,[5] giving a total of 12,300 L/(ha·yr) (1,310 gal/(acre·yr)). Furthermore, other cane varieties, known as *energy cane*, are only 30 percent sugars and 70 percent lignocellulose, but yield about twice the biomass.[6] With lignocellulose conversion technologies, ethanol yields can be as high as 23,400 L/(ha·yr) (2,500 gal/(acre·yr)). In addition, energy crops tend to be easier to grow because they are more rugged, resisting drought and pests. They often have a much lower environmental impact because they do not need as much herbicides, pesticides, and fertilizer, and cause less soil erosion. Other high-yield lignocellulosic crops under investigation include sorghum, switchgrass, miscanthus, and hybrid poplar.[7–10]

2.3 Manufacturing Methods

The only established method for producing ethanol from biomass is yeast fermentation of sugars. The sugar source may vary, but the fermentation is virtually identical in all practiced processes. Figure 2 shows two common ways to obtain sugars follow:

1. *Sugar extraction from sugar-bearing crops:* The juice from sugarcane or sweet sorghum can be fermented directly or from the molasses, after crystallized sugar has been produced.
2. *Starch saccharification from starch-bearing crops:* Grains from corn, sorghum, rice, or wheat are converted into sugars using enzymes (i.e., amylases).

Yeasts, such as *Saccharomyces cerevisiae*, convert sugars into ethanol and carbon dioxide. The ethanol-rich fermentation broth, known as wine or beer, is then distilled to concentrate the ethanol. The distillation yields an azeotrope, which is about 96 percent ethanol by volume. In Brazil, this hydrated or hydrous ethanol is used in flexible fuel vehicles (FFVs) or ethanol-only vehicles. However, water must be removed for blending with gasoline as a gasoline oxygenate (e.g., 5, 10, 20 percent ethanol by volume). The most common method for producing anhydrous ethanol is to use a molecular sieve, such as zeolites. Alternatively, azeotropic distillation or extractive distillation may be employed. In the United States, the anhydrous ethanol must be denatured with gasoline before it can be marketed.

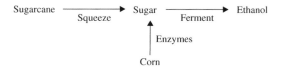

Figure 2 Schematic of established methods for ethanol production.

Ethanol from corn is the most utilized method in United States,[11] whereas ethanol from sugarcane is the basis for Brazil's ethanol industry. Ethanol production from corn differs from sugarcane, because yeasts cannot metabolize starch. Corn starches must be saccharified (hydrolyzed into sugars) using enzymes prior to fermentation. Additionally, the distillation stillage from sugar-derived ethanol, known as *vinasse*, is commonly returned to the fields for fertilization and irrigation. In contrast, the distillation stillage from starch-derived ethanol (i.e., distiller's dry grain) is dried and sold as animal feed. Ethanol from corn is less energy efficient than ethanol from sugarcane because the sugarcane process has leftover fiber (i.e., bagasse), which is used to provide energy for the process.

2.4 Manufacturing Research and Development

Many lignocellulose-to-ethanol processes are under research and development. Some of these processes saccharify lignocellulose into simple sugars using acids or enzymes, known as cellulases.[12,13] The sugars are then converted into ethanol. One process ferments the hydrolyzed sugars into acetic acid, which is then hydrogenated into ethanol.[14] This process produces higher yields, because, unlike ethanol fermentation, mass is not lost as carbon dioxide. The hydrogen can be obtained by gasifying undigested residue; this way, more biomass energy is contained in liquid fuel. The MixAlco process is a similar method under development that produces ethanol. Because it produces a mixture of alcohols, it is discussed in the "Other Fuels" section.[15] Finally, lignocellulose can be gasified to produce, syngas (carbon monoxide and hydrogen), which can be converted into mixed alcohols, mainly ethanol, using a catalyst.[16,17] Alternatively, syngas can be fermented to ethanol using selected microorganisms.[18,19]

2.5 Quality Standards

Ethanol is a well-developed automotive fuel. Table 1 shows the ASTM D 5798 standard specification for denatured fuel ethanol (E75–85) for automotive spark-ignition engines.[20,21] Hydrocarbons used to denature fuel ethanol must meet the requirements of ASTM D 4806.[20,22]

2.6 Vehicle Modifications and Use

Ethanol may be used in spark-ignition engines as a pure fuel, or blended with gasoline. However, to take full advantage of ethanol's unique properties, engines should be specifically designed for it.[23] Vehicles manufactured after 1990 may use E10 or less without modification. Vehicles manufactured before 1990 may require a new carburetor to operate with blends between E5 and E10.[24] Ethanol has poor cold-start properties, so vehicles that run on blends greater than E85 must have a gasoline reservoir for cold starts. Aftermarket conversion of gasoline-powered vehicles to ethanol-fueled vehicles is possible but not recommended because

Table 1 ASTM D 5798 Standard Specification for Denatured Fuel Ethanol (E75–E85) for Automotive Spark-ignition Engines

Property	Value for Class			Test Method
ASTM volatility class	1	2	3	N/A
Ethanol, plus higher alcohols (minimum, volume %)	79	74	70	ASTM D 5501
Hydrocarbons (including denaturant)/(volume %)	17–21	17–26	17–30	ASTM D 4815
Vapor pressure at 37.8°C kPa psi	38–59 5.5–8.5	48–65 7.0–9.5	66–83 9.5–12.0	ASTM D 4953, D 5190, D 5191
Lead (maximum, mg/L)	2.6	2.6	3.9	ASTM D 5059
Phosphorus (maximum, mg/L)	0.3	0.3	0.4	ASTM D 3231
Sulfur (maximum, mg/kg)	210	260	300	ASTM D 3120, D 1266, D 2622
Methanol (maximum, volume %)	0.5			N/A
Higher aliphatic alcohols, C3–C8 (maximum, volume %)	2			N/A
Water (maximum, mass %)	1			ASTM E 203
Acidity as acetic acid (maximum, mg/kg)	50			ASTM D 1613
Inorganic chloride (maximum, mg/kg)	1			ASTM D 512, D7988
Total chlorine as chlorides (maximum, mg/kg)	2			ASTM D 4929
Gum, unwashed (maximum, mg/100 mL)	20			ASTM D 381
Gum, solvent-washed (maximum, mg/100 mL)	5			ASTM D 381
Copper (maximum, mg/L)	0.07			ASTM D 1688
Appearance	Product shall be visibly free of suspended or precipitated contaminats (shall be clear and bright).			Apperance determined at ambient temperature or 21°C (70°F), whichever is higher.

Note: From Refs. 20 to 22.

Table 2 Vehicle Modifications for Vehicles after 1990

Ethanol Content in Fuel	Carburetor	Fuel Injection	Fuel Pump	Fuel Pressure Device	Fuel Filter	Ignition System	Evaporative System	Fuel Tank	Catalytic Converter	Basic Engine	Motor Oil	Intake Manifold	Exhaust System	ColdStart System
≤10%														
10–25%	✓	✓	✓	✓	✓	✓	✓	✓						
25–85%	✓	✓	✓	✓	✓	✓	✓	✓	✓	✓	✓			
≥85%	✓	✓	✓	✓	✓	✓	✓	✓	✓	✓	✓	✓	✓	✓

Note: From Ref. 24.

of the necessary changes in component materials, the high cost, and need for extensive engine recalibration.[25] Table 2 shows component modifications required to operate vehicles manufactured after 1990 on blends higher than E10.

Flexible-fuel vehicles (FFVs) available in the United States may run on gasoline and ethanol blends up to E85. These vehicles use an electronic control unit (ECU) and several sensors to adjust the engine. In Brazil, FFVs operate with E20, hydrous E100 (96 percent ethanol v/v, 4 percent v/v water) and any ratio of the two. In the United States and Brazil, FFVs are different. The Brazilian ECU can tolerate some water to accommodate use of hydrous ethanol.

Water in gasoline/ethanol can be problematic. Large amounts of water will cause an engine to run poorly, or not at all. Additionally, water can cause blends to separate into two phases, which will also cause poor engine operation. There is concern that high ethanol blends could be contaminated with water, yet remain a single phase. If this contaminated E85 were added to a half-full tank of gasoline in a FFV, it could phase separate depending on the percentage of water. Figure 3 is a ternary phase diagram for gasoline/water/ethanol mixtures and illustrates how phase separation might occur.[26] The minimum blend in Brazil is E20, which may be mixed with higher water content without phase separating. In the United States, anhydrous ethanol is used for blending to avoid phase separation problems.

Ethanol is hydroscopic and can be corrosive to common metals used in fuels systems. Additionally, plastics and elastomers may deteriorate or soften in the presence of ethanol.[23] Table 3 lists metallic and nonmetallic materials that are compatible and incompatible with high blends of ethanol.

2.7 Performance

Ethanol (76,000 Btu/gal) contains less energy than gasoline (114,132 Btu/gal).[27] It takes 1.5 gallons of ethanol to equal the energy of 1 gallon of gasoline. Because ethanol and ethanol blends have a lower volumetric energy density than gasoline,

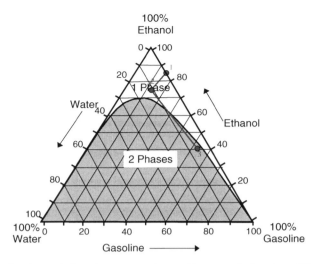

Figure 3 Gasoline/water/ethanol ternary phase diagram. Point 1—E85; Point 2—E85 contaminated with water; Point 3—contaminated E85 blended with pure gasoline (phase seperates). (From Ref. 26.)

Table 3 Incompatible and Compatible Materials with High Ethanol Blends Like E85

	Incompatible	Acceptable Compatibility
Metallic	Zinc	Unplated steel
	Brass	Stainless steel
	Lead	Black iron
	Aluminum	Bronze
	Terne (Lead-tin-alloy) plated steel	
	Lead-based solder	
	Copper	
	Magnesium	
Nonmetallic	Natural rubber	Buna-N
	Polyurethane	Neoprene rubber
	Cork gasket	Polyethylene
	Leather	Nylon
	Polyester-bonded figerglass	Polypropylene
	Polyvinylchloride (PVC)	Nitrile
	Polyamides	Viton
	Metyl-methacrylate plastics	Flourosilicones
	Fiberglass reinforced plastic laminate	Teflon

Note: From Refs. 20 and 32.

fuel economy will be lower. For blends of E10 or less, the difference in fuel economy is generally not noticeable.[28]

Ethanol (E100) is an excellent fuel for internal combustion engines (ICE).[3] It has a higher octane rating than gasoline, 113,[29] which gives it excellent anti-knock properties and allows ethanol (E100) to be used in engines with higher compression ratios. Ethanol burns faster, allows more efficient torque development, and gives a vehicle increased power (5 percent for pure ethanol, 3 to 5 percent for E85).[20,23]

Compared to gasoline, ethanol has poor cold-start properties due to its high heat of vaporization.[30] Gasoline requires less heat to vaporize than ethanol and is blended with ethanol to improve its cold-start properties.[30] E85 has similar cold-start properties as 87 octane gasoline.[20] In the winter months to improve cold starts, fuel that is sold as E85 is, in fact, 70 percent ethanol and 30 percent gasoline by volume.[25]

Ethanol has been tested and used in diesel engines; however, the physical and thermodynamic characteristics of alcohols do not make them particularly suitable fuels for compression-ignition engines; therefore, we have limited out discussion to ethanol's performance in spark-ignition engines.[23]

2.8 Emissions

The blend of gasoline affects emissions. Ethanol does not have sulfur; therefore, it does not contribute to SO_X. Emission trends that are typical of all ethanol/gasoline blends are decreased CO, decreased particulate matter, decreased total hydrocarbon emissions, increased NO_X, and increased aldehydes.[23,31,32] Ethanol has a higher vapor pressure than gasoline; therefore, evaporative emissions tend to increase with increasing percentages of ethanol.

2.9 Fuel Transportation and Distribution

Ethanol/gasoline blends cannot use existing petroleum pipeline systems because of possible contamination by water and residues from other petroleum products or incompatibility with the system due to the corrosive properties of alcohols.[20] In the United States, ethanol is transported either by rail or truck. Generally, rail is used for transmission of anhydrous denatured ethanol from the manufacturing facility to the distribution/blending stations. Trucks are used for distribution to the fueling stations.

Blending occurs at the distribution center, although sometimes it may also occur at the fueling stations. Transportation containers must be free of water and other contaminants. If other fuels have been transported in the same container, it should be washed with solvent and air dried.[25]

The same technologies used for dispensing gasoline and diesel fuel may be employed with ethanol and ethanol blends.[25] However, care should be taken to ensure material compatibility of ethanol and higher blends with containers and transferring equipment. Table 3 lists materials that are incompatible and compatible with high blends of ethanol.

2.10 Fuel Storage

Ethanol is stored and handled in a similar fashion as gasoline. Above-ground and below-ground storage systems are both viable. Larger storage volumes will be required to house the same energy equivalent as gasoline.[23] Materials used may be different due to compatibility issues.[20] For example, metal-plated tanks should not be used. Fiberglass tanks maybe used, but must be coated with a chemical-resistant rubber. Table 3 lists some materials that are incompatible and compatible with high blends of ethanol like E85. Additionally, extra precautions may be necessary to prevent contamination with water and evaporative emissions.

2.11 Safety

The safety standards for handling and storing ethanol and its gasoline blends are the same as those for gasoline.[20] Special attention should be given, however, to material compatibility.

2.12 Subsidy

Currently, ethanol and biodiesel are commercially being used to curb gasoline and diesel use, respectively. According to the Energy Information Agency (EIA), 140 billion gallons of gasoline were used in the United States during 2006. The current U.S. capacity for ethanol production is 5 to 6 billion gallons (3 to 4 percent of the gasoline market).[29] For ethanol to be competitive, government subsidy is required. The Volumetric Ethanol Excise Tax Credit (VEETC) provides ethanol blenders/retailers with $0.51 per pure gallon of ethanol blended, or $0.0051 per percentage point of ethanol blended (i.e., E10 is eligible for $0.051/gal; E85 is eligible for $0.4335/gal). The incentive is available until 2010.[33]

Skeptics criticize subsidies as a waste of tax money; however, biofuel subsidies return more revenue to the U.S. Treasury than they cost.[29] The benefit to taxpayers is that this tax credit is usually passed on to the consumer as lower pump prices for high-octane, ethanol-enriched fuel. According to the Consumer Federation of America, consumers who purchase gasoline with 10 percent ethanol could be saving as much as $0.08 per gallon compared to straight gasoline.[34] Biofuels stimulate the economy by creating jobs, which increases wages and taxes and reduces unemployment benefits and farm program payments.[29]

Many argue that corn-derived ethanol is causing a food shortage because it competes with corn grown for human consumption. According to iowacorn.org, only about 12 percent of the corn grown in the United States is used for human consumption. Figure 4 shows the distribution of corn usage in the United States for

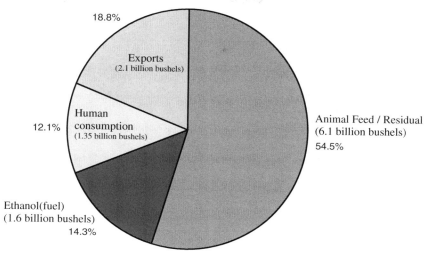

Figure 4 Corn usage in the United States (2005/2006 statistics). (From Ref. 36.)

2005/2006. Ethanol fuel comprises a significant portion of U.S. corn production, but it is not creating a food shortage.

The Energy Policy Act of 2005 included a historic provision—the Renewable Fuels Standard (RFS). The RFS is a directive for the United States to increase its renewable fuel usage each year. The schedule begins with 4 billion gallons per year in 2006, increasing annually to 7.5 billion gallons per year in 2012.[35]

2.13 Current Availability

Ethanol production has increased annually and will continue to grow to meet the Renewable Fuels Standard. Figure 5 shows the annual U.S. production of ethanol. In 2006, 46 percent of America's gasoline was blended with ethanol, and approximately one third of all gasoline in the United States is E10.[25,35] All gasoline vehicles since 1990 are approved to use blends up to 10 percent ethanol (E10).[28,35] As a result, fueling stations that dispense E10 blends may not label the pumps. This rule varies, depending on the state. All E85 pumps are labeled with a bronze E85 pentagon.[25] As of April 2007, the U.S. DOE EERE listed more than 1,100 E85 fueling stations across 40 states serving approximately 6 million FFVs.[27,33] To find an E85 fueling station, the Alternative Fuels Data Center has a station locator at www.eere.energy.gov/afdc/infrastructure/locator.html. Additionally, the National Ethanol Vehicle Coalition provides a flexible fuel vehicle guide on its Web site, e85fuel.com.

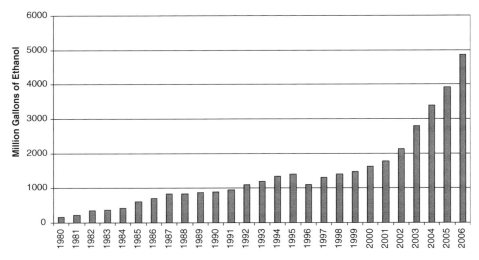

Figure 5 Annual ethanol production in the United States. (From Ref. 29.)

2.14 Ethanol Links

National Ethanol Vehicle Coalition (NEVC)	www.e85fuel.com
Renewable Fuels Association (RFA)	www.ethanolrfa.org
Govenors' Ethanol Coalition	www.ethanol-gec.org
U.S. Dept. of Energy EERE	www.eere.energy.gov
Ethanol Promotion and Information Council (EPIC)	www.drivingethanol.org/
American Coalition for Ethanol	www.ethanol.org
Iowa Corn Growers Association	www.iowacorn.org
Energy Information Administration	www.eia.doe.gov

3 BIODIESEL AND VEGETABLE OIL

3.1 Introduction

Rudolf Diesel first developed his engine to run off peanut oil; however, petroleum became less expensive and the use of vegetable oil as fuel was abandoned. More recently, petroleum reserves are becoming more difficult to find, so exploration and drilling costs are increasing. Diesel was a visionary and knew that vegetable oils, though not considered important in 1912, would "in the course of time become as important as petroleum and coal tar products of the present time."[37] If Rudolf Diesel were alive today, he likely would run his diesel on pure biodiesel.

Biodiesel is an attractive biofuel, made from vegetable oils and fats. It may be used in diesel engines without modification. In many ways, biodiesel is a wonder fuel. Compared to petroleum-derived diesel, biodiesel extends the life of diesel engines, has lower emissions, is biodegradable, nontoxic and has a higher flash point (making it safer to transport). Vegetable oil (VO) can be run directly in a diesel engine, but requires an additional fuel tank and modification to the fuel system.

This section focuses on biodiesel because it is a more developed technology and more practical than vegetable oil. Because these two fuels are very closely related, they are combined into one section.

3.2 Biomass Source

Biodiesel is can be made from virgin and waste vegetable oils and animal fats. There are hundreds of species of oil-producing plants that may be used to produce biodiesel.[37] Those with the greatest yields per acre are preferred for fuel production. Increasing yield per acre is a dominate area in biodiesel research. Rapeseed (canola) and soybean oils are the two most common crops used to produce biodiesel. Rapeseed is the primary oil crop used for biodiesel in Europe. In the United States, soybean is the dominate oil crop. Table 4 lists 48 common oil-producing plants with annual production averages.

3.3 Manufacturing Methods

According to the National Biodiesel Board, "Biodiesel is defined as the mono alkyl ester of long-chain fatty acids derived from VOs or animal fats, for use in compression-ignition (diesel) engines."[38] Oils are triglycerides; a glycerol molecule bonded to three long-chain fatty acids. Transesterification chemically converts oil and alcohol into biodiesel. Fatty acids are hydrolyzed from the triglyceride and transformed into esters (biodiesel).[1,37,38] Figure 6 illustrates this chemical reaction. A strong acid or base may be used as a catalyst. Sodium hydroxide (NaOH) is the most common catalyst because it is the least expensive. Typically, methanol is used due to cost and reaction performance; however, ethanol and higher alcohols could be used.[38]

The result of transesterification is two separate liquid phases. The bottom phase is glycerol and the top phase is alkyl esters (biodiesel). The biodiesel is separated from the glycerol and then washed to remove remaining alcohol, catalyst, and soap that may have formed. Typically, water washing is necessary to meet the American Society of Testing and Materials (ASTM) standards.[37]

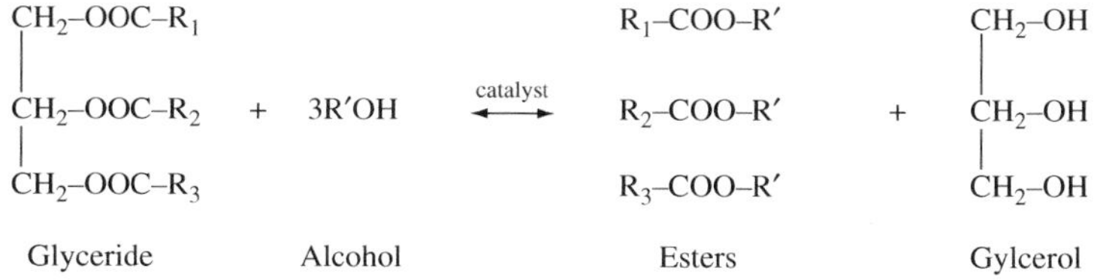

$$\begin{array}{c} CH_2-OOC-R_1 \\ | \\ CH_2-OOC-R_2 \\ | \\ CH_2-OOC-R_3 \end{array} + 3R'OH \xrightleftharpoons{\text{catalyst}} \begin{array}{c} R_1-COO-R' \\ R_2-COO-R' \\ R_3-COO-R' \end{array} + \begin{array}{c} CH_2-OH \\ | \\ CH_2-OH \\ | \\ CH_2-OH \end{array}$$

Glyceride　　　　Alcohol　　　　　　Esters　　　　　Gylcerol

Figure 6 Transesterification of trigylcerides with alcohol. (From Ref. 3.)

Table 4 Oil-producing Plants with Average Production Yields

	Plant	Latin Name	kg oil/hectacre
1	**oil palm**	*Elaeis guineensis*	5000
2	macauba palm	*Acrocomia aculeata*	3775
3	pequi	*Caryocar brasiliense*	3142
4	buriti palm	*Mauritia flexuosa*	2743
5	oiticia	*Licania rigida*	2520
6	**coconut**	*Cocos nucifera*	2260
7	avacado	*Persea americana*	2217
8	brazil nut	*Bertholletia excelsa*	2010
9	macadamia nut	*Macadamia terniflora*	1887
10	**jatropha**	*Jatropha curcas*	1590
11	babassua palm	*Orbignya martiana*	1541
12	jojoba	*Simmondsia chinensis*	1528
13	pecan	*Carya illinoensis*	1505
14	bacuri	*Platonia insignis*	1197
15	castor bean	*Ricinus communis*	1188
16	gopher plant	*Euphorbia lathyris*	1119
17	piassava	*Attalea funifera*	1112
18	olive tree	*Olea europaea*	1019
19	**rapeseed**	*Brassica napus*	1000
20	opium poppy	*Papaver somniferum*	978
21	**peanut**	*Arachis hypogea*	890
22	cocoa	*Theobroma cacao*	863
23	**sunflower**	*Helianthus annus*	800
24	tung oil tree	*Aleurites fordii*	790
25	rice	*Oriza sativa*	696
26	buffalo gourd	*Cucurbita foetidissima*	665
27	**safflower**	*Carthamus tinctorius*	655
28	crambe	*Crambe abyssinica*	589
29	sesame	*Seasmum indicum*	585
30	camelina	*Camelina sativa*	490
31	mustard	*Brassica alba*	481
32	coriander	*Coriandrum sativum*	450
33	pumkin seed	*Cucurbita pepo*	449
34	euphorbia	*Euphorbia lagascae*	440
35	hazelnut	*Corylus avellana*	405
36	linseed	*Linum usitatissimum*	402
37	coffee	*Coffea arabica*	386
38	**soybean**	*Glycine max*	375
39	**hemp**	*Cannabis Sativa*	305
40	cotton	*Gossypium hirsutum*	273
41	calendula	*Calendula officinalis*	256
42	kenaf	*Hibiscus cannabinus L.*	230
43	rubber seed	*Hevea brasiliensis*	217
44	lupine	*Lupinus albus*	195
45	palm	*Erythea salvaorensis*	189
46	oat	*Avena sativa*	183
47	cashew nut	*Anacardium occidentale*	148
48	**corn**	*Zea mays*	145

These figures are based on international averages and may vary with climate, region, and subspecies grown. Those listed in bold are the ten most common plants used. (From Ref. 37.)

Other methods for using VO in diesel engines include direct use of VO, blends of VO with diesel or kerosene, microemulsions of VO with various solvents, as well as other methods of transesterification.[39] The details of these methods are not discussed as base-catalyzed transesterification of VO to biodiesel is superior.

3.4 Quality Standards

Biodiesel is a well-developed biofuel. The ASTM approved quality standards for biodiesel in December 2001 under the designation of D-6751.[1,38] The ASTM

Table 5 Specification for Biodiesel[a] (B100) – ASTM D6751-07b (March 2007)

Property[b]	ASTM Method	Limits	Units
Calcium and magnesium, combined	EN 14538	5 max.	ppm (µg/g)
Flash point (closed cup)	**D 93**	**93 min.**	**degrees C**
Alcohol control (one of the following must be met)			
1. Methanol content	EN14110	0.2 max.	% volume
2. Flash point	D93	130 min.	degrees C
Water and sediment	**D 2709**	**0.05 max.**	**% vol.**
Kinematic viscosity, 40°C	D 445	1.9–6.0	mm^2/sec.
Sulfated ash	D 874	0.02 max.	% mass
Sulfur			
S 15 Grade	**D 5453**	**0.0015 max. (15)**	**% mass (ppm)**
S 500 Grade	**D 5453**	**0.05 max. (500)**	**% mass (ppm)**
Copper strip corrosion	D 130	No. 3 max.	
Cetane	D 613	47 min.	
Cloud point	**D 2500**	**Report**	**degrees C**
Carbon residue 100% sample	D 4530[c]	0.05 max.	% mass
Acid number	**D 664**	**0.50 max.**	**mg KOH/g**
Free glycerin	**D 6584**	**0.020 max.**	**% mass**
Total glycerin	**D 6584**	**0.240 max.**	**% mass**
Phosphorus content	D 4951	0.001 max.	% mass
Distillation, T90 AET	D 1160	360 max.	degrees C
Sodium/potassium, combined	EN 14538	5 max.	ppm
Oxidation Stability	**EN 14112**	**3 min.**	**hours**
Workmanship	Free of undissolved water, sediment, and suspended matter		

[a]Biodiesel is defined as the mono alkyl esters of long chain fatty acids derived from vegetable oils or animal fats, for use in compression-ignition (diesel) engines. This specification is for pure (100%) biodiesel prior to use or blending with diesel fuel. A considerable amount of experience exists in the United States with a 20% blend of biodiesel with 80% diesel fuel (B20). Although biodiesel (B100) can be used, blends of over 20% biodiesel with diesel fuel should be evaluated on a case-by-case basis until further experience is available.
[b]**Boldface = BQ-9000 Critical Specification Testing Once Production Process Under Control**
[c]The carbon residue shall be run on the 100% sample.

Table 6 Advantages and Disadvantages of Biodiesel (B100) Usage

Advantages	Disadvantages
• Nontoxic • Higher flash point (less flammable) • Biodegradable • Extends engine life • Cleans fuel system • Easily made • Small performance gain* • Adds value to agriculture • Safe to handle • Compatible with engines made since 1994** • Sulfur free • Increased cetane number • No offensive odor • Lower hydrocarbon, particulate matter and CO emissions • Improved lubricity • Minimal change in infrastructure and distribution systems	• Slightly higher cost (2006) • Higher cloud point • Small performance loss* • Increased NO_X emissions • Availability is not widespread • Could invalidate warranty of fuel systems • Six-month shelf life • May require hose and gasket replacement • Currently requires government subsidy

*Depends fuel blend and engine used
**Refers to OEM gaskets and hoses

standards are shown in Table 5. Commercially sold biodiesel must meet this standard. BQ-9000 is a quality-assurance program analogous to systems established by the International Organization of Standards (ISO).[40] Both producers and distributors of biodiesel are encouraged to participate in the BQ-9000 program. The maintenance of quality fuel is essential for customer acceptance and industrial growth.

3.5 Vehicle Modifications

Biodiesel can be run in any diesel engine without modification. If the engine is older than 1994, engine gaskets and hoses may need replacement because biodiesel can deteriorate rubber hoses and gaskets.[1] Biodiesel helps maintain fuel systems by dissolving buildup. This may require changing the fuel filter frequently until the system is clean.

For a diesel to run on VO, the kinematic viscosity must be below 20 centistokes (cSt). To achieve this, a second fuel tank must be added and heated by routing the radiator cooling lines through the VO fuel tank. These vehicles must be started with conventional diesel fuel. Once the VO viscosity is below 20 cSt (160°F), the fuel line is switched via a solenoid valve. In a converse manner, the vehicle must be shut off using conventional diesel to ensure proper startup.[37]

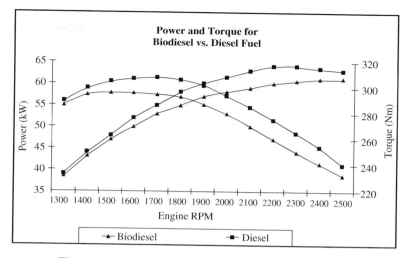

Figure 7 Power torque curve for B100. (From Ref. 1.)

3.6 Performance

Pure biodiesel (B100) has 12 percent less energy per kilogram than conventional diesel (37 vs. 42 MJ/kg) but 7 percent higher combustion efficiency.[37] As a result, performance and fuel efficiency decrease by about 5 percent.[37] In most cases, the difference is hardly noticeable. The power and torque curve for B100 is shown in Figure 7.

In the case of biodiesel blends, there have been reports that fuel efficiency increased. St. Johns Schools in Michigan report that using B20 (20 percent biodiesel by volume) increased fuel mileage from 8.1 to 8.8 miles per gallon.[41] However, this is a singular report and may not be true for all users of B20. The exact performance of biodiesel depends on the specific engine, its condition, and the blend of biodiesel used. In any case, the performance of biodiesel is similar to that of conventional diesel.

Biodiesel has a higher cloud point than conventional diesel, making it less tolerant to cold temperatures ($<32°F$). Below-freezing temperatures can be tolerated when using blends of B20 or less.[42] Figure 8 shows how cloud point varies with blend percentage. Free fatty acids increase the biodiesel cloud point; therefore, biodiesel made from waste vegetable oil will have a higher cloud point than that made from virgin oil.[37]

3.7 Emissions

"Biodiesel is the first and only alternative fuel to have a complete evaluation of emission results and potential health effects submitted to the U.S. Environmental Protection Agency under the Clean Air Act Section 211(b)."[38] Biodiesel contains about 10 percent oxygen by mass, which makes it burn cleaner than

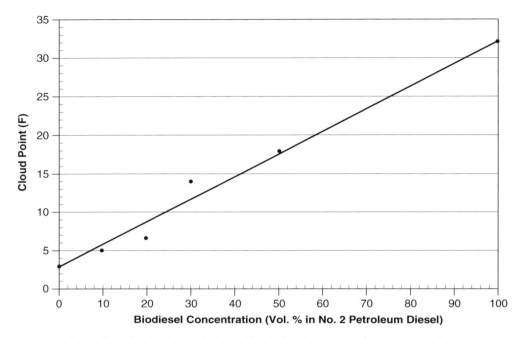

Figure 8 Cloud point variation with biodiesel concentration. (From Ref. 7.)

conventional diesel. Biodiesel has lower hydrocarbon (HC), particulate matter (PM), and carbon monoxide (CO) emissions. Biodiesel does not contain sulfur and does not contribute to SO_X emissions. Figure 9 shows the percent change in emissions from conventional diesel as a function of blend percentage. These changes are not absolute and depend on the feedstock and the specific engine.[43,44] Figure BD-4 is only representative of heavy-duty highway engines on soy biodiesel; however, the general trends are representative of biodiesel usage.[43]

The effect of biodiesel on NO_X emissions is debated. Nitrogen oxides are formed when nitrogen is oxidized. The use of oxygenated fuels increases the production of NO_X emissions. Several factors affect NO_X emissions: blend percentage, degree of saturation, and engine technology/age.[43,44] In 2006, the National Renewable Energy Laboratory (NREL) published a report in which it had tested 43 heavy-duty engines running on B20 and concluded "that B20 has no net impact on NO_X."[43] Some engines decrease NO_X and others increase NO_X, the difference being the technology of each vehicle. Unsaturated fatty acids increase NO_X emissions.[1,44] Therefore, feedstocks with more unsaturated fatty acids will produce more NO_X emissions. Most reports conclude that biodiesel increases NO_X emissions by a small percentage and that future engine technology can correct this increase.[1,37,38,43,44]

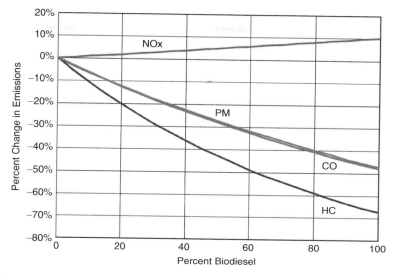

Figure 9 Average emissions impacts of biodiesel for heavy-duty highway engines. (From Ref. 8.)

3.8 Fuel Transportation Issues

Because biodiesel has a higher flash point than conventional diesel (150°C (300°F) and 52°C (125°F), respectively), it is easier and safer to transport. Regular shipping services such as FedEX and UPS can transport biodiesel.[1,37] Shipping methods used for conventional diesel may also be used.

3.9 Storage Issues

Biodiesel can be handled and stored in the same containers and places as conventional diesel.[37] Biodiesel has a shelf life ranging from about one year in warm climates to indefinitely in cold climates.[37] Because biodiesel is derived from VO, it may serve as a carbon source for microorganisms. To prevent microbial growth in warm humid climates, it is recommended to add a biocide or biostat.[37] Additionally, it is important to minimize biodiesel exposure to air and oxidizing agents. Waste vegetable oil should not be stored for extended periods of time because it may spoil.[37]

3.10 Current Availability

Biodiesel is sold just like conventional diesel. Biodiesel is typically blended with conventional diesel and is labeled "B#." The "B" indicates biodiesel and the number is the volume percentage of biodiesel. Figure 10 shows a fueling station that sells biodiesel (B20) and ethanol (E85 and E10).[45] Various retailers carry one or more of the following blends: B2, B5, B10, B11, B50, B98, B99, B100.

Figure 10 Biofuels gas station. (From Ref. 10.)

To find a retailer near you, the National Biodiesel Board has a retailer locator on its Web site, http://www.biodiesel.org/buyingbiodiesel/retailfuelingsites/.[38]

The availability of vegetable oil for the end consumer is very limited. Food-grade oil may be purchased in supermarkets at a premium price. Waste vegetable oil may be available from a local restaurant. Otherwise, vegetable oil is not commercially available to the consumer for use as fuel.

3.11 Production Capacity

Vegetable oil production capacity will limit the growth of the biodiesel industry. According to the Energy Information Agency (EIA), the United States uses 50 to 60 billion gallons of diesel fuel every year.[46] In 2006, The United States produced approximately 250 million gallons of biodiesel (∼0.5 percent of diesel market).[38] As of November 2006, the National Biodiesel Board reported that the *capacity* of U.S. biodiesel plants is 582 million gallons (~1 percent of biodiesel market). If the United States were to use all its waste cooking oil and animal fats, and fallow farm land were planted with high-yield rapeseed to produce biodiesel, the result would displace about 24 percent of the annual diesel fuel usage, by the most generous estimates.[37] For the biodiesel industry to grow past this limit, breakthroughs must be made with a high-yield oil crop.

3.12 Subsidy

Energy Policy Act of 2005 Section 1344 extends the tax credit for biodiesel through 2008. A subsidy of $1.00 per gallon is given for biodiesel made from virgin VO, and $0.50 for biodiesel from waste cooking oils.[47]

3.13 Safety

Comprehensively, biodiesel is the safest fuel available. It is "more biodegradable than sugar and less toxic than salt."[1] It produces fewer carcinogens and has a higher flash point than conventional diesel. Figure 11 compares biodiesel's flash point to other conventional fuels. Because biodiesel is very safe and environmentally friendly, it is excellent for use in marine environments and sensitive areas such as national parks and forests.[37] Because biodiesel has lower emissions, it is an attractive fuel for use in school buses.

3.14 Current Research and Future of Biodiesel

Biodiesel is well established and has been defined with standards. Most engine manufactures have warmed up to biodiesel and have begun covering biodiesel

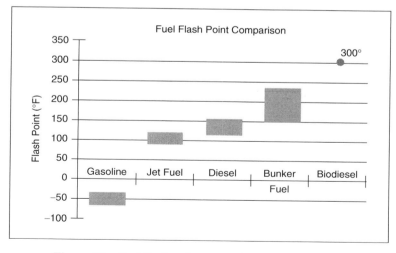

Figure 11 Fuel flash point comparison. (From Ref. 7.)

under vehicle warranty. The main challenges for the future of biodiesel will be production volume, education of the public, maintenance of fuel quality, improvement of government policy and incentives, and, most importantly, economics.

Current biodiesel research is focused on increasing vegetable oil production. Algae show great potential as a mass biooil producer and could provide enough oil to completely replace conventional diesel.[37] Currently, several engineering issues must be addressed before algae can become an economical player as a feedstock for biodiesel. These include issues regarding economy of scale, diffusion of CO_2 into the growth solution, solar efficiency of algae, oil yield along with economical extraction methods, and ecological implications of genetically modified algae.

4 HYDROGEN

4.1 Introduction

On a mass basis, hydrogen has more energy than any other fuel. On a volume basis, hydrogen has the least energy of any fuel. This comparison, in short, is the life story of hydrogen.

Hydrogen is a mammoth. The advantages and prospects of realizing the *hydrogen economy* are enormous. Likewise, the hurdles and engineering issues that must be overcome to get there are equally great. From a theoretical viewpoint, it is easily argued that hydrogen is the perfect fuel, which is why hydrogen research continues to draw entrepreneurs, investors, government support, and environmentalists. From a practical viewpoint, hydrogen is far from perfect. Many experts agree that realizing the hydrogen economy is many decades away, by the most optimistic estimates.

This section will only scratch the surface of information about hydrogen. We will only mention that which pertains to hydrogen used as a transportation fuel. Furthermore, our focus is on hydrogen produced in a "green" fashion.

4.2 Manufacturing Methods

Hydrogen is an energy carrier rather than an energy source. Despite being the most abundant element in the universe and the third most abundant element in Earth's crust, pure hydrogen gas is not readily available. Hydrogen is molecularly bound in many organic and inorganic compounds. The primary resources for hydrogen production are water, biomass, and hydrocarbons. To liberate hydrogen takes energy. Once made, the hydrogen is *stored* energy; thus, hydrogen is an energy carrier.

There are many hydrogen sources and production methods. For hydrogen to be green, it should be produced from water or biomass. Figure 12 illustrates the various pathways that may be used to produce hydrogen from these resources. The following is a brief description of each process.

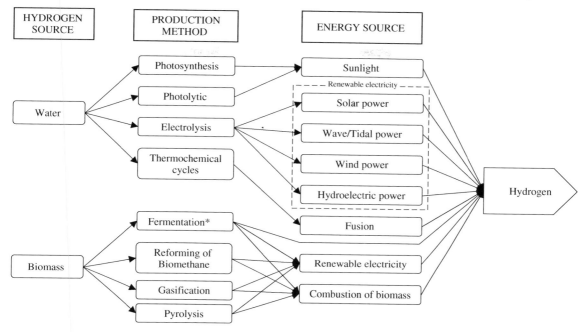

Figure 12 Green routes to hydrogen production. Fermentation may not require a heat source. Gasification and pyrolysis may only require initial heat source. (From Refs. 1 and 4.)

Photolytic/Photoelectrolysis

Photoelectrolysis employs catalysts that absorb photons from sunlight to split water into hydrogen and oxygen.[48,49] Photoelectrolysis is described as running a fuel cell in reverse.[49] This technology is still under research and is considered the Holy Grail of hydrogen production, because it would directly convert sunlight into hydrogen.[48]

Electrolysis

Electrolysis is the decomposition of water using an electrical current. An anode and cathode are placed in water with an electrolyte. When current is applied, hydrogen is formed at the cathode and oxygen at the anode. Equation (1) describes the chemistry of electrolysis. Electrolysis is a well-known technology and produces very pure hydrogen. Of the renewable paths to hydrogen, electrolysis is the only one commercially used.[50,51]

Electrolysis is very energy intensive. In principle, operating at high temperatures (900°C to 1,000°C) can reduced the energy requirement.[52] Commercial electrolysis plants achieve efficiencies of 70 to 75 percent (electrolyzers range from 75 to 90 percent).[50,53] There are other variations of electrolysis such as alkaline electrolysis, polymer electrolyte membrane electrolysis, and high-temperature electrolysis. Regardless of the method, the net reaction is that described by equation (1).[52]

$$H_2O + \text{electricity} \rightarrow H_2 + \frac{1}{2}O_2 \qquad (1)$$

Thermochemical Cycles

Thermochemical cycles produce hydrogen using high-temperature cycling of chemical reactions that split water. Thermochemical cycles operating in conjunction with a high-temperature reactor offer high-efficiency hydrogen generation comparable to high-temperature steam electrolysis. This is well suited for use with nuclear power or future fusion power generation.[55]

Photosynthesis

Photosynthetic production of hydrogen uses modified green algae that produces hydrogenase enzymes to strip hydrogen from water.[53] This technology is still in the initial stages of research, but could prove to be a key player in the direct conversion of light to hydrogen.[55]

Fermentation

The fermentation path to hydrogen production would employ microorganisms that consume biomass. This route to hydrogen is attractive because it does not require sterile conditions or sunlight, and it may use a wide variety of feed stocks. The conversion of glucose to hydrogen by anaerobic fermentation is shown in equation (2).[53]

$$C_6H_{12}O_6 + 2H_2O \rightarrow 2CH_3COOH + 2CO_2 + 4H_2 \qquad (2)$$

This technology is still under research. This route to hydrogen would have a low external energy requirement because the microorganisms consume energy from the biomass.

Reforming of Biomethane

Steam reforming of natural gas accounts for 48 percent of the hydrogen produced in the United States.[51] High-temperature steam is used to strip the hydrogen from methane. This conventional technology could be modified to use methane produced from manure or other biomass resources. The primary hurdle is increasing methane production.

Gasification/Pyrolysis

Gasification is the thermal decomposition of organic material (biomass) in a low-oxygen environment that employs 20 to 40 percent of the stiochiometric oxygen needed for combustion.[52] The production gas (synthesis gas) contains hydrogen, carbon monoxide, carbon dioxide, low-molecular-weight alkanes, as well as other compounds. Gasification temperature ranges from about 700°C with a catalyst, to about 900°C without a catalyst.[52] This production route requires purification if high-purity hydrogen is required. Pyrolysis is similar to gasification, except that the biomass is heated in the absence of oxygen.

"The cost of producing hydrogen ... is one of the biggest barriers on the path toward a hydrogen economy."[50] Furthermore, the cost of producing hydrogen in an environmentally friendly way is an even larger hurdle. Currently, steam reforming of natural gas is the most economical route to hydrogen production. Steam reforming of fossil fuels, primarily methane and naptha, accounts for 96 percent of all hydrogen produced.[51] Breakthroughs must be made for green hydrogen to compete with fossil fuel hydrogen.

4.3 Quality Standards

Fuel cells require high-purity hydrogen to prevent poisoning. Because hydrogen is not a conventional fuel, new standards must be developed for storage, delivery, infrastructure, and end use. Fuelcellstandards.com is a "site is dedicated to assist the worldwide community working to develop and interpret fuel cell codes and standards."[56] "The key players in this area are SAE, ISO TC 22 SC 21, and ISO TC 197."[57] These organizations are actively forming committees to develop and evaluate standards for all aspects of hydrogen. In 2004, ASTM formed a subcommittee (D03) to develop standards on hydrogen and fuel cells.[58] Standards are important to the development of any commercialized product. In the case of hydrogen, standards are essential for public acceptance and safety. Links to hydrogen standards Web sites are listed in the "Hydrogen Links," section 4.11.

4.4 Hydrogen Vehicles

Hydrogen is unique as a transportation fuel. Transportation during the Industrial Revolution was powered by pistons. The steam locomotive, the model-T Ford, and the diesel engine all operate via piston-driven engines. Hydrogen-powered vehicles can employ an internal combustion engine (ICE), too; however, the fuel cell is most efficient way to convert hydrogen to power. Fuel cells do not involve direct combustion nor do they require high operating temperatures. A fuel cell extracts the electrochemical potential from hydrogen to create electricity.

How much hydrogen is equal to a gallon of gasoline? Comparing heat of combustion, 1 lb. (0.45 kg) of hydrogen has the equivalent energy of approximately 19.5 lb. of gasoline (3.1 gallons).[59] Table 7 compares the fuel economy and fuel equivalents to a kilogram of hydrogen for a proton exchange membrane fuel cell vehicle (PEMFCV), ICE gasoline vehicle and a hybrid electric vehicle (HEV) running on gasoline. For a better illustration, the hydrogen contained in a gallon of water (0.92 lb. or 0.42 kg) will provide the same traveling capacity for a PEMFCV as a gallon of gasoline in a typical car.[50,60]

Table 7 Fuel Economy and Fuel Equivalents for PEMFC, ICE, and HEV

	PEMFCV	ICE	HEV
Fuel	H_2	Gasoline	Gasoline
Fuel economy	55 mpkg	24 mpg	34 mpg
Equivalence to 1 kg hydrogen	1.00 kg	2.29 gal	1.62 gal

Note: From Ref. 61.

Fuel Cells (FC)

Hydrogen used in vehicles is likely to employ a proton exchange membrane fuel cell (PEMFC). FCs produce electricity through an electrochemical reaction rather than direct combustion in a heat engine. As shown in Table 8 and Figure 13, there are many types of FCs each classified by the electrolyte used; however, the PEMFC is currently the only practical FC for vehicular applications.[50,61,62] Figure 14 illustrates the internals of a PEMFC. The U.S. Department of Energy describes the PEMFC as follows:

> When hydrogen is fed to a PEM fuel cell and encounters the first catalyst-coated electrode, called the anode, the hydrogen molecules release electrons and protons. The protons migrate through the electrolyte membrane to the second catalyst-coated electrode, called the cathode, where they react with oxygen to form water. The electrons, however, can't pass through the electrolyte membrane to the cathode.

OEM	Honda	GM	Daimler Chrysler
Model	FCX	Equinox	F-Cell A-type car
Year of manufacture	2006	2006	2002
Energy efficiency	60%	----	----
FC power output (max)	134 hp	125 hp	92 hp
Motor power output (max)	127 hp	126 hp	87 hp
Motor torque	201 ft*lb$_f$	236 ft*lb$_f$	----
Range	270^A - 350^B miles	200 mi	93 miles
Fuel tank capacity	171 liters	----	----
Gas pressure	5,100 psi	10,000 psi	5075 psi
Max speed	93 mph	100	87 mph
Acceleration 0–60 mph	----	12 s	16 s
Operating range, low/high	−22°F /	−13°F / 113°F	----
Payload	----	750 lbs	----
Seating	4	4	----

A Determined using EPA calculation method
B When driven in LC4 mode (calculated by Honda)

Figure 13 Comparison of FC vehicle specifications.

Table 8 Fuel-Cell Technologies

Fuel-Cell Type	Temp. Range	Efficiency	Capacities	Primary Application	Notes	
Polymer electrolyte or proton exchange membrane	<100°C (<212°F)	50–60%	Polymer membrane (thin plastic film)	100 W to 250 kW per cell	Transportation, stationary	Fast startup, high power density, rapid response to power demand, relatively rugged
Phosphoric acid	160°–220°C (320–430°F)	37–55%; up to 72–80% with heat recovery	Concentrated phosphoric acid	25–250 kW per cell	Stationary	Fuel of choice is natural gas
Solid oxide	800°–1,000°C (1,500–1,800°F)	45–65%; up to 70–85% with heat recovery	Solid nonporous ceramic materials	200 W per cell; 300 kW to 3 MW per module	Stationary, utility	Typically applied in stacks of hundreds; a plant might produce up to 10 MW
Alkaline	23°–250°C (70–482°F)	50–60%	Potassium hydroxide solution (35–50% KOH)	2.2 kW	Spacecraft	Being developed for other applications
Molten carbonate	650°–660°C (1,200°F)	45–60%; 70–85% with heat recovery	Melted carbonate salt mixture	250 kW to > 1 MW	Stationary, utility	Corrosive electrolyte limits durability

*Without recovery of cogenerated heat, unless otherwise noted
Note: From Ref. 50.

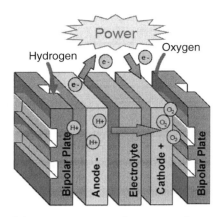

Figure 14 Illustration of how a proton exchange membrane (PEM) fuel cell works. (From Ref. 60.)

Instead they must travel around it—this movement of electrons is an electrical current."[61]

Depending on the area, a single FC will produce an electric potential around 0.7 V.[53] In applications, many individual FCs are connected in a stack to produce the power necessary to propel a vehicle. In Figure 15, the FC stack produces 5kW.

FC vehicles have been developed by GM, Honda, Ford, DaimlerChrsyler, Toyota, Nissan, as well as others [50]. Figure 13 compares three FC vehicles and their specifications. Although hydrogen FC vehicle prototypes have been developed and successfully demonstrated, cost is prohibitive. To put this in perspective, prototype FCs cost $2,000 $3,000/kW with high volume production FCs costing approximately $225/kW,[61] whereas the ICE is on the order of $25 to 35/kW, an order of magnitude less. According to the DOE, FCs must approach $50/kW to be competitive.[61,62]

Advantages of FCs

- Not limited by Carnot efficiency
- Low operating temperature (~80°C)[60]
- Most efficient way to use hydrogen (65–85%)[48]
- &bull High power density
- &bull Reliable power generation[48]
- No harmful emissions[48,50]
- Quiet[48]
- Can serve as remote electrical power[55]

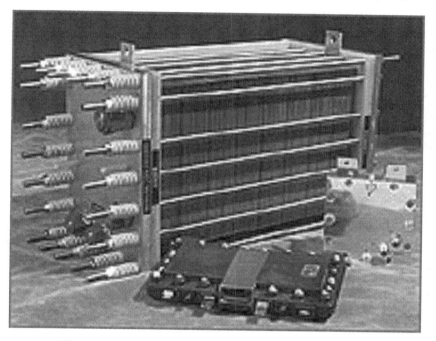

Figure 15 Fuel cell (5 kW) manufactured by Plug Power.

Disadvantages of FCs

- Require high-purity gas; CO poisoning[53]
- Expensive[61]
- Material and component are not durable[48]
- Air, heat, and water management[60]

Internal Combustion Engines (ICEs)

Although FCs are the cleanest, most efficient way to use hydrogen, ICEs offer the potential of a less expensive route to hydrogen use.[69] Hydrogen-powered IC engines have approximately 20 percent greater efficiency than those running on gasoline.[53] Even though hydrogen may increase efficiency, power is lost due to the low volumetric energy density of hydrogen gas. To compensate for the loss of power, engine displacement volumes must increase.[52] ICEs must be redesigned to accommodate the combustion properties of hydrogen, particularly its fast flame speed.[70] Combustion of hydrogen *does* produce NO_X emissions; however, NO_X emissions may be reduced to levels less that gasoline engines using exhaust gas recirculation.[69]

The excellent paper by Karim lists the following advantages of hydrogen-powered ICEs:[71]

- Reduced cyclic variations which reduced emissions, improves efficiency, and allows quieter and smoother operation.
- High-octane number
- Excellent cold-start characteristics
- High-speed engine operation
- Higher efficiency than other IC engine fuels
- Lean mixture operation
- Tolerates impure gas

Karim [71] also describes the following disadvantages:

- Low-volumetric energy density leads to power loss–compressed hydrogen (200 atm) has approximately 5% of the energy as the same volume of gasoline.
- Potential problems with uncontrolled pre-ignition and backfiring into the intake manifold
- Serious limitations to the application of cold exhaust gas recirculation for exhaust emissions control
- Serious limitations to effective turbo charging
- Safety problems
- Material compatibility issues
- Ice formation from exhaust in extreme cold
- Potential durability issues
- A hydrogen-powered engine must be 40-60% larger in size than for gasoline-powered engine with the same power output

4.5 Emissions

Hydrogen is the cleanest fuel on the plant. This hallmark quality motivates hydrogen research and development. FCs have zero emissions producing only water and heat.[60] Only when combusted does hydrogen produce NO_X.[50]

Because hydrogen is an energy carrier, the hydrogen production method must be considered. Hydrogen produced from reformed fossil fuels, produces carbon dioxide. Likewise, hydrogen produced via electrolysis indirectly contributes to carbon dioxide emissions if the electricity is produced from fossil fuels. For this reason, many environmentalists only favor hydrogen produced from renewable sources (water and biomass) using renewable energies (wind, solar, hydropower, etc.).

Emissions from a manufacturing plant are easier to remove than from a tail pipe. Carbon sequestration methods could capture CO_2. The most responsible technologies will avoid emission production. Interestingly, if all passenger vehicles in the United States ran on reformed hydrogen (without carbon sequestration) using FCs, 30 to 50 percent less CO_2 would be emitted than if the fossil fuel were burned in ICEs.[50] This results from the high efficiency of FCs. Even though hydrogen from fossil fuels is not completely clean or green, it is an improvement over direct use of fossil fuels regarding emissions.

4.6 Fuel Transmission and Distribution

Currently, hydrogen is only used as a transportation fuel where municipalities, organizations, or corporations have established pilot programs for research and/or demonstration.[50,72] Distribution and storage are very costly due to the lack of infrastructure and hydrogen's low volumetric density. Hydrogen transportation is arguably the biggest economic hurdle facing the commercialization of hydrogen vehicles. Figures 16, 17, and 18 show that transportation costs can add $0.50 to $3.50 to the cost of a kilogram of hydrogen. Compared to gasoline, this is enormous. According to the EIA, "Distribution, marketing and retail dealer costs and profits combined make up 9 percent ($0.204) of the cost of a gallon of gasoline ($2.27 in 2005)."[73]

Bulk hydrogen must be delivered from a production facility to single point, and then distributed to many smaller refueling stations. Yang and Ogden evaluated transmission and distribution costs by three modes: pipelines, compressed-gas trucks, and cryogenic liquid H_2 trucks.[74] Each mode has its own advantages and disadvantages. Determining which mode is the most economical depends on two key parameters: transportation distance and flow rate.[74] Figure 16 shows

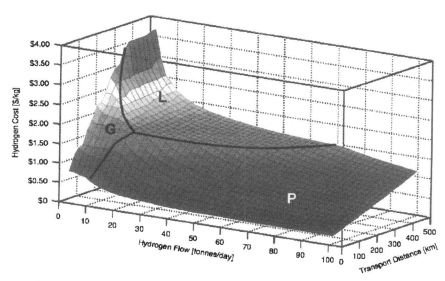

Figure 16 Minimum hydrogen transmission costs as a function of H_2 flow and transport distance. (From Ref. 27.)

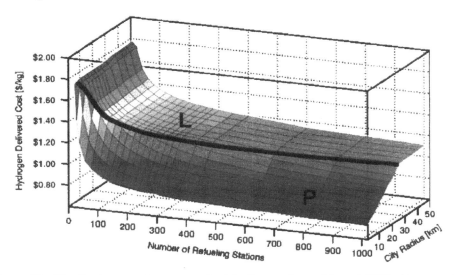

Figure 17 Distribution cost for a network of fueling stations with an 1,800 kg/day capacity as a function of number of refueling station and city radius. (From Ref. 27.)

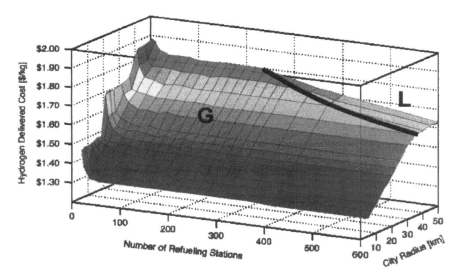

Figure 18 Distribution cost for a network of fueling station with an 1,800 kg/day capacity as a function of number of refueling station and city radius. (From Ref. 27.)

the minimum transmission cost as a function of flow rate and transport distance. From Figure 16, pipelines are favored at high flow rates and trucking is favored at low flow rates; compressed gas for short distances and liquefied hydrogen for long distances. Figures 17 and 18 show distribution costs as a function of the number of refueling stations and the city radius, for 1,800 kg/day and 500 kg/day capacity fueling stations, respectively. Distribution may contribute $0.80 to $1.80 per kilogram of hydrogen.

The models used by Yang and Ogden are based on idealized assumptions; their predictions have not been compared to real scenarios because none exist.[74] Nonetheless, their models provide valuable insights. For Figures 17 and 18, note that the region of many refueling stations within a small radius is very unlikely.[74] In the future, advances in fuel storage (e.g., carbon nanotubes, metal hydrides) may create more cost-efficient distribution modes.

Distributed manufacturing (on-site hydrogen production) is an alternative to hydrogen transport. Although the, transmission and distribution are reduced, the generation cost increases.[52]

4.7 Storage Issues

Hydrogen has a very low volumetric density. Increasing its density consumes energy, thus reducing energy efficiency. This section focuses on hydrogen storage on board vehicles. The fuel tank is central to any vehicle fuel system. Following is a list of key requirements for on-board hydrogen storage systems:

- Refueling time less than 3 minutes[54]
- Range of 300 miles or greater (5 to 13 kg storage capacity)[54,55]
- Low volume; high density[54]
- Safety (fueling, accidents, etc.)[54]
- Hydrogen discharge temperatures below FC temperature[54]
- Operating temperatures range from $-50°C$ to $150°C$[48]
- Fast kinetics of hydrogen uptake and release[48]

There are many on-board storage schemes. Figure 19 compares the volumetric and gravimetric energy densities for hydrogen storage technologies, and Figure 20 compares their cost. The 2010 and 2015 target goals are listed in these figures. None of the current technologies have met the target goals. The following is a brief overview of each hydrogen storage systems:

Pressurized Hydrogen in Composite Tanks

A composite carbon fiber tank is used to store gaseous hydrogen at pressures of 5,000 to 10,000 psi.[54] Compression consumes 2 to 10 percent of the energy content of hydrogen fuel, depending on the starting pressure and compression efficiency.[50,55,74] This is the storage system used in Honda's FCX FC vehicle and is commercially available.[64] According to the International Energy Agency (IEA), composite pressurized tanks cost $500 to $600/kg H_2.[54]

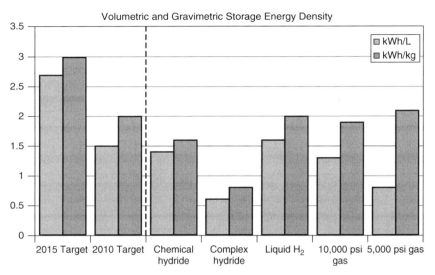

Figure 19 Volumetric and gravimetric energy densities for hydrogen storage technologies. (From Ref. 15.)

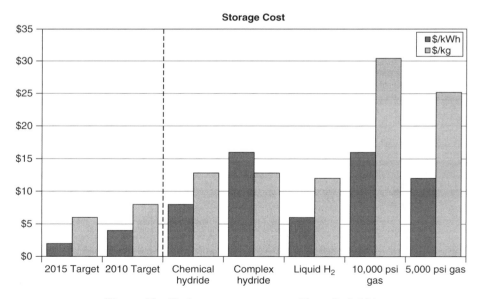

Figure 20 Hydrogen storage costs. (From Ref. 15.)

Glass Microspheres

Microscale glass spheres are filled with hydrogen at high pressure (5,000 to 10,000 psi) and high temperature (300°C).[54] Then, the spheres are cooled and transferred to a low-pressure fuel tank. To release the hydrogen, the microspheres are heated to 200°C to 300°C. The main advantages of this system are low operating pressure and conformable fuel tanks.[54] The liberation temperature is higher than the FC temperature; thus, additional heating is required reducing fuel efficiency. Additionally, there are challenges to prevent breakage while recharging the microspheres.[54]

Cryogenic Liquid Tanks

Liquid hydrogen is the highest energy density storage system (~20% hydrogen of system mass) and may be kept at low pressures.[54] Liquid hydrogen is extremely cold (−253°C, 20K). Liquefaction requires 30 to 40 percent of the energy content of hydrogen.[48,50,54] Boil-off is another problematic issue; 1 to 3 percent is lost per day.[48,53]

Chemical Cycling

Another way to store hydrogen as a liquid is to use a chemical cycling system. A few common chemical cycling systems are borohydride ($NaBH_4$), rechargeable organic liquids, and anhydrous ammonia (NH_3). In each of these systems, a reversible chemical reaction releases hydrogen. The key advantages of chemical cycling are medium-density storage and control. Except for NH_3, spent chemicals must be returned to a filling station for off-board regeneration.[54] Depending on the chemicals used, safety and toxicity may also be a concern.

On-Board Reforming

On-board reforming strips hydrogen-rich compounds (gasoline, methanol, ethanol, etc.) to form hydrogen gas. The advantage is that a liquid fuel would be stored on board and hydrogen would be produced as needed. However, as the NRC points out, "Significant technical barriers exist, such as size, weight, cost and long start-up times."[55]

Solid-State Storage

Solid-state storage of hydrogen would be safe and efficient.[54] These systems fall into two subcategories: carbon nanostructures and metal hydride systems. These systems physically or chemically absorb hydrogen into small structures or interstitial areas, where it is bound in the solid medium. Hydrogen may be liberated for use at elevated temperatures (100°C to 500°C) or by adding water (water-reactive metal hydrides).[54] These systems are still in development. The main challenges are to maximize storage density, lower liberation temperatures, increase absorption kinetics, and increase durability.[54]

4.8 Current Availability

Hydrogen vehicles have been created for demonstrations by many developers.[50,64–67] Various hydrogen production schemes and fueling systems have also been created and tested successfully.[50,54,55] However, in all scenarios, the primary limiting factor to commercial development has been cost. A close second is practical engineering hurdles such as storage and distribution, which also contribute high costs.

Hydrogen fueling stations are few and far between. Only a few key areas regions (e.g., Washington, D.C., California) have public hydrogen fueling stations. Busby includes a table of worldwide hydrogen fueling stations.[50] The U.S. Department of Energy provides an excellent alternative fueling station locator on its Web site, www.eere.energy.gov/afdc/infrastructure/locator.html.

Each year, the United States produces 90 million tons of hydrogen, and the vast majority is used in refining and fertilizer production.[75] Of the hydrogen produced, very little is green. Steam reforming of natural gas is the least expensive route to hydrogen production and accounts for 48 percent of the hydrogen market.[51] The remainder is reformed coal and oil; only 4 percent is from electrolysis. Furthermore, "to replace all the gasoline sold in the United States with hydrogen from electrolysis would require more electricity than is sold in the United States at the present time."[51]

In some ways, a hydrogen economy would be similar to our current gasoline economy. Consumers would purchase hydrogen at fueling stations, and outwardly vehicles will likely look much the same as gasoline vehicles. However, much research, development, and cost reduction must be accomplished for it to become a reality. "Of the three primary markets (transportation, portable applications, and micropower generation) for fuel cells, transportation is the furthest from mass production and widespread use, but hydrogen fuel-cell vehicles seem to generate the most excitement, by virtue of their visibility and lavish publicity."[50]

4.9 Safety

The public perceives hydrogen to be an extremely dangerous fuel; however, the literature generally accepts the hazards of hydrogen to be similar to gasoline.

> "It is now reasonably well established that the hazards of hydrogen use in transportation would be different from, but not necessarily worse than, the hazards of using petroleum fuels."[76]

> "In general, hydrogen is neither more nor less inherently hazardous than gasoline, propane or methane."[77]

> "In general, however, hydrogen is neither more nor less inherently dangerous than conventional fuels. Some of its properties make it safer, while others make it more

dangerous. With proper handling and controls, hydrogen can be as safe as other fuels in common use, or even safer."[50]

"Hydrogen is neither more nor less dangerous than most other energy carriers. In some respects, hydrogen has qualities that make it safer than most. Hydrogen is not poisonous, it burns rapidly with low radiation heat, and due to its low density compared to air, has the ability to spread rapidly in open surroundings."[78]

"It is possible to manufacture and utilize hydrogen just as safely as with today's gasoline systems."[78]

Because hydrogen is a gas, its safety issues are different than for liquid fuels. "The approach to handling safety issues for hydrogen applications ... requires consistent standards in several areas from production to use and also requires a way of handling the unanticipated safety-related events bound to occur in a technology that is significantly different from that currently used."[52] Until hydrogen is a mainstream fuel and used by the public, hydrogen safety will remain a debate. Below are safety advantages of hydrogen:

- Lighter that air, disperses easily[76]
- Burns rapidly with little radiant heat; therefore, risk of secondary fires is low[76,77]
- Nontoxic
- Nonpoisonous[77]
- Noncarcinogenic[77]
- Not an air or water pollutant[77]
- Prove safety track record in industry[77]

However, hydrogen also has the following safety disadvantages:

- Hard to detect—colorless, odorless, tasteless[76,77]
- Very hot clear flames[76]
- Large range of ignitability—4 to 74 percent hydrogen to air by volume[76]
- Requires little energy to ignite[76]

4.10 Current Research

For hydrogen to become mainstream fuel, it must be economical and practical. Hydrogen production is a proven technology; it can be produced through various methods. Many companies have developed fuel cells and have made hydrogen demonstration vehicles. Hydrogen fueling stations have also been created and are available in a few locations. So why is hydrogen not more prevalent? The reason is simple: cost. Every step of hydrogen, from *well to wheel*, is very expensive. Hydrogen research is not about making it work; it is about making it practical. The hydrogen economy will only become a reality when fuel prices have risen and hydrogen costs have decreased.

4.11 Hydrogen Links

Fuel cell/Hydrogen Infrastructure Codes & Standards	www.fuelcellstandards.com
U.S. Fuel Cell Council	www.usfcc.com/
Fuel Cell Today	http://fuelcelltoday.com/index/
National Renewable Energy Laboratory	www.nrel.gov/
National Hydrogen Association	www.hydrogenassociation.org/
Hydrogen Safety Report	www.hydrogensafety.info/
Hydrogen.gov	www.hydrogen.gov/index.html
American Hydrogen Association	www.clean-air.org/
International Electrotechnical Commission	www.iec.ch/
USDOE Energy Eff. and Renewable Energy	www.eere.energy.gov
Union of Concerned Scientists	www.ucsusa.org
American National Standards Institute	www.hcsp.ansi.org
Wikipedia.org	http://en.wikipedia.org/wiki/Hydrogen
Sandia Nat. Lab., Hydride Info. Center	http://hydpark.ca.sandia.gov/

5 OTHER BIOFUELS

Ethanol, biodiesel, and hydrogen are the dominate biofuels; however, there are other potential biofuels for transportation. Below, we briefly describe additional possibilities.

5.1 Methanol

Methanol, like ethanol, is an excellent automobile fuel.[79] Because of its excellent combustion properties (e.g., high octane rating), it is used as a racing fuel.[79] Many characteristics of methanol fuel are similar to ethanol. Methanol is synthesized from gasification or anerobic digester gases.[80] Some advantages of methanol are increased power, increased efficiency, and reduced knock.[79,80] Disadvantages include health hazards, such as blindness and acidosis, poor cold-start properties, corrosion, low lubricity, and formaldehyde production.[23,80]

5.2 Butanol

Butanol is an alcohol that can be produced via fermentation. Historically, butanol was fermented using *Clostridium acetobutylicum* in the ABE fermentation, which produces acetone, butanol and ethanol.[81] Environmental Energy Inc. (EEI) has developed a patented process that produces only butanol and hydrogen.[82] In both fermentations, sugars and starches are feedstocks.[81,82]

Butanol may be blended with gasoline or used directly.[82,83] Compared to ethanol, butanol has more favorable physical properties, such as higher energy content and less susceptibility to water contamination.

5.3 Mixed Alcohols

The MixAlco process anaerobically ferments any biodegradable material into mixed organic acids, which are then chemically converted into mixed alcohols. The MixAlco process is very robust; all process steps have been demonstrated on the laboratory scale and some at the pilot scale. Because the MixAlco process does not require sterility, costs are very low.[15] Another route to mixed alcohols involves gasifying lignocellulose to syngas (carbon monoxide and hydrogen), which can be converted into mixed alcohols using a catalyst.[16,17]

5.4 Petroleumlike Biofuels

Petroleumlike compounds may be extracted from hydrocarbon-producing plants and used as fuel. For example, *Euphorbiaceae* and *Asclepiadaceae* have several species that produce these compounds.[84] Biocrude extracted from *E. lathyris* "contained ethylene (10%), propylene (10%), toluene (20%), xylene (15%), C5–C20 non-aromatics (21%), coke (5%), C1–C4 alkenes (10%) and fuel oil (10%)."[84] Biocrude will be processed much like petroleum crude to produce fuel.

5.5 Synthesis Hydrocarbons

Fisher-Tropsch is a well-known synthesis process that converts syngas to gasoline, diesel, and waxes.[85] Unfortunately, Fisher-Tropsch is an expensive technology and is practiced only at large scales where economies can be realized. Another gas-to-liquids (GTL) technology can oligimerize methane or alcohols into gasoline, diesel, or jet fuel. SynFuels International Inc. is developing the technology and estimates production costs of $25 to $28 per barrel from low-cost feedstocks.[86]

5.6 Methane

Anaerobic digesters are a common route to methane production. Typically, these systems do not generate enough methane to run more than a small generator.[87] Biomethane could be used as a transportation fuel.

6 CLOSING REMARKS

Currently, none of the biofuels mentioned in this chapter significantly displace petroleum. Some require subsidy, some are too expensive, and others are still under research and development. Displacing petroleum will require parallel use of renewable energy and biofuels. Renewable experts agree that there is no single silver bullet. In recent years, biofuels have gained momentum and will continue to grow as long as the sun shines.

REFERENCES

1. Greg Pahl, Biodiesel: Growing a New Economy, Chelsea Green, 2005.
2. Energy Information Agency, EIA Kids Page Ethanol Energy Timeline—milestone in ethanol energy history, www.eia.doe.gov/kids/history/timelines/ethanol.html. Accessed on January 15, 2007.
3. Stephen W. Mathewson, *The Manual for the Home and Farm Production of Alcohol Fuel*, Ten Speed Press, 1980.
4. J. Finguerut, Ethanol Production in a Cane Sugar Mill, Workshop—Louisiana State University, Baton Rouge, 2005.
5. U.S. Department of Energy (DOE), "Energy Efficiency and Renewable Energy, Biomass Program," http://www1.eere.energy.gov/biomass/ethanol_yield_calculator.html. Accessed on December 7, 2007.
6. A.G. Alexander, *The Energy Cane Alternative* (Sugar Series 6), Elsevier, Amsterdam, 1985.
7. M. H. Nguyen and R. G. H. Prince, "A Simple Rule for Bioenergy Conversion Plant Size Optimization: Bioethanol from Sugar Cane and Sweet Sorghum," *Biomass and Bioenergy*, **10**(5/6), 361–365 (1996).
8. May Wu, Ye Wu, and Michael Wang, "Energy and Emission Benefits of Alternative Transportation Liquid Fuels Derived from Switchgrass: A Fuel Life Cycle Assessment," *Biotechnology Progress* **22** (4), 1012–1024 (2006).
9. H. Hartmann, "Environmental aspects of energy crop use—a system comparison," In Philippe Chartier, A. A. C. M. Beenackers, and G. Grassi, ed., *Biomass for Energy, Environment, Agriculture and Industry*, Proceedings of the 8th European Biomass Conference, Editor(s): Vienna, Oct. 3–5, 1994 (1995).
10. C. Luo, David L. Brink, and H. W. Blanch, "Identification of Potential Fermentation Inhibitors in Conversion of Hybrid Poplar Hydrolyzate to Ethanol," *Biomass and Bioenergy*, **22** (2), 125–138 (2002).
11. Ethanol Industry Outlook, Renewable Fuels Association, www.ethanolrfa.org. 2006.
12. M. M. Rabbani, "Acid Hydrolysis Processes for Cellulosic Biomass," Esc. Univ. Ing. Tec. Agric., Villava, Spain. *Ingenieria Quimica* (Madrid, Spain), **21** (245), 139–46 (1989).
13. Lizbeth Laureano-Perez, Farzaneh Teymouri, Hasan Alizadeh, and Bruce E. Dale, "Understanding Factors that Limit Enzymatic Hydrolysis of Biomass. Characterization of Pretreated Corn Stover," *Applied Biochemistry and Biotechnology* 121–124 (2005).
14. T. Eggeman, and D. Verser, "The Importance of Utility Systems in Today's Biorefineries and a Vision for Tomorrow," *Applied Biochemistry and Biotechnology*. 129–132, 361–381 (2006).
15. Mark T. Holtzapple, Advanced Biomass Refinery: Third-Generation Technology, Abstracts, 62nd Southwest Regional Meeting of the American Chemical Society, Houston, October 19–22, 2006, SRM-156. Publisher: American Chemical Society, Washington, D.C.
16. M. Xiang, D. Li, W. Li, B. Zhong and Y. Sun, "Potassium and nickel doped β-Mo$_2$C catalysts for mixed alcohols synthesis via syngas," *Catalysis Communications* **8** (3), 513–518 (2007).

17. J. M. Campos-Martin, A. Guerrero-Ruiz, J. L. G. Fierro, "Structural and Surface Properties of CuO-ZnO-Cr$_2$O$_3$ Catalysts and Their Relationship with Selectivity to Higher Alcohol Synthesis," *Journal of Catalysis*, **156**, 208–218 (1995).
18. G. Najafpour, and H. Younesi, "Ethanol and Acetate Synthesis from Waste Gas Using Batch Culture of *Clostridium ljungdahlii*," *Enzyme & Microbial Technology*, **38** (1–2), 223–228 (2006).
19. K. T. Klasson, M. D. Ackerson, E. C. Clausen, and J. L. Gaddy, "Bioconversion of Synthesis Gas into Liquid or Gaseous Fuels," *Enzyme and Microbial Technology*, **14** (8), 602–608 (1992).
20. *Guidebook for Handling, Storing, & Dispensing Fuel Ethanol*, Prepared for the U.S.D.O.E. by the Center for Transportation Research Energy Systems Division at Argonne National Laboratory.
21. Renewable Fuels Association, *Guidelines for Establishing Ethanol Plant Quality Assurance and Quality Control Programs*, RFA Publication #040301, www.ethanolrfa.org. Accessed on August 1, 2007.
22. D4806-98 for Denatured fuel ethanol for blending with gasoline, http://www.distill.com/specs/US-1.html. Accessed on May 11, 2007.
23. Keith Owen and Trevor Coley, *Automotive Fuels Reference Book*, 2nd ed., Society of Automotive Engineers, Warrendale, Pennsylvania, 1995.
24. H. Joseph, ANFAVEA, anfavea@anfavea.com.br, henry.joseph@volkswagen.com.br. Accessed on March 5, 2007.
25. U.S. Dept. of Energy EERE, *Handbook for Handling, Storing, and Dispensing E85*, www.e85fuel.com/pdf/e85_technical_booklet.pdf. Accessed May 12, 2007.
26. E de Oliviera, Ph.D. Dissertation, University of Waterloo, Ontario, 1997.
27. National Ethanol Vehicle Association, http://www.e85fuel.com. Accessed on April 30, 2007.
28. Low-level ethanol blends, http://www.nrel.gov/docs/fy05osti/37135.pdf. Accessed on April 30, 2007.
29. Renewable Fuels Association, http://www.ethanolrfa.org. Accessed on April 30, 2007.
30. M. Gautam and D.W. Martin II, "Combustion Characteristics of Higher-alcohol/Gasoline Blends," *Proc. Instn. Mech. Engrs.*, **214** Part A (2000).
31. Robert K. Niven, "Ethanol in Gasoline: Environmental Impacts and Sustainability," *Renewable & Sustainable Energy Reviews* **9** 535–555 (2005).
32. Richard L. Bechtold, *Alternative Fuel Guidebook: Properties, Storage, Dispensing and Vehicle Facility Modifications*, Society of Automotive Engineers, 1997.
33. U.S. Department of Energy, Efficiency and Renewable Energy (EERE), http://www.eere.energy.gov/. Accessed on April 30, 2007.
34. Mark Cooper, Over a Barrel: Why Aren't Oil Companies Using Ethanol to Lower Gasoline Prices? Confederation of America, May 2005.
35. American Coalition for Ethanol, www.ethanol.org. Accessed on May 1, 2007.
36. Iowa Corn Growers Association, www.iowacorn.org. Accessed on May 1, 2007.
37. Joshua Tickell, *From Fryer to Fuel Tank: The Complete Guide to Using Vegetable Oil as an Alternative Fuel*, 3rd ed., Ticknell Energy, Hollywood, California, 2003.
38. The National Biodiesel Board., www.biodiesel.org. Accessed on July 21, 2007.

39. Fangrui Ma and Milford A. Hanna., "Biodiesel Production: A Review," *Bioresource Technology*, **70** (1999).
40. BQ 9000 Quality Management Program, www.bq-9000.org. Accessed on January 15, 2007.
41. Gail R. Frahm, St. Johns School Buses Rolling 1 Million Miles on Biodiesel, Michigan Soybean Promotion Committee press release, April 27, 2004.
42. Steve Howell, "Rigorous Standards Ensure Biodiesel Performance," *Biodiesel: On The Road to Fueling the Future*, National Biodiesel Board Biodiesel Report, 2001.
43. R. L. McCormick et al., "Effects of Biodiesel Blends on Vehicle Emissions," NREL Milestone Report NREL/MP-540-40554, October 2006.
44. K. Shaine Tyson, 2004 Biodiesel Handling and Use Guide, DOE/GO-102004-1999, November 2004.
45. National Renewable Energy Laboratory. Image Library. Picture of biofuels fuel pump, http://www.nrel.gov/data/pix/Jpegs/13531.jpg. Accessed on March 20, 2007.
46. Energy Information Administration., www.eia.doe.gov. Accessed on March 20, 2007.
47. U.S. Congress, Safe, Accountable, Flexible, and Efficient Transportation Act of 2004, Section 5102.
48. P.P. Edwards et al., "Hydrogen Energy," *Phil. Trans. R. Soc. A*, **365**, 1043–1056(2007).
49. Robert F. Service, "Profile: Daniel Nocera, Hydrogen Economy? Let Sunlight Do the Work.," *Science*, **315** (February 9 2007). www.sciencemag.org.
50. Rebecca L. Busby, *Hydrogen and Fuel Cells: A Comprehensive Guide*, PennWell, Tulsa, Oklahoma, 2005.
51. Michael F. Hordeski, P.E., *Alternative Fuels: The Future of Hydrogen.*, The Fairmont Press, Lilburn, Georgia, 2007.
52. Bent. Sorenson, *Hydrogen and Fuel Cells*, Elsevier, Amsterdam, 2005.
53. A. G. Dutton, Hydrogen Energy Technology, Tyndall Working Paper TWP 17. Tyndall Centre for Climate Change, p. 30. Available from http://www.tyndall.ac.uk/publications/working_papers/wp17.pdf. Accessed on March 12, 2007.
54. International Energy Agency, 2006 Hydrogen production and storage, R&D priorities and gaps, http://www.iea.org/Textbase/papers/2006/hydrogen.pdf. Accessed on March 12, 2007.
55. National Research Council, *The Hydrogen Economy: Opportunities, Costs, Barriers, and R&D Needs*, The National Academies Press, 2004 (394 pages), www.nap.edu.
56. www.fuelcellstandards.com. Accessed on December 7, 2007.
57. Karen. Hall, Personal e-mail communication. March 23, 2007., khall@ttcorp.com.
58. Committee on Gaseous Fuels Forms, "Subcommittee on Hydrogen and Fuel Cells," *Standardization News*, December 2004. www.astm.org.
59. Energy Information Administration (EIA), Alternatives to Traditional Transportation Fuels: An Overview, June 1994. ftp://ftp.eia.doe.gov/pub/solar.renewables/alt_over.pdf #page=58.
60. U.S. Department of Energy, Energy Efficiency and Renewable Energy, http://www1.eere.energy.gov/hydrogenandfuelcells/education/pdfs/fuel_cell_facts.pdf. Accessed on March 20, 2007.

61. Robert. Rose, Questions and Answers about Hydrogen and Fuel Cells, www.fuelcells.org. Accessed on April 10, 2007.
62. U.S. Department of Energy, Energy Efficiency and Renewable Energy, http://www1.eere.energy.gov/hydrogenandfuelcells/pdfs/01_paster_hydrogen_prog.pdf.
63. Ibrahim. Dincer, "Environmental and Sustainability Aspects of Hydrogen and Fuel Cells," *International Journal of Energy Research*, **31**: 29–55 (2007).
64. Honda Worldwide, http://world.honda.com/news/2007/4070108FCXConcept/. Accessed 2007.
65. General Motors Corporation., Chevrolet Equinox specifications, http://www.gm.com/company/gmability/adv_tech/400_fcv/fact_sheets.html.
66. Tokyo Gas Co., Ltd., Corporate Communications Dept. Introduction of DaimlerChrysler F-Cell Fuel Cell Vehicle, October 16, 2003, http://www.tokyo-gas.co.jp/Press_e/20031016-1e.pdf.
67. Stefan, Geiger, "Opening Doors to Fuel Cell Commercialisation: DaimlerChryler F-Cell," *Fuel Cell Today*, April 5, 2004, http://www.fuelcelltoday.com/FuelCellToday/FCTFiles/FCTArticleFiles/Article_788_DCF-Cell0404.pdf.
68. National Renewable Energy Lab (NREL), Fuel Cell, http://www.nrel.gov/data/pix/Jpegs/12508.jpg. Accessed on April 10, 2007.
69. James W. Heffel, "NO_x Emission and Performance Data for a Hydrogen Fueled Internal Combustion Engine at 1,500rpm Using Exhaust Gas Recirculation," *International Journal or Energy Research* **28** 901–908 (2003).
70. Jehad A. A. Yamin, "Comparative Study Using Hydrogen and Gasoline as Fuels: Combustion Duration Effect," *International Journal or Energy Research*, **30**: 1175–1187 (2006).
71. Ghazi A. Karim, "Hydrogen as a Spark Ignition Engine Fuel," *International Journal of Hydrogen Energy*, **28** 569–577 (2003). www.elsevier.com/locate/ijhydene.
72. Arizona Public Service, Alternative Fuel Pilot Plant & Hydrogen Internal Combustion Engine Vehicle Testing, http://avt.inl.gov/pdf/hydrogen/h2factsheet.pdf. Accessed on April 11, 2007.
73. Energy Information Agency (EIA), A Primer on Gasoline Prices, http://www.eia.doe.gov/pub/oil_gas/petroleum/analysis_publications/primer_on_gasoline_prices/html/petbro.html. Accessed on December 7, 2007.
74. Christopher Yang and Joan Ogden, "Determining the Lowest-cost Hydrogen Delivery Mode," *International Journal of Hydrogen Energy*, **32** 268–286 (2007). www.elsevier.com/locate/ijhydene.
75. U.S. Dept. of Energy Hydrogen Program, Hydrogen Production, www.hydrogen.energy.gov. Accessed on December 7, 2007.
76. Daniel Sperling (editor), *Alternative Transportation Fuels: An Environmental and Energy Solution*, Greenwood Press, Westport, CT, 1989.
77. U.S. Dept. of Energy, EERE., Regulators' Guide to Permitting Hydrogen Technologies. Version 1.0 PNNL-14518, Released January 12, 2004.
78. Bellona, Hydrogen Report, 2002, http://www.bellona.org/reports/hydrogen.
79. John H. Perry and Christiana P. Perry, *Methanol: Bridge to a Renewable Energy Future*, University Press of America. 1990.

80. Mark A. Deluchi, Daniel Sperling, and Robert A. Johnston, A Comparative Analysis of Future Transportation Fuels, Institute of Transportation Studies, University of California Berkley. 1987. Research Report UCB-ITS-RR-87-13.
81. Wikipedia. Clostridium acetobutylicum. http://en.wikipedia.org/wiki/Clostridium_acetobutylicum. Accessed on December 7, 2007.
82. Environmental Energy Inc, http://www.butanol.com/. Accessed on December 7, 2007.
83. Dupont, Biobutanol Fact Sheet, http://www.dupont.com/ag/news/releases/BP_DuPont_Fact_Sheet_Biobutanol.pdf. Accessed on December 7, 2007.
84. Dipul Kalita, Hydrocarbon Plant—New Source of Energy for Future, Renewable & Sustainable Energy Reviews. 2006. www.elsevier/locate/rser.
85. Ayhan Demirbas, "Progress and Recent Trends in Biofuels," *Progress in Energy and Combustion Science*, **33** 1–18 (2007).
86. Kenneth R. Hall, *A New Gas to Liquids (GTL) or Gas to Ethylene (GTE) Technology*. Texas A&M University, College Station. (Unpublished Manuscript). 2007.
87. Fredrick T. Varani and John J. Buford Jr., *Fuels from Waste Chapter VI: The Conversion of Feedlot Wastes into Pipeline Gas*, Academic Press, 1977.

CHAPTER 9

LIFE-CYCLE ASSESSMENT AS A TOOL FOR SUSTAINABLE TRANSPORTATION INFRASTRUCTURE MANAGEMENT

Gerardo W. Flintsch
Virginia Polytechnic Institute and State University
Blacksburg, Virginia

1	INTRODUCTION	257	
2	TRANSPORTATION INFRASTRUCTURE ASSET MANAGEMENT	258	
3	SUSTAINABLE TRANSPORTATION INFRASTRUCTURE MANAGEMENT	261	
4	PERFORMANCE MEASURES	262	
5	LIFE-CYCLE COSTING	264	
	5.1 Deterministic LCCA	267	
	5.2 Probabilistic LCCA	267	
	5.3 Perceived Problems and Solutions	268	
	5.4 Tools for Life-Cycle Cost Analysis	269	
6	LIFE-CYCLE ASSESSMENT	270	
	6.1 LCA Methodology	271	
	6.2 Environmental Impacts of Transportation Projects	274	
7	OTHER IMPACTS OF TRANSPORTATION PROJECTS	277	
8	CONCLUSIONS AND RECOMMENDATIONS	278	

1 INTRODUCTION

Efficient and effective asset management is critical for the long-term sustainability of transportation infrastructure systems and networks. On one hand, sound transportation infrastructure systems (e.g., highway, rail, aviation, and maritime) play a vital role in encouraging a more productive and competitive national economy.[1] On the other hand, increasing demands, shrinking financial and human resources, and increased infrastructure deterioration have made the task of maintaining our transportation infrastructure systems more challenging than ever before.

Many of the nation's transportation infrastructure systems are reaching the end of their service lives. Infrastructure systems have gradually deteriorated with age as a result of environmental action and use that, in many cases, significantly exceeds the design expectations. Shortfalls in funding and changing population patterns have placed an even larger burden on our aging power plants, water systems, airports, bridges, highways, and school facilities. For example, in a recent survey conducted by the American Society of Civil Engineers (ASCE), America's transportation infrastructure, the nation's critically important foundation for economic prosperity, only received a cumulative grade of D.[2]

2 TRANSPORTATION INFRASTRUCTURE ASSET MANAGEMENT

Under these circumstances, decision makers are faced with competing investment demands and must distribute limited resources so that the transportation infrastructure systems are maintained in the best possible condition. Infrastructure management systems have emerged as tools to support these decisions and, as such, they may help bridge the gap between infrastructure condition and user expectations. Public and private agencies have always tried to maintain their infrastructure assets in good and serviceable condition at a minimum cost, and thus they practiced infrastructure management. However, as most of the nation's infrastructure systems reached maturity and the demands started to rapidly increase in the mid-1960s, infrastructure agencies started to focus on a systems approach for infrastructure management. This process has led to today's asset management concept, as illustrated in Figure 1. The process started with the development of pavement management systems, continued with bridge management systems and infrastructure management systems, and has recently evolved into asset management.[3–5]

Asset management combines engineering principles with business practice and economic theory, and it has been proposed as a solution for balancing growing demands, aging infrastructure, and constrained resources in the transportation sector.[6] Good asset management implies a systematic integrated approach to

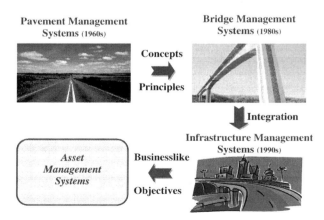

Figure 1 Engineering management system evolution.

project selection, analysis of tradeoffs, resource optimization, programming, and budgeting. The objectives of asset management include building and preserving facilities more cost effectively and with more satisfying performance, delivering the best value for the available resources to customers and enhancing the accountability of the agencies.[7]

As with all engineering management systems, efficient asset management relies on accurate asset inventory, condition, and system performance information; considers the entire life-cycle cost of the asset; and combines engineering principles with economic methods, thus seeking economic efficiently and cost-effectiveness (Figure 2). Effective asset performance modeling is necessary to find the best timing for the maintenance and rehabilitation actions, as well as to assess the impact of the decisions on the overall performance of the infrastructure systems being managed. The asset inventory, condition, and performance data are used to develop feasible alternatives based on the agency goals and policies, user expectations, and available resources. Alternative investments and funding scenarios are then evaluated to determine their impact on system performance and their compliance with current and future user expectations and agencies' financial constraints. Decision makers use this information to prepare short-term and long-term plans that are more systematic, broader in scope, and more supportable by field data than those determined using traditional approaches. Once the plans are implemented, the performance results are monitored to verify the assumptions and predictions made at the alternative evaluation and planning stages.[6]

Figure 3 presents an example of a general functional framework for an infrastructure asset management system.[8] This framework builds on the scheme presented in Figures 2 and consists of modules (tools and methods) that can be used for various types of infrastructure assets, individually or holistically. The foundation of the system is a database that includes asset inventory, condition, usage, and treatment information. The information is integrated and analyzed though a series of modular applications. Strategic decision-support tools allow

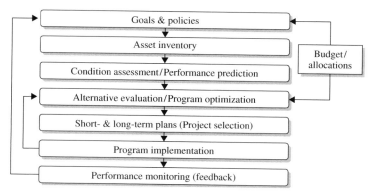

Figure 2 Asset management framework. (From Ref. 6.)

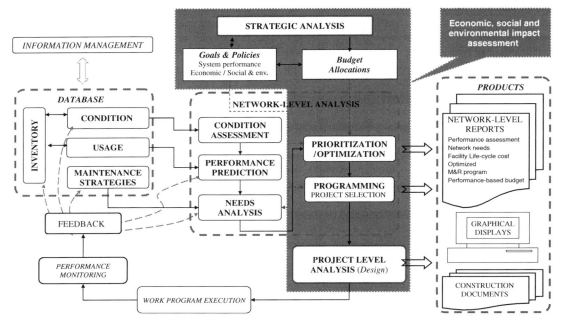

Figure 3 Intelligent infrastructure asset management system framework. (From Ref. 8.)

the overall goals of system performance and the policies of the agency to be set by analyzing tradeoffs among competing infrastructure classes and programs. Network- or program-level tools are used to evaluate and predict asset performance over time; to identify appropriate maintenance, rehabilitation, replacement, or expansion investment strategies for each asset; to evaluate the different alternatives; to prioritize or optimize the allocation of resources; and to generate plans, programs, and budgets. These tools produce reports and graphical displays tailored to different organizational levels of management and executive levels, as well as to the public. Projects included in the work program are designed using project-level analysis tools. The infrastructure management cycle continues with the execution of the specified work. Changes in the infrastructure assets resulting from the work conducted, as well as from normal deterioration, are periodically monitored, preferably by means of nondestructive techniques, and input into the system.

These strategic, network and project-level transportation infrastructure management decisions have many technical, economic, social and environmental impacts that should be considered in the decision-making process. The assessment of these impacts over the life-cycle of the infrastructure systems is necessary for making informed decisions on how to define policies, allocate resources, select projects, and/or design and construct these projects, as illustrated by the overlapping shadowed area in Figure 3. Current practice typically includes the consideration of technical and economic issues through life-cycle cost analysis (LCCA). LCCA techniques are used by most state departments of transportation (DOTs),[9] but they are generally limited to project-level pavement analysis, using just agency costs.[10] Some management systems use LCCA for network-level analysis; however, the technique does not enjoy widespread application.[6] In addition, the National Environmental Policy Act (NEPA) requires the assessment of the social, economic, and environmental impact of federally funded projects.[11] However, the application often has focused mainly on environmental impact affecting human heath and natural ecosystems. Various levels of analysis (environmental impact statement, environmental assessment, or categorical exception) are required, depending on the nature and significance of a specific impact.

3 SUSTAINABLE TRANSPORTATION INFRASTRUCTURE MANAGEMENT

Life-cycle costing has been used extensively in infrastructure asset management to consider all the costs of a project over its service life. However, the approach to evaluate potential transportation projects, programs, and strategic plans on the basis of a combination of engineering and economic criteria is no longer sufficient in the context of sustainable transportation infrastructure. Sustainable development has been defined as "the development that meets the needs of the present without compromising the ability of future generations to meet their own needs."[12] The effects of transportation projects on the environment should be fully considered because (1) these effects can be lasting and substantial and (2) they often are important to the quality of life. In the last two decades this topic has attracted enormous attention.[12–19] Furthermore, social issues are also starting to be considered in this process, in addition to the environmental and economic dimensions when developing engineering solutions to societal needs.

ASCE states that *"the demand on natural resources is fast exceeding supply in the developed and developing world. Environmental, economic, social and technological development must be seen as interdependent and complementary concepts, where economic competitiveness and ecological sustainability are complementary aspects of the common goal of improving the quality of life."*[20] To achieve sustainable development transportation investments (and technological advances) should help achieve (or balance) economic, environmental, and social goals. Therefore, these decisions should be placed in a context of economic development, ecological sustainability, and social desirability, manifested in a mix of cultural and social values.[21] The simultaneous optimization of these three objectives is being referred to as the triple bottom line of sustainable development.

4 PERFORMANCE MEASURES

The triple bottom line of sustainable development is also reflected in the parameters or performance measures that transportation stakeholders use to evaluate the performance of transportation in general and the transportation infrastructure systems and networks in particular. Asset management requires the monitoring of the transportation systems using performance measures that reflect the agency's goals and objectives. The National Cooperative Highway Research Program (NCHRP) Report 551 defined performance measurement as *"a way of monitoring progress toward a result or goal, more specifically it helps to keep track and forecast the impacts within and outside the system."*

Existing performance measures and analytical approaches are used to evaluate the effects of highway investment on economic productivity, user benefits (that include reductions in fatalities and serious injuries), environmental impacts, and societal benefits, among other factors. The analytical approaches used for quantifying these measures are grounded in financial, economic, and transportation theory and involve classical financial criteria such as benefit/cost ratios, as well as user (e.g., reduction in fatalities and serious injuries), environmental (pollution reduction) and societal benefits (e.g., reduction in traffic congestion). The report identified numerous performance measurements and organized them in the following 10 categories: preservation, accessibility, mobility, operations and maintenance, safety, environmental impacts, economic development, social impacts, security, and delivery. Examples of performance measures in the various categories are presented in Table 1.[22]

The report also found that typical performance measures are expressed as output or outcome measures. *Outputs* refer to the quantities that the agency produced or used (e.g., the number of staff hours spent or the tons of asphalt used). *Outcomes* refer to the resulting improvements in performance or condition as a result of those outputs (e.g., the increased percent of pavement in good condition, improved safety, etc.).

The current thinking in asset management emphasizes outcome measures or results as the most desirable, since they emphasize results and accountability for them. This has been reflected in the increasing use of performance-based contracts and warranty clauses in transportation infrastructure construction contracts. However, most state agencies currently utilize outputs. The primary reasons for this are because outputs are easier and less expensive to measure, may be easier to communicate to nontechnical audiences, and provide an immediate indication of accomplishment, outcomes of an investment or a program, by contrast, may be long term. Therefore, NCHRP Report 551 recommends the use of a blend of output and outcomes in life-cycle analysis. Both types of measures will be used in this investigation to support better quantification of life-cycle costs and benefits at different levels of investment and the resulting level of transportation service.[22]

Table 1 Examples of Performance Measures

Group	Properties Measured	Examples
Preservation	Condition of the transportation system and actions to keep the system in a state of good repair; often specific to the type of asset	• Physical condition (e.g., extent or severity of distress, deviations from nominal track gauge) • Indices that combine a number of condition measurements or that relate to user perceptions of condition (e.g., pavement condition index or bridge health index) • Nontechnical measures (e.g., financial asset value)
Accessibility	Ability of people and goods to access transportation services, often expressed from a user's perspective	• Density of opportunities enabled by transportation services (e.g., number of households within a 30-minute drive of a center, or number of jobs within a 10-minute walk of transit stops) • Ability of a facility to serve a particular user group (e.g., a particular segment of population or type of freight) • Availability of modes and modal choice
Mobility	Time and cost of making a trip and the relative ease or difficulty with which a trip is made, essentially congestion	• Measures of offer that reflect a supplier (offer) perspective (e.g., volume-capacity ratio, capacity-related level of service) • Measures that reflect a user (demand) perspective (e.g., speed, travel time, delay, trip reliability, and user cost)
Operations and Maintenance	Effectiveness of the transportation system in terms of throughput and travel costs and revenues from a system perspective and level of service (customer experience)	• Measures of vehicle occupancy or freight capacity • Cost efficiency (e.g., average cost per mile or per vehicle-mile traveled (VMT)) • Systemwide fuel efficiency
Safety	Quality of transportation service in terms of crashes or incidents that are harmful to people and damaging to freight, vehicles, and transportation infrastructure	• Asset conditions that contribute to or detract from safety • "True" safety measures gauged by the number, frequency, severity, and cost of accidents • Harm to agency personnel as well as drivers and passengers, particularly in work zones • Risk of future safety problems at candidate locations
Environmental Impacts	Protection of the environment	• Measures associated with key impact areas, including air quality, groundwater, protected species, noise, and natural vistas • Output-based performance measures for actions critical to mitigating the above impacts (e.g., protecting wetlands, using snow and ice chemicals that protect groundwater and air quality)

(continues)

Table 1 (*continued*)

Group	Properties Measured	Examples
Economic Development	Direct and indirect impacts of transportation on the economy	• Cost of transportation experienced by users and shippers and are expressed in measures such as economic output (e.g., gross state product), employment (e.g., jobs created), and income • Indirect measures (transportation's contribution to the general economy), such as manufacturers/shippers/employers who have relocated for transportation purposes, and measures of truck travel per unit of regional economic activity
Social Impacts	Effects on the broader society	• Measures of the effects on neighborhoods adjacent to transportation facilities or on disadvantaged population groups
Security	Protection of travelers, freight, vehicles, and system infrastructure from terrorist actions	• Risk exposure measures
Delivery	Measures focused on the delivery of transportation projects and services to the customer	• Output-oriented accomplishment measures that complement outcome-oriented measures in the other categories • Measures of efficiency and effectiveness in use of resources, and impacts on customers that need to be considered in evaluation of alternative delivery strategies

Note: From Ref. 22.

The following sections provide guidelines on current practices for life-cycle cost analysis and life-cycle assessment of transportation infrastructure projects, programs, and strategic plans, and discuss how these analysis tools can be used to support asset management decisions.

5 LIFE-CYCLE COSTING

Life-cycle costing, or whole life costing, is an economic analysis tool that is used extensively in highway project decision making and asset management. The concept of LCCA was originally developed by the U.S. Department of Defense in the early 1960s to increase the effectiveness of government procurement. Since then, the concept of LCCA has spread from defense-related issues to a variety of areas. From the very beginning, LCCA has been closely related to design and development because it allows costs to be eliminated before they are incurred as opposed to cutting costs afterward.[23]

LCCA accounts for all costs associated with infrastructure investment, including construction, operation, maintenance, renewal, retirement, and any other requirement over the expected service life of the facility. The process of incorporating into the design all the "true costs" has also been referred to as Life-cycle Engineering. For transportation projects, the Transportation Equity Act for the 21st Century defined LCCA as "... a process for evaluating the total economic worth of a usable project segment by analyzing initial costs and discounted future costs, such as maintenance, user, reconstruction, rehabilitation, restoring, and resurfacing costs, over the life of the project segment." A usable project segment is defined as a portion of a highway that, when completed, could be open to traffic independent of some larger overall project.[24] The Federal Highway Administration (FHWA) has encouraged and, in some cases, mandated the use of LCCA as a decision support tool when analyzing major investment decisions; however, the agency emphasizes that the results are not decisions in and of themselves. The FHWA Life-Cycle Cost Analysis Primer provides the basic background for transportation officials to investigate the use of LCCA to evaluate alternative infrastructure investment options and demonstrates the value of such analysis in making economically-sound decisions.[25] This primer proposes the following five-step framework:

1. Establish design alternatives.
2. Determine activity timing.
3. Estimate costs (agency and user).
4. Compute life-cycle costs.
5. Analyze the results.

Often, the logical-analytical framework implied in such analyses is as important as the LCCA results themselves.[27]

Before any analysis can take place, several basic assumptions are needed, such as the same-benefit assumption, analysis period length, and inclusion or exclusion of user costs. These issues directly affect the reliability of final results. For example, if the same-benefit assumption does not hold, the LCCA method should not be used in the selection process. Other methods, like benefit-cost analysis, are more appropriate in such conditions.[25]

The process starts with the definition of the general parameter, such as analysis period and discount rate. The analysis period should be sufficiently long to reflect the performance difference among alternative projects. FHWA recommends a period of over 35 years but, because most highway segments are required to be in service for a longer time, some DOTs require a longer analysis period (e.g., 50 years for Virginia DOT). To evaluate the sustainability of the transportation infrastructure, this analysis period should be even longer; such as 100 years. Second, the discount rate must be defined. In the United States, the recommended rates are published by the Office of Management and Budget.

Once the general parameters are determined, the following five categories of costs should be evaluated and calculated for the project's life cycle:[26]

- Agency costs (construction, rehabilitation, maintenance, salvage return or disposal, engineering and administration, and investment)
- Vehicle operating costs
- Travel time costs (dollar value of time spent on the roadway)
- Accident costs
- Environmental costs

Project costs are generally regarded as agency costs; they include initial construction, rehabilitations and repairs, maintenance, engineering, administration, supervision and inspection, and traffic control. For many agencies, existing management systems could provide pavement condition and usage information as well as pavement deterioration models and appropriate maintenance and rehabilitation options and costs. The other four categories of costs are user costs. Many agencies only consider project costs in their LCCA procedures because of the difficulties associated with the quantification of user costs. However, user cost can account for up to 95 percent of the total highway transportation cost and should not be ignored. Several models for estimating direct user costs as a function of pavement condition and user delay costs as a function of lane closure practices have been proposed.[27-30] Other costs that are harder to quantify include air pollution, noise, neighborhood disruption, effects on business and property value, and vulnerability costs.

A common characteristic shared by most projects to which LCCA is applied is that the studied systems are dynamic, which means their properties evolve over time and change with their environments. This characteristic introduces a variety of uncertainties and risk assessment requirements into the LCCA process. In terms of uncertainty consideration, there are two traditional LCCA approaches: deterministic and probabilistic (risk analysis). Deterministic methods are simpler and easier to implement, but they do not assess any risk that might be incurred. Probabilistic methods can account for the uncertainty in variables, parameters, and results, but it is relatively difficult to collect all of the information needed for applying these models. Both methods require rehabilitation timings and costs as input variables to their models. The following sections briefly discuss these two approaches. In addition, soft computing-based approaches have been proposed to account for possible ambiguities in the input variables.[31]

5.1 Deterministic LCCA

In the deterministic methods, all input variables (costs) are assumed known and given a single, fixed value. The primary formula used to calculate the net present value over the life cycle of the facility under investigation is the following:

$$\text{LCC} = \text{Initial cost} + \sum_{k=1}^{n} \text{Future cost} \times \left[\frac{1}{(1+i)^k}\right] \qquad (1)$$

where

LCC = net present value of life-cycle cost
Initial cost = project costs in the first year
Future cost = project costs in year k
k = year number
n = analysis period
i = discount rate

Interpretation of deterministic LCCA results is simple and straightforward. The alternative with lowest expected life-cycle costs is favored. However, deterministic LCCA does not evaluate any uncertainties surrounding the input variables of economic analysis. Thus, this approach often ignores information that could improve the decision. A limited sensitivity analysis may be conducted with various combinations of inputs, but it still often conceals uncertainty that may be crucial to the decision-making process and may lead to debate over the validity of the results.

5.2 Probabilistic LCCA

The uncertainty in engineering economic analysis has been partially addressed by adopting probabilistic methods and simulation techniques. Probabilistic LCCA methods allow the model to consider the variability associated with input parameters over the life cycle of a project. Each input variable is associated with a probability density function (PDF). Then, a simulation model is run for a certain number of iterations, or until some criteria are met. The statistical characteristics of output results, such as mean and variance, represent the risk associated with future outcomes. A variety of computer-based tools can realize such functions. Probabilistic LCCA assesses three basic questions about risk:[27]

1. What can happen?
2. What is the likelihood of it happening?
3. What are the consequences of it happening?

The computational capability of today's computers has made possible the use of simulation for risk analysis. Monte Carlo simulation is widely used in risk analysis. The statistical characteristics of output variables can be captured with sufficient accuracy by running the simulation thousands or even tens of thousands of times. For probabilistic LCCA, the final outputs would be the full range of possible life-cycle costs and the relative probability of any particular total cost actually occurring. Walls and Smith proposed a methodology for using Monte Carlo simulation and risk analysis Excel Add-in tools a probabilistic for pavement LCCA.[27] StratBencost uses a similar approach and provides default median and ranges for all variables relevant to the user costs.[26]

5.3 Perceived Problems and Solutions

FHWA identified the following technical issues as the main concerns surrounding LCCA implementation: selecting an appropriate discount rate; quantifying nonagency costs such as user costs; securing creditable supporting data, including traffic data; projecting costs and travel demand throughout the analysis period; estimating salvage value and useful life; estimating maintenance costs and effectiveness; and modeling asset deterioration.[6]

Many of these technical issues are being addressed. The use of real discount rates and real dollars has been recommended.[32,33] For many agencies, existing management systems could provide asset condition and usage information as well as asset deterioration models and appropriate maintenance and rehabilitation options and costs.

Several researchers have provided models for estimating direct user costs as a function of pavement condition and user delay costs as a function of lane closure and work-zone design practices.[27-30,34,35] However, most of these models have not been extensively calibrated. Furthermore, the user costs incurred during highway maintenance operations, in many cases, become the overriding cost factor in an engineering economic analysis.[34] Thus, some recently developed LCCA tools are more focused on this aspect of the pavement life cycle. These tools allow engineers to fully design the length and daily scheduling of work-zone operations to provide the lowest total costs over the life of a project. Both agency costs and user costs are considered in the LCCA of the proposed rehabilitation plan. For example, a methodology for two-lane highways has been proposed.[34] The method involves selecting the optimal work-zone length and schedules by minimizing the total project cost, including both agency and user costs. The model accounts for the variation of traffic demand over time and between locations.

Another issue that has come to the forefront in the life-cycle cost field is how to deal with preventive maintenance. Preventive maintenance techniques have been shown to be cost-effective in comparison to corrective maintenance. Thus, tools are needed that allow users to consider the effects of preventive maintenance. This is not just a simple programming issue, however, because debate still exists on the effectiveness of preventive maintenance in extending the service life of the treated pavements rather than simply causing a short-term change in the slope of the deterioration curve. Another question is how exactly to incorporate preventive maintenance into an LCCA procedure. Time-dependent treatments (e.g., joint sealing every two years) would be easy to implement, assuming the software tools allow the user to establish any number of maintenance activities for their design alternative. However, treatments based on pavement condition thresholds would be much more difficult to integrate into typical LCCA software. Pavement condition prediction would have to become a required piece of any software package, and the users would need to be able to establish their own deterioration functions, based on their historical pavement management data. Research is ongoing to determine pavement condition thresholds for effective preventive maintenance treatments. The full inclusion of these alternatives into LCCA software tools requires additional work as well.

5.4 Tools for Life-Cycle Cost Analysis

In addition to engineering economic analysis procedures included in *stovepipe* management systems, there are many specific tools for economic analysis of infrastructure investments. These include models for optimizing highway investments (given funding constraints or performance objectives) such as HERS/ST;[36] highway life-cycle cost analysis software, such as MicroBencost[28] and StratBencost;[26] and multicriteria multimodal tradeoff analysis tools, such as TransDec.[37] In addition, software packages have been developed for life-cycle cost analysis of specific transportation infrastructure assets, such as pavements,

Table 2 Comparison of LCCA Tools for PMS

Software Package	Operating System		User Costs		Risk Analysis	Pavement Condition Prediction	Analysis Level		Economic Indicator		
	Win.	DOS	Calculated Operating	Work-Zone			PL	NL	NPV	B/C	IRR
MicroBENCOST		✓	✓				✓		✓	✓	✓
StratBENCOST	✓		✓		✓		✓	✓	✓	✓	✓
HERS/ST	✓(1)		✓			✓	✓		✓	✓	
HDM-4	✓		✓			✓	✓	✓	✓	✓	✓
PID	✓		✓	✓		✓	✓	✓		✓	
FHWA	✓			✓	✓		✓		✓		

(1)Windows version was not available when the paper was published.
PL = Project level
NL = Network level
NPV = Net present value
B/C = Benefit cost ratio
IRR = Internal rate of return
Note: From 44.

bridges, rail infrastructure, and buildings, among others.[29,30,38–42] Fuller and Paterson recommended the use of LCCA for the evaluation of building design alternatives and provided a detailed methodology for conducting the analysis.[43] Furthermore, several off-the-shelf commercial analysis tools for supporting life-cycle analysis are available from a number of vendors.

For illustration purposes only, some of the most frequently used engineering economic tools for highway investments are compared in Table 2. These tools are reviewed in detail elsewhere.[44] This is only a partial list, included to provide examples of the available software and methods; additional tools are discussed elsewhere in NCHRP Report 545: Analytical Tools for Asset Management.[45]

6 LIFE-CYCLE ASSESSMENT

Although some LCCA methodologies and software consider some social and environmental costs, sustainable transportation infrastructure development and management requires methods and tools to measure and compare the full environmental impacts of human activities for the provision of goods and services (both of which are often referred to as *products*). In this context, transportation infrastructure projects, networks and system are analyzed using life-cycle assessment (LCA), a method for evaluating the impact on the environment—such as climate change, acidification, toxicological stress on human health and ecosystems, the depletion of resources, water use, land use, and noise—caused by all human activities throughout the whole life cycle of transportation infrastructure. Although traditionally the LCA has been a product-centered approach, recent efforts have expanded the approach to include process design, where economic technological and environmental constraints control the decision making. This latter, broader approach should be applied to transportation infrastructure analysis.

LCA is a relatively new and still evolving application,[46] with its roots in research related to energy requirements in the 1960s[47] and pollution prevention, which was formally initiated in the 1970s.[48] Among various definitions, two of the most widely accepted definitions about LCA come from the Society of Environmental Toxicology and Chemistry (SETAC) and the International Organization for Standardization (ISO).

SETAC defines LCA as follows:

An objective process to evaluate the environmental burdens associated with a product, process, or activity by identifying and quantifying energy and material usage and environmental releases, to assess the impacts of those energy and material uses and releases to the environment, and to evaluate and implement opportunities to effect environmental improvements. The assessment includes the entire life cycle of the product, process, or activity, encompassing extracting and processing raw materials; manufacturing transportation and distribution; use/re-use/maintenance; recycling, and final disposal.[49]

Similarly, ISO 14040 indicates that LCA is a technique for assessing the potential environmental aspects and impacts associated with a product (process or service) by compiling an inventory of relevant inputs and outputs of the process, evaluating the potential environmental impacts associated with those inputs and outputs, and interpreting the results of the inventory and impact phases in relation to the objectives of the study.[50] ISO groups the environmental impacts over the product's life cycle (i.e., cradle to grave) from raw materials acquisition through production, use, and disposal into three general categories: resource use, human health, and ecological consequences.

6.1 LCA Methodology

Figure 4 summarizes the consensus technical framework for the LCA recommended in ISO 14040.[50,51] According to this standard, LCA consists of four phases: (1) goal and scope definition, (2) inventory analysis, (3) impact assessment, and (4) interpretation. Separate companion standards focus on these steps. ISO 14041 provides guidance for the definition of goal and scope and inventory analysis.[52] ISO 14042 focuses on the life-cycle impact assessment,[53] and

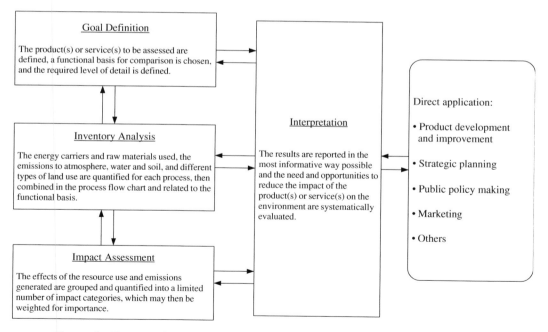

Figure 4 Framework and application of life-cycle assessment. (From Refs. 50, 51.)

ISO 14043 provides guidance for the interpretation of results from an LCA study.[54] This framework is slightly different from that of the SETAC "Code of Practice" in the fourth step.[55] SETAC indicates a life-cycle improvement assessment instead of ISO's life-cycle interpretation, arguing that the life cycle improvement assessment is not actually a phase on its own, but it influences the whole LCA methodology.[49]

The first step, goal definition, objectively defines the scope and boundaries of the study. This phase identifies all relevant elements of the systems, from the natural state of raw materials to the final disposal, and sets boundaries in terms of input, outputs, and routes. The phases of inventory analysis and impact assessment indicated in Figure 4 are often called life-cycle inventory (LCI) and life-cycle impact assessment (LCIA), respectively. The LCI is a methodology for estimating the consumption of resources (materials and energy) and the quantities of waste flows and emissions caused by or otherwise attributable to a product's life cycle, including acquisition of raw materials, processing or manufacturing, transport, use, reuse, and recycling or disposal. The LCI is used to collect data and qualify the various inputs (materials and energy) and outputs (potentially hazardous emissions and waste) throughout the process. The LCIA provides indicators and the basis for analyzing the potential contributions of the resource extractions and wastes/emissions identified in the inventory to a number of potential impacts. This process typically includes the following steps: classification, characterization, and valuation or quantification. In the classification step, the analysis aggregates the results from the inventory into relatively homogeneous categories in relation to the impact on the affected ecosystems. The LCIA starts with the selection of impact categories, category indicators, and characterization models, and the LCI results are assigned to each category.

A particular input or output may affect more than one category. Common general categories used include ozone depletion (OD), acid rain potential (AP), photochemical oxidant impact (POI), global warming potential impact (GWPI), controlled toxic water mass (CTWM), solid mass disposal (SMD), safety risk (SR), human health risk (HHR), ecological risk (ER), and natural resource depletion (RD). The magnitude of the impacts for each of the defined categories is then assessed in the characterization process. For example, the GWPI can be determined by multiplying the mass of pollutant by the global warming potential in kg of CO_2 equivalent. This step yields an environmental profile consisting of a series of environmental indicators (or performance measures). In many cases, the total impact (valuation) is determined by normalizing the various indicators, assigning relative weights to the different impacts, economic, and performance measures, and combining them in a ranking scheme.

Life-cycle interpretation occurs at every stage in an LCA; if two product alternatives are compared and one alternative shows higher consumption of each material/resource than the other, an interpretation purely based on the LCI can be conclusive.[46]

In the interpretation phase, the analyst utilizes the results from the LCI and the LCIA to identify significant issues; evaluates the results and verifies their completeness, consistency, and sensitivity; and prepares a report with conclusions and recommendations. The interpretation often includes a sensitivity and uncertainty analysis. Some researches have proposed the use of risk-based approaches, and multicriteria decision making.[55] Their proposed methodology considers uncertainty and combines LCA with multicriteria decision-making tools into a systematic approach to estimate environmental risks/impacts associated with life-cycle of products, processes and services. The proposed methodology establishes a link between the environmental risks/impacts, cost, and technical feasibility of processes, and thus provides a more comprehensive methodology that is in line with the holistic focus of asset management.

The methodological framework of LCA in Figure 4 is comprehensive and detailed, with the potential of providing insights into the possible environmental effects of any system associated with the provision of goods and services. In the context of transportation infrastructure asset management, however, the time and costs for such a detailed LCA study may not be feasible (or cost-effective) for many of the asset management decision-support functions. Therefore, in order to provide efficient and reliable decision support, simplified LCA approaches, e.g., using only LCI, are often utilized.

There are three approaches of LCI/ LCA used in transportation infrastructure:

1. The direct application (of sometimes simplified) process-oriented modeling, such as the SETAC–EPA and ISO techniques based on a process flow diagram[50,57]
2. Economic input–output–based LCA (EIO–LCA), which estimates the impacts based on the interdependencies among economic sectors in the whole economy
3. Hybrid methods that combines elements of process LCA with input–output approaches, such as the tiered hybrid method[58]

These hybrid models combine the advantages of both approaches. Bilec et al. reviewed hybrid approaches to conduct an LCA, developed a methodology (and software) for assessing the construction phase of building construction, and demonstrated the advantages of combining both methods.[59] The study suggested that transportation, equipment activity, and support functions have the largest effects on the environment.

Several examples of application in the transportation infrastructure field are available in the literature. For example, Arskog et al. conducted an LCA of the repair and maintenance systems for concrete structures and concluded that from an ecological point of view, it appears to be a good strategy to carry out preventive maintenance of a concrete structure before repairs may be necessary.[60]

Stripple conducted an LCA of a road in the Swedish National Road Administration network.[61]

6.2 Environmental Impacts of Transportation Projects

The U.S. Environmental Protection Agency (USEPA) developed a comprehensive review of indicators of the environmental impacts of transportation.[51] The term *indicators* in the report refers to quantitative estimates of the magnitude or severity of environmental impacts of transportation. These indicators, referred to as performance measures in the asset management terminology, may be based on either measurements or modeling and may refer to either historical or projected estimates. A comprehensive impact should consider all impacts, from construction of infrastructure and manufacture of vehicles to disposal of vehicles and parts. The report presents a framework for developing various types of indicators and categorizing transportation activities affecting the environment in all environmental media—air, water, and land resources. Figure 5 summarizes the causes and effects (environmental and social) of transportation activities.

Table 3 summarizes the main environmental impacts on air and water quality, land use, and other categories identified for each transportation mode. The main activities considered in the USEPA report and their respective potential negative impacts by transportation mode are summarized in Table 4.[51] The reader is

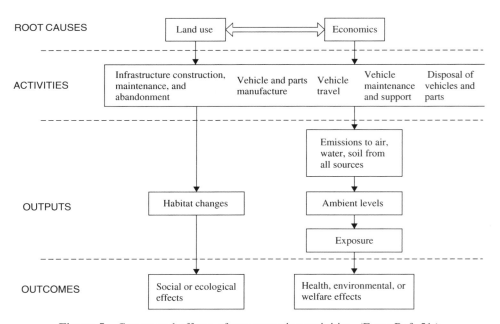

Figure 5 Causes and effects of transportation activities. (From Ref. 51.)

Table 3 Examples of Environmental Impacts by Transportation Mode

Media / Mode	Air	Water Resources	Land Resources	Other Impacts
Road/Highway • Automobiles, trucks, buses • Streets, highways	• Engine and evaporative emissions of CO, HC, NO_X, PM, lead • Emissions of CO_2 from fossil fuel combustion • CFCs released during vehicle manufacture and disposal	• Surface and ground water pollution from runoff (lubricants, coolants, vehicle deposits, roadsalt) • Modification of water systems from road building	• Land taken for infrastructure • Extraction of road-building materials • Abandone drubble from roadworks • Road vehicles withdrawn from service • Waste oil, tires, and batteries	• Local noise • Congestion
Railroad • Freight, intercity passenger, transit rail • Railwaytrack		• Oil and grease • Creosote from track beds	Land taken for rights-of-way and terminals. Dereliction of obsolete facilities. Abandoned lines, equipment, and stock.	• Local noise
Aviation • Aircraft • Airports		• Modification of watertables, river courses, and field drainage in airport construction • Deicing chemicals and degreasers on runways	• Land taken for infrastructure. • Dereliction of obsolete facilities. • Aircraft withdrawn from service • Buffer zones for noise abatement	• Local noise
Maritime • Marine vessels, ferries • Port facilities, canals		• Discharge of ballast wash, oil, spills • Modification of water systems during port construction and canal cutting and dredging • Sanitation device discharge	Land taken for infrastructure Dereliction of obsolete port facilities and canals. Vessels and craft withdrawn from service. Land disposal of dredged material.	• Plastic wastes at sea

Note: From 51.

Table 4 Examples of Activities and Their Environmental Impacts

Mode / Activity	Highway Transportation	Rail Transportation	Aviation Transportation	Maritime Transportation
Infrastructure construction and maintenance, expansion, improvement, and abandonment	• Habitat disruption and land taken for roads and right-of-way • Emissions during construction and maintenance • Releases of deicing compounds • Highway runoff	• Habitat disruption and land taken • Emissions during construction and maintenance	• Habitat disruption and land taken • Emissions during construction and maintenance • Releases of deicing compounds • Airport runoff	• Direct deterioration of habitats and water quality from dredging or other navigation improvements • Habitat disruption and contamination from disposal of dredged material • Habitat disruption and land taken for ports and marinas
Vehicle and parts manufacture	• Toxic releases and other emissions	• Toxic releases	• Toxic releases	• Toxic releases during manufacture of maritime vessels and parts
Vehicle travel	• Tail pipe and evaporative emissions • Fugitive dust emissions from roads • Emissions of refrigerant agents from vehicle air conditioners • Noise • Hazardous materials incidents during transport • Roadkill	• Exhaust emissions • Noise • Hazardous materials incidents during transport	• High altitude emissions • Low altitude/ground level • Emissions • Noise • Hazardous materials incidents during transport	• Air pollutant emissions • Habitat disruption caused by wakes and anchors • Introduction of non native species • Hazardous material incidents during transport • Wildlife collisions • Overboard dumping of solid waste • Sewage dumping
Vehicle maintenance and support	• Releases during terminal operations: tank truck cleaning, maintenance, repair, and refueling • Releases during passenger vehicle cleaning, maintenance, repair, and refueling • Leaking under ground storage tanks(USTs) containing fuel	• Releases during terminal operations: car cleaning, maintenance, repair, and refueling • Emissions from utilities powering rail	• Emissions from ground support equipment	• Releases of pollutants during terminal operations
Disposal of vehicles and parts	• Scrappage of vehicles • Motor oil disposal • Tire disposal • Lead-acid batteries disposal	• Rail car and parts disposal	• Airplane and parts disposal	• Scrappage of old vessels and dilapidated parts

Note: From Ref. 51.
Source: VHB, 1992, p.1117. Adapted from Ref. 51.

referred to the full report for details of the various impacts and a detailed list of possible indicators for each type of effect considered.

7 OTHER IMPACTS OF TRANSPORTATION PROJECTS

As indicated previously in this chapter, a comprehensive evaluation of transportation infrastructure investments requires the consideration of not only the direct economic effects of the projects (LCCA) and the environmental impacts (LCA) but also broader social and indirect economic impacts. Although the assessment of these broader impacts is often subjective, their consideration into the decision-making process is in line with the customer focus of asset management. There are several resources available for assessing these broader impacts. For example, NCHRP Report 456, *Guidebook for Assessing the Societal and Economic Effects of Transportation Projects*, provides a detailed framework for assessing transportation systems performance and economic effects (Figure 6).[62] This guidebook provides a wide range of quantitative and qualitative (often subjective) methods for evaluating the various social and economic impacts. The 52 methods covered are structured in a scientific approach: information collection,

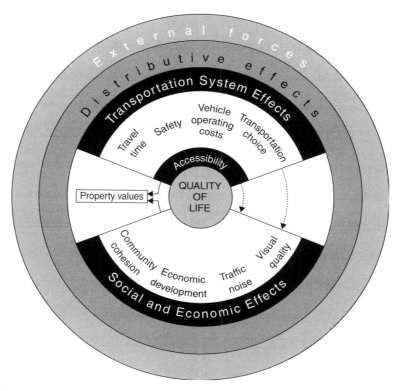

Figure 6 Broader social and economic effects of transportation investments. (From Ref. 62.)

analysis, measurement and presentation, and assessment. Selected resources are provided when available.

The transportation system effects include three traditional system performance effects often included in LCCA: changes in travel time, safety, and operating costs, as well as transportation choice and accessibility. The main broader social and economic effects considered include community cohesion, economic development, traffic noise, and visual quality, which are clearly interrelated with accessibility. Additional information on existing empirical studies, currently available data sources, and prototype study designs at regional, corridor, and local levels are available elsewhere.[63,64]

8 CONCLUSIONS AND RECOMMENDATIONS

The strategic, network, and project-level decisions supported by transportation infrastructure asset management have many technical, economic, social, and environmental impacts over the life cycle of the transportation infrastructure. It is necessary to assess these impacts over the whole life of the infrastructure systems to make informed decisions on how to define policies, allocate resources, select projects, or design and construct these projects in a sustainable manner. To support sustainable development, transportation asset management decisions should be constructed over three pillars: economic development, ecological sustainability, and social desirability. Current practice the consideration of technical and economic issues through LCCA. Environmental aspects are increasingly being considered though LCA, and although the inclusion of social aspects is in its infancy, they should not be neglected. The consideration of environmental and social goals in conjunction with economic considerations promises to enhance transportation infrastructure asset management while promoting sustainable transportation infrastructure systems.

ACKNOWLEDGMENTS

The author would like to acknowledge the input provided by several of his graduate students during the preparation of this chapter. In particular, Chen Chen provided significant input in the LCCA section and Zheng Wu in the LCA section.

REFERENCES

1. GAO, *U.S. Infrastructure, Funding Trends and opportunities to Improve Investment Decisions*, Report to the Congress GAO/RCE/AIMD-00-35, U.S. General Accounting Office, Washington, DC., 2000.
2. ASCE, "Report Card for America's Infrastructure," 2005 Progress Report. American Society of Civil Engineers. http://www.asce.org/reportcard/2005/index.cfm. Accessed April 2007.

3. W. R. Hudson, F. N. Finn, F. H. McCullough, K. Nair, and B. A. Vallerga, *Systems Approach Applied to Pavement System Formulation, Performance Definition and Materials Characterization*, Final Report National Cooperative Highway Research Program Project 1-10, 1968.
4. S.W. Hudson, R. F. Charmichael, L. O. Moser, W. R. Hudson, and W. J. Wilkers, *Bridge Management Systems*, National Cooperative Highway Research Project Report 300, Transportation Research Board, National Research Council, Washington, D.C., 1987.
5. W. R. Hudson, R. C. Haas, and W. Uddin, *Infrastructure Management*. McGraw-Hill, New York, 1997.
6. FHWA, *Asset Management Primer,* Office of Asset Management, Federal Highway Administration, Washington, D.C., 1999.
7. NCHRP, *Transportation Asset Management Guide*, National Cooperative Highway Research Program, Transportation Research Board, Washington, D.C., 2002. http://downloads.transportation.org/ amguide.pdf. Accessed April 2007.
8. G. W. Flintsch and C. Chen, "Soft Computing Applications in Infrastructure Management," *Journal of Infrastructure Systems, ASCE*, **10** (4), 157–166 (2004).
9. M. L. Tischer, Asset Management as a Tool for Statewide Planning. *Transportation Research E-Circular E-C015: Statewide Transportation Planning*, Transportation Research Board, NAS-NRC, Washington, D.C., 2000.
10. K. A. Zimmerman, *Guidelines for Using Economic Factors and Maintenance Costs in Life Cycle Cost Analysis*. Final Report, SD96-08-F. South Dakota Department of Transportation, 1997.
11. NEPA, "National Environmental Policy Act," U.S. Environmental Protection Agency http://www.epa.gov/compliance/nepa/obtaineis/index.html. Accessed April 2007.
12. WCED, *Our Common Journey*, World Commission on Environment and Development, Oxford Univ. Press, Oxford, England, 1987.
13. C. M. Jeon and A. Amekudzi, "Addressing Sustainability in Transportation Systems: Definitions, Indicators, and Metrics," *Journal of Infrastructure Systems, ASCE*, **11**, 31–50 (2005).
14. T. Litman and D. Burwell, "Issues in Sustainable Transportation," Victoria Transport Policy Institute (VTPI), Victoria, Canada. http://www.vtpi.org/sus_iss.pdf. 2003.
15. A. R. Pearce, and J. A. Vanegas, "Defining Sustainability for Built Environment Systems," *International Journal of Environmental Technology Management*, **2** (1), 94–113 (2002).
16. J. A. Black, A. Paez, and P. A. Suthanaya, "Sustainable Urban Transportation: Performance Indicators and Some Analytical Approaches," *Journal of Urban Planning and Development*, **128** (4), 184–209 (2002).
17. A. J. Balkema, H. A. Preisig, R. Otterpohl, and F. J. D. Lambert, "Indicators for the Sustainability Assessment of Wastewater Treatment Systems," *Urban Water*, **4**, 153–161 (2002).
18. M. A. Rijsberman, and F. H. M. van de Ven, "Different Approaches to Assessment of Design and Management of Sustainable Urban Water Systems," *Environmental Impact Assessment Review*, **20** (3), 333–45 (2000).
19. M. D. Meyer, and L. J. Jacobs, "A Civil Engineering Curriculum for the Future: The Georgia Tech Case," *Journal of Professional Issues in Engineering Education and Practice*, **126** (2), 74–78 (2000).

20. *American Society of Civil Engineers*, "The Role of the Civil Engineer in Sustainable Development," ASCE Policy Statement 418, http://www.asce.org/pressroom/news/policy_details.cfm?hdlid=60. Accessed April 2007.
21. D. Richards, "Harnessing Ingenuity for Sustainable Outcomes," *The Bridge*, National Academy of Engineering, **29**, (1) (Spring 1999). http://www.asce.org/professional/sustainability/nae_hiso.cfm Accessed April 2007.
22. NCHRP, *Report 551: Performance Measures and Targets for Transportation Asset Management*, T. R. Board, National Cooperative Highway Research Program, Washington, D.C., 2006.
23. J. Emblemsvag, *Life-cycle Costing: Using Activity-based Costing and Monte Carlo Methods to Manage Future Costs and Risks*, John Wiley & Sons, New York, 2003.
24. National Highway Designation Act of 1995, Public Law 104-59, 109 Stat. 568 (1995), Subsection 106(e)(1) of Title 23.
25. FHWA, *Life-Cycle Cost Analysis Primer*, Office of Asset Management, Federal Highway Administration, Washington, D.C., 2002.
26. NCHRP, *Development and Demonstration of StratBENCOST Procedure*, Research Result Digest no 252, Summary of NCHRP 2-18 (3) and 2-18 (4), National Cooperative Highway Research Program, Transportation Research Board, NAS-NRC, Washington, D.C., 2001.
27. J. Walls III and M. R. Smith, *Life-Cycle Cost Analysis in Pavement Design*, Technical Bulletin, FHWA-98-079, Federal Highway Administration, Washington, D.C., 1998.
28. J. L. Memmott, M. Richter, A. Castano-Pardo, and A. Widenthal, *MicroBENCOST User Manual*, Prepared for NCHRP Project 7-12 (2): Metrification and Enhancement of MicroBENCOST Software Package. Transportation Research Board, NAS-NRC, Washington, D.C., 1999.
29. T. Pappagiannakis and M. Delwar, "Computer Model for Life-Cycle Cost Analysis of Roadway Pavements," *Journal of Computing in Civil Engineering*, **15** (2), 152–156 (2001).
30. H. Keraly, *HDM-4 Highway Design & Management, Overview, Pre-Release*. The Highway Development and Management Series: The World Road Association (PIARC), 1999.
31. C. Chen and G. W. Flintsch, "Fuzzy Logic Pavement Maintenance and Rehabiliation Triggering Approach for Probabilistic Life-Cycle Cost Analysis," *Journal of the Transportation Research Board*, TRR No. 1990, 80–91 (2007).
32. S. J. Kirk and A. J. Dell'Isola, *Life Cycle Costing for Design Professionals*. McGraw-Hill, New York, 1995.
33. K.T. Hall, C.E. Correa, S.H. Carpenter, and R.P. Elliot, "Guidelines for Life-Cycle Cost Analysis of Pavement Rehabilitation Strategies," 82nd Transportation Research Board Annual Meeting, Washington, D.C., 2003.
34. S. Chien, Y. Tang, and P. Schonfeld, "Optimizing Work Zones for Two-Lane Highway Maintenance Projects," *Journal of Transportation Engineering*, **128** (2), 145–155 (1995).
35. R. I. Carr, "Construction Congestion Cost (CO^3) Basic Model," *Journal of Construction Engineering and Management*, **126** (2), 105–113 (2000).
36. FHWA, *HERS/ST Highway Economic Requirements System/State Version, Overview*, Draft Final Report, Federal Highway Administration, Washington, D.C., 2001.

37. NHCRP, "Development of a Computer Model for Multimodal, Multicriteria Transportation Investment Analysis," *Research Result Digest*, **258**, Summary of NCHRP 20-29 (2) (S. S. Roop, and S. K. Mathur, eds.), National Cooperative Highway Research Program, Transportation Research Board, NAS-NRC, Washington, D.C., 2001.
38. M. A. Ehlen, "Life-Cycle of Fiber-Reinforced Polymer Bridge Decks," *Journal of Materials in Civil Engineering*, **11** (3), 224–230 (1999).
39. NCHRP, *Bridge Life-Cycle Cost Analysis Guidance Manual*. Draft Report for Project 12-43, (H. Hawk, ed.), National Cooperative Highway Research Program, Transportation Research Board, NAS-NRC, Washington, D.C., 2001.
40. National Institute of Standards and Technologies, "BridgeLCC: Life-cycle Costing Software for Preliminary Bridge Design," http://www.bfrl.nist.gov/bridgelcc/welcome.html. Accessed on December 14, 2007.
41. C. D. Martland, M. B. Hargrove, and A. R. Auzmendi, "TRACS: A Tool for Managing Change," *Railway Track & Structures*, **9** (10), 27–29 (1994).
42. Federal Energy Management Program, *Building Life-Cycle Cost (BLCC) Programs*, Available at http://www.eren.doe.gov/femp/techassist/softwaretools/softwaretools.html#blcc. Accessed December 2002.
43. S. K. Fuller and S. R. Petersen, *Life-Cycle Costing Manual for the Federal Energy Management Program*, Federal Energy Management Program, U.S. Department of Commerce, Washington, D.C., 1995.
44. G.W. Flintsch and J. Kuttesch, "Application of Life-Cycle Cost Analysis for Pavement Management," Paper 04-4557, 83rd Transportation Research Board Meeting, Washington, D.C. (CD-ROM), 2004.
45. NCHRP, Report 545: Analytical Tools for Asset Management, Transportation Research Board, NAS-NRC, Washington, D.C. http://onlinepubs.trb.org/onlinepubs/nchrp/nchrp_rpt_545.pdf. Accessed April 2007.
46. G. Rebitzer, T. Ekvallb, R. Frischknechtc, D. Hunkelerd, G. Norrise, T. Rydbergf, W.-P. Schmidtg, S. Suhh, B. P. Weidemai, and D. W. Pennington, "Life-cycle Assessment Part 1: Framework, Goal and Scope Definition, Inventory Analysis, and Applications," *Environment International*, **30**, 701–720 (2004).
47. M. A. Curran, *The History of LCA*, McGraw-Hill, New York, 1996, pp. 1.1–9.
48. M. G. Royston, *Pollution Prevention Pays*, Pergamon, Oxford, UK, 1979.
49. Society of Environmental Toxicology and Chemistry (SETAC), *Guidelines for Life-Cycle Assessment: A Code of Practice*, Brussels, 1993.
50. International Standard ISO 14040, *Environmental Management—Life-cycle Assessment—Principles and Framework*, 1997.
51. U.S. Environmental Protection Agency (USEPA), Indicators of the Environmental Impacts of Transportation. EPA 230-R-96-009, 1996.
52. International Standard ISO 14041, *Environmental Management—Life-cycle Assessment—Goal and Scope Definition and Inventory Analysis*, 1998.
53. International Standard ISO 14042, *Environmental Management—Life-cycle Assessment—Life-cycle Impact Assessment*, 2000.
54. International Standard ISO 14043, *Environmental Management—Life-cycle Assessment—Life-cycle Interpretation*, 2000.

55. F. Consoli, D. Allen, I. Boustead, J. Fava, W. Franklin, A. A. Jensen, *A Code of Practice. Guidelines for Life-cycle Assessment,* SETAC, Pensacola, Florida, 1993.
56. R. Sadiq and F. L. Khan, "An Integrated Approach for Risk-based Life Cycle Assessment and Multi-criteria Decision-Making," *Business Process Management Journal,* **12** (6), 770–792 (2006).
57. A. Horvath and C. Hendrickson, "Comparison of Environmental Implications of Asphalt and Steel-reinforced Pavements," TRR 1626, Transportation Research Board, Washington, D.C. pp 105–113, 1998.
58. C. W. Bullard, P. S. Penner, and D.A. Pilati, "Net Energy Analysis—Handbook for Combining Process and Input–output Analysis," *Resource and Energy Economics,* **1**, 267–313 (1978).
59. M. Bilec, R. Ries, S. Matthews, and A. Sharrard, "Example of a Hybrid Life-Cycle Assessment of Construction Processes," *Journal of Infrastructure Systems, ASCE* **12**(4), 207–15 (2006).
60. V. Arskog, S. Fossdal, and O. Gjorv, "Life-Cycle Assessment of Repair and Maintenance Systems for Concrete Structures," *International Workshop on Sustainable Development and Concrete Technology,* http://www.cptechcenter.org/publications/sustainable/harskogassessment.pdf. Accessed April 2007.
61. H. Stripple, "Assessment of Roads: An Inventory Model Study," Report IVL-B-1210, Swedish Environmental Research Institute, 1995.
62. Forkenbrock and Weisbrod, *NCHRP Report 456: Guidebook for Assessing the Societal and Economic Effects of Transportation Projects,* National Cooperative Highway Research Program: Washington, D.C., 2001.
63. Economic Development Research Group, Inc. and Cambridge Systematics, Inc., *Using Empirical Information to Measure the Economic Impact of Highway Investments: Volume 1: Review of Literature, Data Sources, and Information Needs,* http://www.edrgroup.com/hpages/pdf/fhwa-wy-impact-vol-1.pdf Accessed April 2007.
64. Economic Development Research Group, Inc. and Cambridge Systematics, Inc., *Using Empirical Information to Measure the Economic Impact of Highway Investments: Volume 2: Guidelines for Data Collection and Analysis,* http://www.edrgroup.com/pages/pdf/fhwa-hwy-impact-vol-1.pdf. Accessed April 2007.

CHAPTER 10

PAVEMENT AND BRIDGE MANAGEMENT AND MAINTENANCE

Sue McNeil
Department of Civil and Environmental Engineering
University of Delaware
Newark, Deleware

1 INTRODUCTION	**283**	
1.1 Defining Asset Management	284	
1.2 Framework for Asset Management	286	
1.3 Historical Background	287	
1.4 Relevant Environmental Issues	289	
2 TOOLS USED IN ASSET MANAGEMENT	**291**	
2.1 Types of Tools	291	
2.2 Investment Level and Trade-off Analysis Tools	292	
2.3 Tools That Evaluate and Compare Options	293	
3 ASSET MANAGEMENT STRATEGIES	**294**	
3.1 Performance Measurement	294	
3.2 Life-Cycle Analysis	294	
4 ASSET MANAGEMENT AND SUSTAINABILITY	**294**	
4.1 Definitions, Indicators, and Metrics	295	
4.2 Case Study: New Zealand	295	
5 CHALLENGES FACING IMPLEMENTATION	**296**	

1 INTRODUCTION

In the United States, there are approximately 4 million miles of roads and 600,000 bridges.[1,2] These transportation assets support almost 3,000,000 million vehicle miles of travel each year.[1] Road and bridge repair, maintenance, rehabilitation, and replacement are significant users of engineered materials and energy to support the transportation of goods and people. In 2004, approximately $35 billion was spent on maintenance and services.[1] At the same time, we are rarely conscious of the environmental issues in the maintenance and management of our pavements and bridges.

For decades, we painted bridges with lead-based paints; we closed lanes on urban freeways at rush hour to do repairs causing massive delays, consuming excess energy and creating air pollution; and we used large quantities of energy in mining, processing, and hauling raw materials to be engineered into a composite material. Fundamentally, maintenance, rehabilitation and reconstruction is a noisy

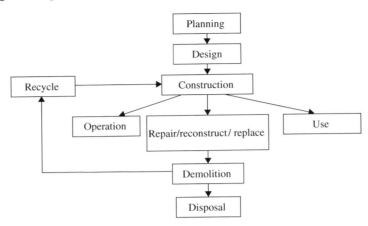

Figure 1 Facility life cycle.

and disruptive process. As our roads get more congested, we are causing more delays to drivers, consuming more energy and emitting more pollutants as drivers are stuck in stop-and-go traffic due to work zones.

The maintenance, repair, reconstruction, and rehabilitation phase is just one stage in the facility life cycle. However, costs and impacts of activities in this phase are heavily influenced by other stages in the life cycle, as shown in Figure 1. Recognizing the life-cycle impact of pavement and bridge maintenance and rehabilitation decisions has become an integral part of asset management—"a systematic process of maintaining, upgrading, and operating physical assets cost-effectively."[3] This chapter introduces the concept of asset management and then presents an historical perspective on asset management as well as applicable tools and techniques. The chapter then reviews opportunities for integrating environmental issues into the asset management framework, as well as work on asset management and sustainability. Figure 2 shows the components of a basic asset management system.

The chapter concludes with a brief case study, the integration of environmental and sustainability issues into the infrastructure decision-making process in New Zealand, followed by a discussion of implementation challenges.

1.1 Defining Asset Management

Highway assets are economic resources that provide services to the public.[4] These highway assets can be divided into two types: physical highway assets and other operational types.[3] Physical highway assets are composed of pavement, structures, tunnels, and hardware (i.e., guardrail, signs, lighting, barriers, impact attenuators, electronic surveillance and monitoring equipment, and operating facilities). Other operational highway assets include construction and maintenance equipment, vehicles, real estate, materials, human resources, and corporate

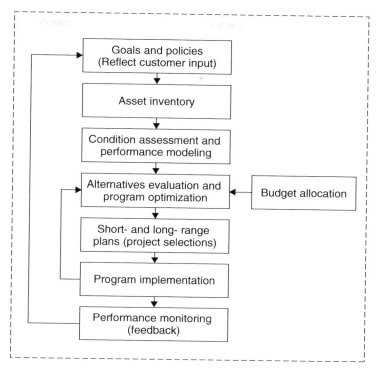

Figure 2 Generic asset management system components. (Modified from Ref. 3.)

data. In this chapter, we focus on physical assets in general, and pavement and bridges in particular.

Ongoing significant investments are required to maintain the physical and operational quality of the public highway assets to ensure safety to the public and to maintain an overall condition above the minimally acceptable level. Asset management is a tool that can be used for managing transportation assets. The Federal Highway Administration (FHWA) has promoted asset management since the mid-1990s to assist transportation agencies in managing different types of assets. In 1999, the FHWA formed the Office of Asset Management. Also in the United States, professional organizations have formed task forces and committees to promote the concept of transportation asset management. These include the AASHTO Task Force on Asset Management formed in 1997, the American Public Works Association (APWA) task force on asset management formed in 1998, and the Transportation Research Board (TRB) task force on asset management established in 2000.[4]

Asset management is described as a strategic enterprise that combines engineering practices and analysis with sound business practices and economic theory.[3] The transportation community is continually refining the definition of asset management to meet the needs of specific organizations. As a result, various definitions have been developed. With each definition, the overall concepts remain the same. The following definitions illustrate this point:

"A systematic process of maintaining, upgrading, and operating physical assets cost-effectively."[3]

"A methodology needed by those who are responsible for efficiently allocating generally insufficient funds amongst valid and competing needs."[5]

"A comprehensive and structured approach to the long-termed management of assets as tools for the efficient and effective delivery of community benefits."[6]

"A comprehensive business strategy employing people, information, and technology to improve the allocation of available funds amongst valid and competing asset needs."[7]

"Asset Management ... goes beyond the traditional management practice of examining singular systems within the road networks, i.e., pavements, bridges, etc., and looks at the universal system of a network of roads and all of its components to allow comprehensive management of limited resources."[8]

"Asset management [is] ... a set of concepts, principles, and techniques leading to a strategic approach to managing transportation infrastructure. Transportation asset management enables more effective resource allocation and utilization, based upon quality information and analyses, to address facility preservation, operation, and improvement."[9]

"The combination of management, financial, economic, engineering, operation and other practices applied to physical assets with the objective of providing the required level of service in the most cost-effective manner."[10]

"A dynamic management system that combines and integrates engineering, business, and technological aspects to optimize infrastructure management under budget constraints. The system collects asset inventory and condition, analyzes the impacts and the asset deterioration process to further establish alternative maintenance strategies, estimates costs and benefits received from maintenance actions, determines the trade-offs among investments in different infrastructure assets, and establishes the prioritization and optimization programs for asset maintenance planning."[11]

1.2 Framework for Asset Management

An overall framework of asset management suggested by the U.S. Department of Transportation consists of an asset inventory assessment, condition assessment and performance modeling, maintenance alternative selection and evaluation, methods of evaluating the effectiveness of each strategy, project implementation, and performance monitoring.[3] System complexity depends substantially on the types of assets that are being managed and the available budgets or resources.

The general concept of transportation asset management is designed to be policy or goal driven. Goals and performance indicators are the key elements that drive the decision-making process of the asset management system and help establish different investment levels.[3] These performance measures include condition data. This condition assessment is used to evaluate the performance of an asset, which can be represented in many forms. Examples of asset condition assessment are safety index, overall structural condition, user cost distance, user satisfaction index, and consumption of road transport, freight, and fuel indicators. The data are stored in databases that include a record of changes over time. Performance measures may also include environmental measures. The performance measures are linked to assets through the inventory that includes a spatial referencing system to locate the data.[12]

The data also support performance modeling that relates the changes in the performance of an asset with the contributing factor(s) that affect(s) the degradation process of such asset or the change in the level of service of an asset. An analysis module supports the development of short- and long-term plans, which involve optimizing the management planning in the most cost-effective manner subject to budget constraints. The optimization module of highway asset management is a decision support tool that balances the objective function, (i.e., minimizes maintenance costs or maximizes overall benefits) and selects the optimum set of maintenance, rehabilitation, and replacement (MR&R) actions for the entire highway network. Its determination is subject to user-defined budget constraints and other considerations. Optimization provides a bundle of projects that satisfy a set of criteria including budget and other constraints over an analysis period. Once the short- and long-term management plans are determined, the overall performance of the developed management plans is monitored before program implementation is executed.

Highway asset management systems are designed to support and enhance various decision-making processes, ranging from the detailed technical aspect of each maintenance project to broader societal goals such as sustainability, green roads, context-sensitive solutions, and environmental improvement. Customizing the asset management framework to a particular organization is essential to the successful implementation and application of an asset management system, as there are many ways to look at the key components or elements of an asset management system.

1.3 Historical Background

After World War II, highway agencies focused on the construction of infrastructure networks to support the growing economy, the increasing popularity of automobiles and improved quality of life. Road and rail networks were developed to provide mobility to users and efficiently move goods. Since the late 1970s, focus has shifted from expansion to preservation. At the same time, emphases in federal, state, and local policies relative to infrastructure management and expansion, budgeting decisions, and staff resource allocations have impacted transportation investment decisions. Gakenheimer captures the reasons why infrastructure was neglected and this shift in emphasis occurred.[13]

Highway agencies continue to face tough challenges in simultaneously expanding and maintaining the infrastructure network. To assist, different technologies have been incorporated to maintain the infrastructure network to prevent catastrophic failure while providing a safe network for users. Among the technologies introduced were the concepts of single-asset-type management systems, such as the pavement management system and bridge management system that were introduced and implemented in the late 1970s to 1990s.

Federal transportation policy and legislation have seen significant shifts in response to constrained budgets and shifting priorities at all levels. Specifically, the Intermodal Surface Transportation Efficiency Act (ISTEA), adopted in 1991, implemented a national intermodal transportation system approach intended to "link highway, rail, air, and marine transportation. Prior to ISTEA, the Federal-Aid Highway Program had been directed primarily toward the construction and improvement of four federal-aid systems: the Interstate, primary, secondary, and urban highways."[14] ISTEA required public sector agencies to shift their focus away from capacity expansion and emphasize the preservation and operation of the country's $1 trillion investment in its highways and bridges.[3]

The Transportation Equity Act of the 21st Century (TEA-21) continued to build on the policies of ISTEA and offered greater spending flexibility to fund highway safety and transit programs. In response to shifting priorities and the need to assess trade-off analyses using constrained funding sources, the principles of asset management began gaining acceptance in the transportation industry. The 2005 legislation—Safe, Accountable, Flexible, Efficient Transportation Equity Act: A Legacy for Users (SAFETEA-LU)—supports principles of asset management through the use of performance measures, the emphasis on preservation and recognition of the challenges of providing transportation for the twenty-first century.

Prior to 1995, the view of asset management in the United States was that it was something private sector companies did; transportation agencies in Australia and New Zealand said they practiced,[15] and state departments of transportation (DOTs) thought they should be practicing. In September 1996, the American Association of State Highway and Transportation Officials (AASHTO) and the Federal Highway Administration (FHWA) held the first asset management workshop focused on sharing experiences in both the public and private sectors.[16] Since 1996, a series of activities has helped to advance the state of the art and state of the practice of asset management.[17,18]

Asset management has long been an important component of the management practices in the private sector. Recently, asset management has been receiving significant interest in the public sector around the world. Many agencies are implementing asset management concepts as a way to expand their infrastructure management practices. Examples can be found in the United States, Australia, and Canada. Several factors motivated the different agencies to include asset management strategies in their agency's objectives. The following are objectives defined by several agencies:[3,19]

- To improve the highway management efficiency and capability
- To support the paradigm shift from new construction to maintenance management
- To reinforce budget demands by providing rational justification for investment in infrastructure when competing with other publicly supported programs
- To increase public acceptance and accountability
- To support trade-off decisions as demand continues to grow, causing increased congestion and wear and tear on the system
- To overcome personnel constraints due to downsizing problems and competition in the employment market
- To improve communication with customers, owners, and elected officials

1.4 Relevant Environmental Issues

Key environment issues should be considered in the maintenance of our roads and bridges:

- Emissions (both from the processes and the delays to the existing traffic)
- Consumption of energy (including the processes, the embodied energy in producing the materials, and the fuel consumption by delayed motorists)
- Impacts of the ecology from erosion and silt during construction; the use of salt in winter to enable all-weather operations; construction noise, and nighttime construction; construction and reconstruction waste; and the use of recycled material

Although there is little research or literature that addresses environmental issues in maintenance and operations directly, the literature includes several relevant concepts and discussions of these issues in the context of the life cycle of the infrastructure, or highway construction in general. However, the relationship between infrastructure management and maintenance, and the environment is significant. Jonsson suggests that for an infrasystem, "the environmental impacts related to the use of the system generally overshadow the impacts from construction and maintenance."[20] However, Vesikar and Söderqvist look at the life cycle management of concrete infrastructures and find that 35 percent of all energy consumed and 30 percent of environmental impacts generated are associated with the operation, maintenance, repair, rehabilitation, and renewal of concrete infrastructure.[21]

Other research integrates environmental factors into different phases of the facility life cycle. For example, Jeon et al. focus on the planning portion of the life cycle and include environmental sustainability indices to assess the sustainability of projects in the Atlanta region.[22] Mobile sources emissions and vehicle hours of travel per capita are used as sustainability indicators. Hendrickson and Horvath analyze the role of the construction sector in terms of materials, energy, and other resource consumption; environmental emissions; and wastes.[23]

At a higher level of abstraction, Jeon and Amekudzi, and Forman and Alexander, look at societal and agency goals related to the assessment of sustainability of transportation systems, and the ecological effects of roads respectively.[24,25] Jeon and Amekudzi list various mission statements of transportation agencies that include direct or indirect reference to environmental and quality of life issues. For example, Montana Department of Transportation's mission statement is, "MDT's mission is to serve the public by providing a transportation system and services that emphasize quality, safety, cost, effectiveness, economic vitality and sensitivity to the environment."[24] In contrast, Forman and Alexander focus on ecology and the impacts of roads on roadsides; populations; water, sediment, chemicals and streams; and policy and planning.[25]

Exploring the impact of bridge and highway maintenance on the environment in a systematic manner is challenging. In terms of stakeholders, there are several different roles:

- *Owner-agencies.* Authorities, state departments of transportation, and private sector owner-operators are the stewards of the environment.
- *Regulators.* In the United States, federal agencies such as the Environmental Protection Agency and the Federal Highway Administration are the gatekeepers.
- *Users.* The users demand better or more convenient travel without really understanding the environmental implications.
- *Society.* The society as a whole sees transportation as supporting quality of life with little regard for the externalities.

Alternatively, we can examine the environmental impacts in terms of different types of materials and construction processes in a similar manner to Horvath and Hendrickson in comparing asphalt and steel-reinforced concrete pavements,[26] or Keoleian et al. in comparing cementitious composite link slabs and conventional steel expansion joints for concrete bridges.[27] Horvath and Hendrickson included consumption of electricity, fuels, ores and fertilizers; ozone depletion potential of chemical releases; toxic chemical discharges to the air, water and land; hazard waste generation; and conventional air pollutants. Keolian et al. consider energy consumption, solid waste generation, raw material construction and construction-related traffic congestion.[27]

Finally, another important area is the use of recycled materials in maintenance and rehabilitation. Road bases, hotmix asphalt, and other applications use recycled pavements, tires, roof shingles, and foundry sand.[28–31] Materials that are engineered to use recycled products can reduce consumption of virgin materials. Equally important is the safe and appropriate disposal of waste materials.[32]

In the remainder of this chapter, we provide an overview of the various types of tools available and then present examples of how environmental measures are integrated into them.

2 TOOLS USED IN ASSET MANAGEMENT

Asset management is a complex process that relies on data and information to support decision making. A variety of tools are available to support this process, and many agencies have developed custom systems to support their specific needs. The tools described here are illustrative. They are not intended to be mutually exclusive, nor are they intended to represent a comprehensive or exhaustive catalogue of tools.

2.1 Types of Tools

Databases
Databases are basic tools for organizing, compiling, accessing, and reporting information on particular topics such as the asset inventory or highway performance standards by functional classification. Some state DOTs have a number of databases designed to collect inventory; performance and historical information such as construction history, pavement, and bridge condition; bridge inventory; and traffic and accident data. Steps have been taken to make this data user friendly with outputs delivered in text and graphic formats.

Management Systems
In the 1990s, in response to ISTEA mandates that were later rescinded, many agencies built asset-specific management systems, for pavements and bridges.[33,34] These concepts have been generalized to other types of assets such as signals, slopes, grade crossings, and rights of way. In addition, congestion, safety, public transit, and intermodal management systems have been developed by some state DOTs to assess trade-off analyses and investment options and identify system needs.

Management systems are used to inform agencies of the status, needs, and trends of infrastructure assets. They also enable staff to evaluate infrastructure deterioration to generate and rank candidate projects for action. As the AASHTO Asset Management Guide states, "Effective management systems and complete, current and accurate information on transportation infrastructure are practical necessities in meeting the policy and process requirements of asset management."[9] Central to these management systems are an inventory of the assets and some assessment of their condition and performance. These systems and data are fundamental building blocks for other types of assets.

Simulation Models
Simulation models provide detailed analyses of performance, costs, and impacts of decisions regarding transportation systems. Although useful for analyzing

complex problems with many interactive elements, they require considerable data input and either a well-structured set of decision rules or repetitive runs to analyze different options.

Other Tools
Other tools include the following:

- *Sketch planning tools.* These provide "analyses of performance, costs, and impacts of transportation decisions" allowing the user to explore a range of options quickly and effectively.[35]
- *What-if tools.* These tools offer simple and easy-to-use analytic procedures that may be used to explore the impact of specific scenarios.
- *Mapping and visualization tools.* Geographic Information Systems (GIS) play an important role as they preserve the spatial relationships among various assets and provide a visual representation in the form of maps of the location of assets and related maintenance, rehabilitation, and improvement activities.

2.2 Investment Level and Trade-off Analysis Tools

A number of investment and trade-off analysis tools have been developed that aid transportation officials in assessing infrastructure needs and measures of effectiveness over multiple years for a range of budget levels within and across investment categories. For example, the National Bridge Investment Analysis System (NBIAS) allows users to analyze impacts to the national bridge network while applying different budget levels to the bridge network.[35] Similarly, the Highway Economic Requirements System (HERS) reports on the condition and performance of the nation's transportation system.[35]

The Highway Economic Requirements System (HERS) is a simulation tool created by FHWA to provide an estimated highway budget or develop an understanding of the impact on performance of a specific budget level. HERS-ST is a modified version of HERS for use by state departments of transportation. The tool includes capacity analysis, condition assessment and safety analyses to assess the cost of expanding, upgrading, and maintaining the highway system.[35] It is designed to analyze trade-offs between preservation and mobility programs for state DOTs.[35]

Three questions can be addressed using the HERS/ST model:

1. What is the impact of a change in budget over a specified time frame?
2. What investment is required to maintain current conditions?
3. What is the change in budget to improve the network to a specified level?

The model functions by simulating network segments relative to the cost to implement a specific strategy and its impact in travel time, safety, vehicle operating costs, emissions, and improvement costs. Using an incremental benefit/cost analysis in HERS-ST, users can assemble a series of projects that address overall policy goals.

The Oregon Department of Transportation (ODOT) began using a customized version of HERS to supports its investment decision-making process in the late 1990s.[36] Since 1991, ODOT had been using the Highway Performance Monitoring System Analytical Process that focuses on engineering criteria. HERS provided the tools to recognize the impact of investments on highway users. HERS now supports needs analyses, the development of the State Transportation Improvement Program (STIP) and special studies such as value of time.[37]

HERS-ST includes two environmental measures: (1) the costs of damages from vehicular emissions of air pollutants and (2) vehicle operating costs including fuel and oil consumption and tire wear. The estimates are derived from look-up tables and models of vehicle usage. As these measures are presented in terms of monetary values, they are integrated into the economic analysis that is used to select the maintenance and rehabilitation strategies.

2.3 Tools That Evaluate and Compare Options

There are a number of benefit/cost analysis tools that can be applied at the project level. The wide range enables the user to evaluate the benefits and costs associated with specific highway and bridge projects, ranging from adding capacity and changing alignment to rehabilitation and maintenance activities. Among the benefit/cost analysis tools available are MicroBENCOST (analyzes highway projects ranging from added capacity, to rehabilitation work, to bridge projects), StratBENCOST (highway improvement analysis tool that compares several projects during concept development), STEAM (analyzes the costs, benefits and impacts of multimodal investments), and IDAS (analyzes the benefits and costs of Intelligent Transportation System investments).

One such tool, AssetManager NT, was recently developed under National Cooperative Highway Research Project 20-57, "Analytical Tools for Asset Management." The prototype Asset Manager NT focuses on network-level trade-offs between different types of assets. It uses the outputs from individual management systems such as the Pavement Management System (PMS) and the Bridge Management System (BMS) as inputs to the trade-off analysis. The tool provides graphical analysis of the trade-offs. Asset Manager NT is designed to assist users in understanding how "different patterns of investment in transportation assets will affect the performance of the system over the long term ... pavement versus bridge, geographic areas, or system subnetworks."[35] The tool uses performance measures from the individual management systems as selected by the users. The tool has the flexibility to include environmentally focused performance measures.

3 ASSET MANAGEMENT STRATEGIES

3.1 Performance Measurement

Performance measures are a mechanism for relating actions to goals and objectives. Measures related to preservation, mobility, and safety are commonly associated with asset management, although measures reflecting environmental and societal impacts are also considered. These performance measures are useful in assisting decision makers in setting priorities, identifying potential financial resources, and allocating funds. They may be used to assess needs, evaluate system performance, and communicate actions and results with customers. Consistent evaluation of the performance of transportation assets requires the use of performance measures. "Performance measures allow decision makers to compare actual performance with desired performance, as well as to provide the basis for making investment decisions to improve performance of the transportation system."[38]

Performance measures also serve as a common language between planning and programming, and between asset management and strategic decision making.[39,40] They are widely used in many countries to enhance accountability to the general public and to improve communications with decision makers and management.[41]

Environmental performance measures include air quality, groundwater, protected species, noise, and natural vistas. They can also include output-based performance measures that benchmark actions that mitigate environmental impacts, such as protecting wetlands, constructing wildlife passages across transportation facilities, using snow and ice chemicals that protect groundwater and air quality, and monitoring and controlling hazardous materials.[42]

3.2 Life-Cycle Analysis

Life-cycle analysis is a widely used strategy for asset management in general and for the analysis of environmental impacts.[19] Life-cycle analysis recognizes the various stages of the facility life cycle and recognizes the impacts of decisions in all phases of the life cycle, including environmental impacts. Life-cycle analysis is used by several researchers in the selection of materials and processes.[21,26,27]

4 ASSET MANAGEMENT AND SUSTAINABILITY

Interest in sustainability has grown over the past two decades, since the Brundtland Commission defined sustainable development as "development that meets the needs of the present without compromising the ability of future generations to meet their own needs."[43] This interest focuses on environmental, economic, and social sustainability—the three-legged stool. Transportation infrastructure management has also started paying attention to sustainability.[24]

At the same time, all public-sector agencies practice asset management. The real question is whether these agencies are systematically and cost effectively maintaining, upgrading, and operating their transportation assets. Asset management clearly becomes an appropriate vehicle for addressing sustainability in general and environmental sustainability in particular.

Experiences in agencies in the United States are documented in conference papers and presentations, a series of case studies developed by the Federal Highway Administration's Office of Asset Management, and material on the Transportation Asset Management Today (TAMT) Web site (http://assetmanagement.transportation.org). In April 2005, an International Scanning Tour made up of representatives from federal, state, and local government and from academia completed a tour of Canada, Australia, New Zealand, and England to review asset management practices and applications.[44]

This section provides two examples of how agencies are integrating asset management and sustainability.

4.1 Definitions, Indicators, and Metrics

Jeon and Amekudzi present a thorough literature review of links among asset management and sustainability, beginning with the role that sustainability plays in the mission and overall goals of agencies (the strategic element of asset management) through to the specific measures used (the link to the tactical element of asset management).[24] They found that many agencies included sustainability in their mission statement and then included measures ranging from CO_2 emissions by mode, through to hazardous materials incidents.

4.2 Case Study: New Zealand

In New Zealand, the Land Transport Management Act requires Transit New Zealand to "exhibit a sense of social and environmental responsibility." This includes the objective: "To operate the state highway system in a way that contributes to the integrated, safe, responsive and sustainable land transport system."[44]

This is put into practice in the context of several government policies, including the New Zealand Waste Strategy, National Energy Efficiency and Conservation Strategy, New Zealand Biodiversity Strategy, Confirmed Climate Change Policy and Package, and the Sustainable Development Programme of Action; and support by legislation in the form of the Resource Management Act of 1991, which requires Transit New Zealand to avoid, remedy, or mitigate the adverse effects of its activities on the environment.

Transit New Zealand responded with a strategic goal to "improve the contribution of state highways to the environmental and social well-being of New Zealand including energy efficiency and public health" and an environmental plan. The environmental plan includes recycling and reusing resources to reduce waste; adoption of noise standards for routine maintenance works; using noise-reducing surfaces; noise, water, and landscaping retrofitting; trialing the use of low-growth roadside grass; reporting on environmental performance; and working withsuppliers.

5 CHALLENGES FACING IMPLEMENTATION

As a paradigm shift for most public agencies, asset management implementation is a challenge. For example, resource allocation across an entire state with limited resources offers significant challenges to both management and agency staff working to reflect the needs-based analysis while promoting economic development activities. Asset management says that agencies need to make resource allocations among programs and across geographic regions based on societal goals and performance expectations, not by historical splits or formulas that do not correlate to objective indications of system condition.[33]

The 2004 Transportation Research Board Asset Management Peer Exchange Meeting identified six barriers to asset management implementation:

1. Lack of integration using more sophisticated analytical tools to evaluate and prioritize maintenance and rehabilitation (M&R) projects
2. Database issues, such as the inability of existing legacy systems to conduct predictive analyses and related costs associated with data collection
3. Lack of adequate communication tools and methods to relate to different audiences
4. Jurisdictional issues such as gaps in asset management approaches between agencies at different levels of government (in some states, the majority of roadway infrastructure is under local or county jurisdiction)
5. Institutional issues such as lack of coordinated and consistent asset management implementation
6. Costs.[46]

In addition to the findings of the TRB peer exchange, the Asset Management Guide[9] included a synthesis of asset management practice. The synthesis categorized challenges facing state DOTs into two groups: technical and institutional (which are identified as more significant and difficult to overcome).

Including environmental impacts in asset management presents a challenge. However, this challenge must be met if public sector agencies are going to serve as the stewards of the infrastructure.

REFERENCES

1. U.S. Department of Transportation (US DOT), Highway Statistics 2004, http://www.fhwa.dot.gov/policy/ohim/hs04/index.htm. Accessed October 2005.
2. U.S. Department of Transportation (US DOT), Status of the Nation's Highways, Bridges, and Transit: 2004 Conditions and Performance, Report to Congress, http://www.fhwa.dot.gov/policy/2004cpr/chap2c.htm#body. Accessed February 2006.

3. U.S. Department of Transportation (US DOT), Federal Highway Administration (FHWA), *Asset Management Primer* 1999.
4. S. McNeil, M. L. Tischer, and A. J. DeBlasio, "Asset Management: What Is the Fuss?" *Transportation Research Record*, **1729**, 21–25 (2000).
5. N. H. Danylo and A. Lemer, Asset Management for the Public Work Manager Challenges and Strategies, Findings of the APWA Task Force on Asset Management, American Public Works Association, Kansas City, MO, 1998.
6. Austroads, *Strategy for Improving Asset Management Practices*, Association of Australian and New Zealand Road Transport and Traffic Authorities, New South Wales, Australia, 1997.
7. Transportation Association of Canada (TAC), *Primer on Highway Asset Management Systems*, Transportation Association of Canada, Ottawa 1996.
8. Organization for European Cooperation and Development (OECD), Asset Management System, Project Description, Paris, France 1999.
9. Cambridge Systematics, Incorporated, *Transportation Asset Management Guide*, American Association of State Highway & Transportation Officials, Washington, DC and Federal Highway Administration, Washington, D.C., 2002, http://downloads.transportation.org/amguide.pdf. Accessed October 2, 2005.
10. Mildura Rural City Council, Policy Number 9.13: Asset Management Policy (Infrastructures), http://www.mildura.vic.gov.au/Files/AssetManagementPolicyInfrastructure9.13.pdf. Accessed on December 8, 2007.
11. P. Herabat and S. McNeil, "Highway Asset Management System," in T.F. Fwa (ed.), *The Handbook of Highway Engineering*, CRC Press, 2005.
12. James P. Hall, (ed.). *Geospatial Information Technologies for Asset Management*, Transportation Research Circular E-C108, Transportation Research Board, Washington, D.C., 2006.
13. R. Gakenheimer, "Infrastructure Shortfall—The Institutional Problems," *Journal of the American Planning Association*, **55** (1) (1989).
14. E. Schweppe, "Do Better Roads Mean More Jobs?" *Public Roads*, **65** (6), 19–22 (2000).
15. G. Norwell and G. Youdale, Managing the Road Asset, PIARC Special Report on Concessions 1997.
16. U.S. Department of Transportation, Federal Highway Administration (FHWA), Asset Management, Advancing the State of the Art into the 21st Century through Public-Private Dialogue, FHWA-RD-97-046, U.S. Department of Transportation, Washington, D.C., September 1996.
17. W. Oberman, J. Bittner, and E. Wittwer, Synthesis of National Efforts in Transportation Asset Management, Midwest Regional University Transportation Center, May 2002. Available: http://www.mrutc.org/research/0101/report0101.pdf. Accessed December 20, 2003.
18. A. Switzer and S. McNeil, "A Road Map for Transportation Asset Management Research," *Public Works Management and Policy*, **8** (3), 162–165 (January 2004).
19. U.S. Department of Transportation (US DOT), Federal Highway Administration, *Life-Cycle Cost Analysis Primer*, U.S. Department of Transportation, August 2002.
20. D. K. Jonsson, "The Nature of Infrasystem Services," *Journal of Infrastructure Systems*, **11** (1), 2–7 (2005).

21. Vesikar and Söderqvist, *Life-Cycle Management of Concrete Infrastructures for Improved Sustainability*, Transportation Research Circular E-C049: 9th International Bridge Management Conference, Transportation Research Board, Washington, D.C., 2003 http://onlinepubs.trb.org/onlinepubs/circulars/ec049.pdf.
22. C. M. Jeon, A. A. Amekudzi, and R.L. Guensler, *Evaluating Transportation System Sustainability: Atlanta Metropolitan Region*, TRB Annual Meeting CD-ROM, Transportation Research Board, Washington, D.C., 2007.
23. C. Hendrickson and A. Horvath, "Resource Use and Environmental Emissions of U.S. Construction Sectors," *Journal of Construction Engineering and Management*, **126** (1), 38–44 (January/February 2000).
24. C. M. Jeon and A. Amekudzi, "Addressing Sustainability in Transportation Systems: Definitions, Indicators, and Metrics," *ASCE Journal of Infrastructure Systems.* **11** (1), 31–50 (March 2005).
25. R. T. T. Forman and L. E. Alexander, "Roads and Their Major Ecological Effects," *Annual Review of Ecology and Systematics*, **29**, 207–231 (1998).
26. A. Horvath and C. Hendrickson, "Comparison of Environmental Implications of Asphalt and Steel Reinforced Concrete Pavements," *Transportation Research Record*, **1626** (1998).
27. G. A. Keoleian, A. Kendall, J. E. Dettling, V. M. Smith, R. F. Chandler, M. D. Lepech and V. C. Li, "Life Cycle Modeling of Concrete Bridge Design: Comparison of Engineered Cementitious Composite Link Slabs and Conventional Steel Expansion Joints," *Journal of Infrastructure Systems*, **11** (1), (March 2005).
28. "Two New England States Revamp RAP Use in Recycling Specs," *Roads & Bridges*, **31** (10), (1993).
29. R. G. Packard, "Concrete Solutions," *Roads & Bridges*, **30** (9) (1992).
30. J. Harrington, "Recycled Roadways," *Public Roads*, **68** (4), (2005), http://www.tfhrc.gov/pubrds/05jan/02.htm.
31. National Research Council of Canada, Reuse and Recycling of Road Construction and Maintenance Materials, 2005, http://sustainablecommunites.fcm.ca/files/Infraguide/Roads_and_Sidewalks/reuse_recycl_rd_constr_maint_materls.pdf. Accessed on December 8, 2007.
32. N. Z. Siddiki, D. Kim, and R. Salgado, "Use of Recycled and Waste Materials in Indiana," *Transportation Research Record* **1874** (2004).
33. American Association of State Highway & Transportation Officials (AASHTO), *Pavement Management Guide*, American Association of State Highway & Transportation Officials, Washington, DC, 2001.
34. W.E. Robert, A.R. Marshall, R.W. Shepard, and J. Aldayuz, *Pontis Bridge Management System: State of the Practice in Implementation and Development*, Transportation Research Board E-Circular, 9th International Bridge Management Conference, Orlando, Florida, 2003, Transportation Research Board; Washington D.C., pp. 49–60, http://gulliver.trb.org/publications /circulars/ec049.pdf. Accessed October 13, 2005.
35. Cambridge Systematics, Incorporated, *Analytical Tools for Asset Management,* NCHRP Report 545, Transportation Research Board, Washington, D.C., 2005.

36. U.S. Department of Transportation (US DOT), Federal Highway Administration, Highway Economic Requirements System: The Oregon Experience, Transportation Asset Management Case Studies, Federal Highway Administration, Washington DC, FHWA-IF-03-037, 2003.
37. U.S. Department of Transportation (US DOT), Federal Highway Administration, Highway Economic Requirements System—State Version, HERS-ST, http://www.fhwa.dot.gov/infrastructure/asstmgmt/hersindex.htm. Accessed October 2, 2005.
38. J. C. Falcocchio, "Performance Measures for Evaluating Transportation Systems: Stakeholder Perspective," *Transportation Research Record*, **1895**, 220–227 (2004).
39. R. Barolsky, Performance Measures to Improve Transportation Planning Practice: A Peer Exchange, Transportation Research Board E-Circular, E-C073, May 2005. http://trb.org/publications/circulars/ec073.pdf. Accessed October 13, 2005
40. Anthony M. Pagano, S. McNeil, and E. Ogard, "Linking Asset Management to Strategic Planning Processes: Best Practices from State DOT's", *Transportation Research Record*, **1848**, 29–36 (2003).
41. International Scanning Study Team, Transportation Performance Measures in Australia, Canada, Japan and New Zealand, Federal Highway Administration, December 2004, http://assetmanagement.transportation.org/tam/aashto.nsf/All+Documents/ 38F1D8E2D42 9FBD985257042003F764C/$FILE/2004%20-%20Transportation%20Performance%20 Measure.pdf. Accessed October 13, 2005.
42. Cambridge Systematics, Incorporated, *Performance Measures and Targets for Transportation Asset Management*, National Cooperative Highway Research Program (NCHRP) Report 551, Transportation Research Board, Washington D.C., 2006.
43. World Commission on Environment and Development (WCED), *Our Common Journey*, Oxford University Press, Oxford, England, 1987.
44. P. Bugas-Schramm, The Tipping Point? Trip Summary on the International Scan on Transportation Asset Management, APWA Reporter, American Public Works Association, September 2005, http://www.apwa.net/Publications/Reporter/ReporterOnline/index.asp?DISPLAY=ISSUE&ISSUE_DATE=092005&ARTICLE_NUMBER=1113. Accessed October 13, 2005.
45. American Association of State Highway and Transportation Officials (AASHTO), Asset Management Peer Exchange: Using Past Experience to Shape Future Practice, Proceedings of a Workshop Held in Scottsdale, Arizona, American Association of State Highway and Transportation Officials, Washington, D.C., 2000.
46. P. Hendren, *Transportation Research Board. Asset Management in Planning and Operations: A Peer Exchange*, Transportation Research Circular, No. E-C076, 2005, http://trb.org/publications/circulars/ec076.pdf. Accessed October 13, 2005.

CHAPTER 11

IMPACTS OF THE AVIATION SECTOR ON THE ENVIRONMENT

Victoria Williams
Centre for Transport Studies
Imperial College
London, United Kingdom

1	**INTRODUCTION**	**301**
2	**IMPACTS OF AIRCRAFT IN FLIGHT**	**302**
2.1	Global Impacts	303
2.2	Landing and Take-off—Local Impacts	307
3	**STAKEHOLDERS AND AIRCRAFT IMPACTS**	**311**
3.1	Airspace: Air Traffic Management and Its Contribution to Impacts and to Mitigation	311
3.2	Airport Operators and the Mitigation of Aircraft Impacts	313
3.3	Airlines: Operating Practices and Their Environmental Impact	314
3.4	The Passengers: Aviation, Travel Behavior, and Equity	315
3.5	Beyond Civil Passenger Transport—Other Airspace Users	317
4	**BEYOND AIRCRAFT IN FLIGHT—THE WIDER IMPACTS OF AVIATION**	**318**
4.1	Airports	318
4.2	Aircraft Lifecycle: Manufacture, Maintenance, and Disposal	318
5	**GOVERNING AVIATION: REGULATING GLOBAL AND LOCAL IMPACTS**	**319**
5.1	Aircraft Legislation	319
5.2	Controlling Fuel	321
5.3	Controlling Airports	321
5.4	Controlling Total Emissions	322
6	**FUTURE IMPACTS—GROWTH, TECHNOLOGY, AND THE ENVIRONMENTAL CHALLENGE**	**323**
6.1	Market Growth	323
6.2	New Technologies	324
6.3	Research	325
6.4	Challenges	326
7	**CONCLUSION**	**327**

1 INTRODUCTION

Sustainability issues have influenced the air transport industry throughout its development. Safety has driven improvements in aircraft design and in air traffic management. Economic factors have also played a role, both within the industry—shaping airline operations, networks, and alliances—and beyond it, driving demand for aviation growth. In societal terms, aviation has opened gateways to international trade and tourism, offering new opportunities for interaction, economic growth, and employment. These facets of sustainability each interact increasingly with growing environmental concerns. Improvements in efficiency and reductions in noise and emissions have not been enough to counter the growth in impacts driven by rising traffic demand.

This chapter discusses the impacts of the aviation sector on the environment. The three largest impacts of aircraft—contributions to climate change, noise nuisance, and poor air quality—are introduced, with a review of their effects on a local and global scale and a discussion of how those effects can be quantified. The issues are then explored in more detail from the perspective of key stakeholders, illustrating how the decisions taken by air space managers, airport operators, airlines, passengers and noncommercial airspace users contribute to the environmental impacts and could contribute to their mitigation.

These discussions highlight the importance of building environmental consciousness into each element of the air traffic system to control aircraft impacts. Beyond aircraft in flight, the wider environmental issues associated with the air industry are considered, including the impacts of airport infrastructure and access and the processes relating to the manufacture and maintenance of aircraft.

As an internationally operating industry, aviation presents challenges for policies to restrict its impacts. This chapter includes a review of legislation, outlining its successes and limitations in controlling aircraft, fuel, airport growth, and the rise in total impacts. Finally, the future is addressed, with a discussion of forecast traffic growth, expected technological advances, and the ambitious research targets being set. As environmental pressure on aviation grows, we raise the important issue of prioritization of impacts and the challenge of objectively quantifying and comparing the full environmental costs of flight, arguing that decisions trading off one environmental impact for another should not be left to the commercial decision making of airlines and aircraft manufacturers.

These are complicated issues, both scientifically and politically, and decisions regarding aviation's future should be carefully weighed, including taking into account the inequitable distribution of aviation's benefits and its environmental costs.

2 IMPACTS OF AIRCRAFT IN FLIGHT

Aircraft in flight have a wide range of environmental impacts. These can cover spatial scales from the local—confined to the airport or its close vicinity—to the global. They have lifetimes from hours to centuries. Many of these impacts and their mitigation options interact, requiring difficult trade-off decisions when evaluating new technologies or operational strategies. This section describes the global and local impacts, discussing their causes and implications and the ways in which they can be quantified.

2.1 Global Impacts

Changes in human behavior, and particularly in industrial, agricultural, and social practices, have changed our atmosphere. Concentrations of carbon dioxide (CO_2), methane, and other greenhouse gases have risen, changing the radiative balance of the atmosphere. The environmental and social consequences will grow unless measures are taken to stabilize concentrations, but stabilization would require emissions to be cut far below their current levels. However, exploiting technological and other opportunities to reduce our impact on climate could reduce expected climate changes and slow the rate at which they occur.

Aircraft play a small but growing part in anthropogenic CO_2 emissions. Most emissions from aviation take place in the upper troposphere, close to the tropopause, giving rise to additional (non-CO_2) impacts. These emissions can change the radiative balance of the atmosphere by reflecting incoming solar radiation (reducing heating at the surface) and/or by absorbing outgoing terrestrial radiation (reducing the energy escaping to space). The ice crystals that make up contrails and cirrus clouds also have a net warming impact, although the magnitude of the effect remains uncertain. There are mitigation opportunities, but the trade-offs between different processes present serious challenges to reducing the net climate impact of aviation, particularly with current rates of growth.

The climate impacts of atmospheric change can be described using radiative forcing, a measure of the additional energy trapped in the atmosphere (in units of watts per square meter). Aviation accounted for 3.5 percent of the total anthropogenic radiative forcing of climate in 1992.[1] This share is growing. Greenhouse gas emissions from many large industries such as manufacturing and power generation have been constrained, whether by the Kyoto protocol, national and local emission targets or by changing political and industrial circumstances. Globally, aviation has escaped these processes, exhibiting rapid growth that has overwhelmed improvements in efficiency.

The net radiative forcing by aviation describes the combined impacts of a range of mechanisms that each disrupts the atmosphere, either warming or cooling. Each possesses its own characteristic patterns of impact.

Carbon Dioxide (CO_2)

CO_2 has been the main focus for policies to control climate change. Combustion of fossil fuels releases stores of carbon, sequestered over millennia, into the atmosphere as CO_2, which acts to trap the heat energy radiated by Earth, warming the surface. By 1999, the concentration of CO_2 in the atmosphere had risen to 367 parts per million (ppm) from its preindustrial value of 280 ppm.[2] Of this increase, about one part per million can be attributed to aviation.[1]

Atmospheric CO_2 has a long lifetime, allowing thorough mixing through the atmosphere. The impact does not depend on the altitude or location of emission, and aviation can be compared with other sources. In 1990, aviation accounted

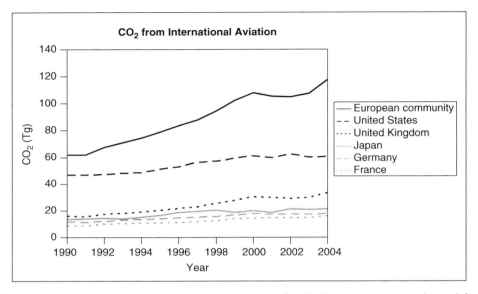

Figure 1 Carbon dioxide from international aviation for the European community and for the five highest-emitting countries.

for less than 3 percent of CO_2 emissions from fossil fuels. The share is rising as aviation grows, while emissions from other industry sectors are constrained by the Kyoto protocol and by other social and political processes.

Figure 1 shows the CO_2 from international aviation for the European community, which now accounts for more than half of the total reported emissions from fuel bunkers for international aviation. The five countries with the highest emissions from international aviation (United States, United Kingdom, Japan,

Table 1 Increases in CO_2 from International Aviation and from National Total Emissions Excluding International Aviation

Country	% Increase CO_2 from International Aviation (1990–2004)	% Increase National Total CO_2 Emissions (1990–2004)
United Kingdom	111	−5
Germany	53	−14
France	83	6
United States	30	20
Japan	61	12
European Community	87	4

Note: Based on Ref. 4.

Germany, and France) are also shown.* This figure shows values reported to the United Nations Framework Convention on Climate Change (UNFCCC) based on fuel sold and represents international departures only.

For most countries, rises in CO_2 emissions from international aviation, shown in Figure 1, are in sharp contrast to changes in the national totals, which exclude international aviation (Table 1).[4]

Nitrogen Oxides (NO_x)

Nitrogen oxides (NO_x) produced in the combustion process contribute both to poor air quality (see local impacts) and to global climate change. At cruise altitudes, NO_x increases ozone. In the mid-latitude Northern Hemisphere, the ozone increase was 6 percent for 1992 and is predicted to rise to 13 percent by 2050. The effect is strongest along major flight routes. The warming that occurs from this increased ozone is partly offset by a decrease in the methane lifetime. Global methane levels are 2 percent lower than they would be without aviation.[1] Methane is a powerful greenhouse gas, so this reduction due to aviation has a cooling effect. However, methane emissions from other sources have driven atmospheric concentrations to more than double their preindustrial level.

Unlike CO_2, the impacts of NO_x emission depend strongly on the altitude and location of emission.

Contrails and Contrail-cirrus

When hot, moist air emerges from the turbine, it expands rapidly, mixing with the cold surrounding air. If condensation occurs in this mixing, the liquid rapidly freezes to form the ice crystals that make up a visible contrail. These ice crystals can rapidly evaporate, but in the right atmospheric conditions they can persist for several hours (Figure 2). Contrails cover about 0.1 percent of Earth's surface, with this expected to reach 0.5 percent by 2050. In the right atmospheric conditions, contrails can grow (condensing more water vapor from the ambient air) and can spread to form large ice clouds that are indistinguishable from natural cirrus. The coverage of these clouds could be much larger than that of linear contrails.

Contrails can prevent solar radiation from reaching the surface and can act as an insulating layer, similar to a greenhouse gas. The net impact is the balance of these two effects and depends on many factors, including time of day, location, and the size and shape of the ice crystals that make up the contrail. On average, the warming impact is greater than the cooling.

Contrail formation depends on the humidity and temperature of background air, and varies with season. Contrails form at lower altitudes in winter than in

---

* Emissions from China are not reported. China is ranked second in the world for total passenger-km when domestic operations are included, but international operations are ranked fifteenth[3] and the CO_2 from international aviation from China is likely to be similar to that of Italy or the Republic of Ireland.

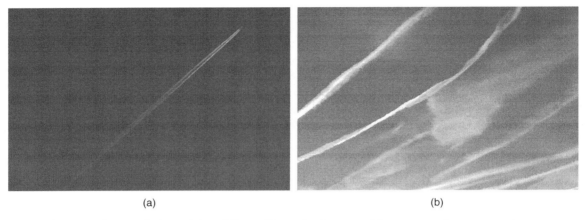

Figure 2 Contrails can dissipate (*a*) rapidly or (*b*) persist for several hours, spreading to cover larger areas.

summer, and also have a greater radiative impact, with winter flights contributing one-fifth of the traffic, but half of the annual radiative forcing.[5]

Other Contributors

Other mechanisms contributing to aviation's climate impact include soot, sulphate, and water vapor. The direct effect of soot and sulphate aerosol is small. For water vapor, most emissions take place in the troposphere, where they are rapidly removed by precipitation. If supersonic aircraft were widely adopted, high-altitude emissions of water vapor would dominate impacts.

Comparing Emissions

The radiative forcing associated with each of the contributors to aviation's climate impact is shown in Figure 3. This measure expresses the additional energy trapped in the troposphere as a global, annual average. It is approximately proportional to the change in global average surface temperature; positive radiative forcing indicates warming, negative indicates cooling. The estimates shown, from the European TRADEOFF study,[6] indicate that the largest contribution to aviation radiative forcing may come from contrail cirrus, although considerable uncertainties remain.

Radiative forcing is just one approach for comparing climate impacts. Although it provides a useful indicator of the global impact of the cumulative effects of aviation, it does not fully compare the spatial and temporal patterns of impacts. Changes in ozone and cirrus clouds have localized impacts along the main flight corridors, principally in the mid-latitude Northern Hemisphere. By contrast, CO_2 and methane are well mixed throughout the atmosphere and have global impacts. Time scales also differ. The lifetime of a cirrus cloud is hours, while CO_2 may remain in the atmosphere for up to 200 years.

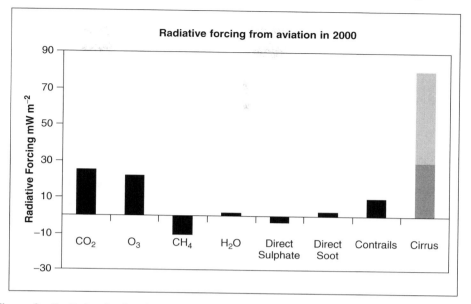

Figure 3 Radiative forcing from aircraft in 2000 (From Ref. 6.) For aviation-induced cirrus, considerable uncertainty remains. The dark gray bar shows a mean estimate, with the light gray showing the upper bound.

Impacts and Implications

Small changes in global average surface temperature are associated with much larger changes on a local and regional scale, and with changes in the frequency and severity of extreme weather events.

The economic consequences of climate change are potentially huge, but efforts to weigh the costs require subjective judgments, which can be distorting. The social benefits of aviation, with short journey times opening up leisure and business opportunities, are largely enjoyed by affluent nations but the impacts are global, with their consequences likely to be worst for those countries least equipped for adaptation.

2.2 Landing and Take-off—Local Impacts

Near airports, the environmental impacts of flying can be readily apparent. The most conspicuous is noise, but air-quality concerns are also significant. Environmental concerns are increasingly a challenge to airport growth, typically expressed through public and political objections to planning applications to increase terminal or runway capacity.

Noise

Noise can disrupt communities near airports (Figure 4). Aircraft operations are the main source. During take-off and climb, aircraft noise comes mainly from the

Figure 4 Aircraft noise impacts can be significant for (*a*) residential areas or (*b*) leisure environments.

engine; for approach and landing the airframe is also significant. On the ground, auxiliary power units used to supply power for lighting, air conditioning, and other electrical systems contribute to airport noise. Other noise sources on the airport site include ground support vehicles and systems, while site access traffic for both passengers and staff also contributes. The combined noise from the airport has a range of social implications.

Exposure to aircraft noise has been linked to health impacts, including an increase in vitality-related conditions like headache and tiredness and increased use of medication and sedatives.[7] For night flights, potential sleep disturbance raises particular concerns. Even outside the night-time hours, time of day can influence the nuisance associated with aircraft noise. Some residents express a willingness to pay to reduce air traffic movements in early morning and the evening on weekdays.[8]

Educational impacts can also be significant. Aircraft noise can be a barrier to communication and can disrupt educational development. In schools exposed to high levels of noise, children's development of reading comprehension can be impaired.[9,10]

There have also been studies linking exposure to aircraft noise to reduced property prices, although it can be difficult to separate noise exposure concerns from other factors and to offset the economic impacts of the employment and other benefits the airport provides.[11]

Over particular environments, aircraft noise impacts can be more strongly felt, either by disturbing the tranquility and so lowering the recreational value of an area or by disrupting animal populations, threatening important and fragile ecosystems.

Quantifying Impacts

The nuisance of aircraft noise can be readily apparent near to airports, but some of its impacts can be difficult to quantify. Although sound intensity is a direct and measurable physical property, translating that to a measure of the nuisance experienced requires subjective judgments about noise perception. This can be particularly challenging for aircraft noise. To allow comparison with other, continuous sources, such as roads or industrial sites, the sound intensity of intermittent aircraft noise is expressed as the level of steady sound that, over the period of measurement, would deliver the same noise energy.

Maps of surface noise exposure can be generated by tracking the trajectories of aircraft arriving and departing from an airport and matching those trajectories to known aircraft types and their noise emissions. There is no global standard time period used to calculate this noise equivalence. In the United States, the Federal Aviation Authority requires the use of day-night average sound level (DNL), weighting sounds between 10:00 P.M. and 7:00 A.M., to reflect the increased nuisance of night noise exposure. The European Environmental Noise Directive weights evening and night-time noise by adjusting by +5 dB for flights during a four-hour evening period and by +10 dB for flights during an eight-hour night period. In the United Kingdom, separate measures are used to restrict night noise, so the standard noise measure is based on the equivalent exposure averaged over a 16-hour day.

Each of these approaches leads to evaluation of the noise exposure at specific locations—usually modeled to shows contour maps for the affected areas. From these maps, noise performance indicators for the airport as a whole can be derived. These include the area and population enclosed by a contour of equivalent noise level. These indicators can track changes in the population exposed to aircraft noise, whether those changes are derived from changes in aircraft type, number of movements, operational changes, or changes in land use.

Trends in noise depend on local airport circumstances. For two London airports, the past 15 years have been very different. At capacity-constrained Heathrow, the population and area enclosed by the 57 dB Leq contour have fallen, as the increase in traffic movements has not been large enough to offset the noise reductions achieved by the use of progressively quieter aircraft.[12] At Stansted, the population within the 57 dB Leq contour in 2004 was double that in the late 1980s and early 1990s, but half that at its peak in 1998. The area enclosed by the 57 dB Leq contour was double at the peak in 1998, but returned to its 1988 value in 2004. Traffic during the period grew fourfold, but the transition to quieter aircraft types allowed the increased movements to be accommodated within the same noise exposure area.[13]

Continuous noise equivalence may not reflect the full health and social impacts and the nuisance of noise exposure. Perceived nuisance varies considerably. Frequency of flights and the distribution of traffic through the day will both affect the nuisance experienced. Seasonal effects can also play a role, as sound insulation can be ineffective if windows are opened for ventilation. Tolerance of noise disturbance also reduces with increasing affluence.[11]

Air Quality
Local air quality is also a significant component of the total environmental impact of an airport in the community.

Aircraft engines produce a range of combustion products. These include the standard products of clean combustion of a hydrocarbon fuel, CO_2 and water vapor, but also sulphur dioxide and other oxides of sulphur, nitrogen dioxide, carbon monoxide, soot, and unburned hydrocarbon. For CO_2 and water vapor, emissions are proportional to fuel mass, so the rate of emission is highest at take-off when the engine operates at maximum thrust. Emissions of NO_x are also highest in the take-off phase. By contrast, carbon monoxide and hydrocarbon emissions are highest when the engine operates at a small fraction of its maximum thrust, so most emissions occur when the aircraft is maneuvering on the ground.

Particles can also impact on human health. Engine advances have substantially reduced the amount of particulate matter in the engine exhaust.

The total impact of an airport on the air quality of the local environment is not confined to aircraft emissions. Other transport modes used for airport access and for aircraft support activities will also contribute, as will stationary sources. For NO_2, total transport emissions are dominated by road, but the aviation share is growing. In 1990, aviation NO_x emissions in the United Kingdom were just 6 percent of the road value, but by 2004 this had risen to 28 percent. This reflects both the rapid growth in air transport and a 60 percent fall in road NO_x associated with the introduction of catalytic converters.[14]

Poor air quality has been associated with health problems, including an increase in the frequency or severity of symptoms of respiratory complaints like asthma. NO_2 can irritate lungs and reduce resistance to illnesses like bronchitis and pneumonia. SO_2 aggravates respiratory illnesses and contributes to 'acid rain', damaging buildings and vegetation.

Quantifying Emissions and Impacts
For air quality, the most informative measures describe concentrations, not amounts emited. This accounts for different meteorological conditions (which govern dispersion), background concentrations, and the characteristics of the built environment, all of which will affect exposure. Observations can combine sampling at key sites with modeling of the physical and chemical processes to build a more complete map of concentrations. The time scale used for measurements and targets can vary, with parallel indicators for different time scales sometimes used—for example, a peak value and an annual mean. This allows different health and environmental impacts to be considered.

3 STAKEHOLDERS AND AIRCRAFT IMPACTS

There are many key actors in the air space system, each with their own role. This section reviews those roles, considering both the contributions to current impacts and the future mitigation opportunities.

3.1 Airspace: Air Traffic Management and Its Contribution to Impacts and to Mitigation

The organization and management of traffic is essential to the safety of air transport. Air traffic control infrastructure has developed to ensure that aircraft can complete their requested route while maintaining safe separation from other traffic. These systems and structures can contribute to environmental inefficiencies in the air traffic system. New technologies and approaches offer significant opportunities to reduce environmental impacts.

One source of inefficiency arises from dependence on ground-based navigation devices. Flight plans are typically defined as a series of waypoints, each coinciding with a ground-based navigation aid. This can result in a route length significantly longer than the direct (great circle) distance between the departure and destination airports.

Diversions away from the direct route can occur for other reasons. Traffic can be grouped into fewer common route segments in order to facilitate the management of traffic. Figure 5 illustrates typical flight routes flown for short and long haul routes into London's Heathrow airport. Both groups of flights show diversions from the shortest route and grouping into a small number of approaches to the airport. Where airport capacity is constrained, the use of hold-stacks to

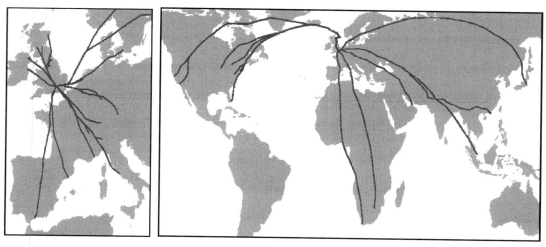

Figure 5 Example flight paths for key (*a*) short and (*b*) long haul routes serving London Heathrow, showing diversions from the straight (great circle) route and grouping into key routes through airspace.

sequence aircraft for landing can also significantly increase journey time, fuel use, and emissions.

Rationalization of airspace could improve direct routing. Currently, many air space boundaries follow national borders. For Europe, particularly, this results in a complex routing network and a costly and inefficient system. Progress is now being made towards the vision of a *single European sky*, with innovative approaches to air traffic management to efficiently control traffic throughout the region.

One option is to divide airspace according to its function, allowing dedicated 'highways' separating high altitude, long distance traffic from other routes to avoid en-route delays associated with crossing traffic.[15]

Of the operational opportunities to reduce fuel consumption identified by the International Civil Aviation Organization (ICAO), improved communication, navigation and surveillance (CNS) is expected to offer the most significant fuel savings.[16] Improving CNS is one step toward allowing pilots to take more responsibility for selecting their own flight trajectories and maintaining their separation from other aircraft.

Maintaining safe separation can be a source of environmental inefficiency if an aircraft has to be diverted. Diversions can be reduced if conflict situations are predicted and avoided early. The type of diversion maneuver selected will also govern the emission penalties, with little-used measures such as speed controls and lateral offsets of flight paths, offering potentially lower-emitting alternatives to conventional vectoring or altitude changes.[17]

Air traffic management could also contribute to the mitigation of climate impacts by controlling cruise altitudes to reduce or eliminate contrail formation. In general, cruise altitude reductions would reduce contrail, but a policy targeting altitude changes to avoid regions supersaturated with respect to ice would deliver more effective contrail reductions.[18] These regions have an average thickness of only 0.5 km, so only a small change in flight level would be required.[19] Typically, reducing cruise altitude increases fuel burn, so there may be a trade-off between the elimination of contrail and the impacts of a slight increase in CO_2 emission.

Cruise altitude emissions are most relevant to global climate impacts, but air traffic management can also contribute to management and mitigation of local impacts, particularly noise. Arrival and departure procedures determine the footprint exposed to aircraft noise. This can be used to avoid sensitive areas.

There are operational opportunities to reduce the noise impact of take-off and landing. Many reduce the time spent at low altitudes. One is *continuous descent approach*, avoiding long, continuous sections of level flight at low altitude. This can be operationally more difficult for pilots and controllers, but offers reductions in noise, emissions, and fuel burn compared with a conventional stepped descent. A steeper approach angle can also keep aircraft higher for longer, but its use is restricted to certain aircraft types, and currently an angle greater than 3.5 degrees is only permitted if required for obstacle avoidance.

Noise preferential routes divert low altitude aircraft away from highly populated areas. Many of these routes are long established and may be difficult to follow accurately using more modern, faster aircraft. Forthcoming changes in aviation navigation technologies are likely to improve the accuracy with which these routes can be flown.

The options for noise mitigation through aircraft operations are limited by the airport and runway configuration. In low or head wind conditions, a preferred runway direction can be used to avoid densely populated areas. This preference can be retained in cross- or tail-wind conditions, provided safety is not compromised. In some cases, there may also be potential to displace the landing threshold along the runway, using the shorter runway requirements of modern aircraft. This could move much of the noise footprint onto the airport site, reducing the impact on the local community. Changes in automated navigation systems would be required, and there are safety concerns as the buffer zone for landing is reduced.

Operational measures can be used to mitigate air quality impacts, but not all measures will reduce all emissions as the dependence on engine operating characteristics varies between pollutants. In general, measures to reduce noise exposure by reducing sections of low-level flight over populated areas could have some air quality benefits, though this will depend on local meteorological conditions. Sites where aircraft pass over a busy road can be particularly problematic for achieving air quality targets.

3.2 Airport Operators and the Mitigation of Aircraft Impacts

Efficient operations can reduce noise and emissions by avoiding delays after pushback from the gate. Optimized traffic sequencing can help. At the planning stage, designing taxiways and aircraft stands to minimize taxiing can reduce emissions, particularly those related to low-power engine operation, such as carbon monoxide and hydrocarbon.

On the ground, the use of auxiliary power units for aircraft systems like lighting and ventilation can be limited either by setting time restrictions or by offering an alternative power supply. This lowers emissions and noise. Tugs can position aircraft before take-off, further reducing the time the aircraft turbines are operated.

Noise can constrain growth before the airport reaches its terminal or runway capacity. For these airports, mitigation measures can allow more operations, if the noise footprint of the traffic (measured as equivalent continuous noise) remains the same.[11] One option for airports is to apply landing charges differentiated by the noise classification of an aircraft to encourage airlines to invest in newer, quieter aircraft types. There are ICAO recommendations for such charges, calling for them to be used only where the problem is severe and to cover only appropriate alleviation costs such as investment in insulation. Alternative approaches include the UK quota count scheme for night noise, which caps total night-time aircraft movements using a measure weighted for louder aircraft.

3.3 Airlines: Operating Practices and Their Environmental Impact

Airline decision making influences noise, air quality, and climate change. The choice of business model is significant. For traditional carriers, the choice between hub-and-spoke or point-to-point operations will be based on the business case. For hub-and-spoke, using small aircraft on feeder routes can increase the catchment area for key (often intercontinental) routes, allowing larger, more efficient aircraft to be used for the long-haul section. Environmentally, this may offer both advantages and disadvantages. Increased fuel efficiency on the long haul hub-to-hub section of the route may be canceled out if there is a substantial increase in the total distance flown compared with a direct route between the departure and destination airports. The hub-and-spoke model also relies on the availability of a number of large and intensively used airports with associated noise and air quality concerns for the community.

Some characteristics typical of low-cost operations are linked with good environmental practice. High seat densities and passenger load factors contribute to low fuel and emissions per passenger-kilometer. Limits on catering and baggage allowances to decrease costs and turn-around time reduce the take-off mass, giving a further efficiency benefit. As the utilization of aircraft in the low-cost business model is high, the service lifetime of the aircraft is reduced, potentially allowing earlier introduction of efficiency improvements into the fleet.

Routes are short, to maximize the number of aircraft rotations per day—typically below two hours for maximum profitability.[20] This relatively short average trip length also has environmental consequences. Short routes tend to be less fuel-efficient per passenger-kilometer,[21] although at very long distances the fuel requirements increase again due to the high take-off mass of the aircraft. This reduced fuel efficiency for short trips may be partly offset by lower probability of contrail formation, as shorter routes do not reach high cruise altitudes.[22] Local environmental impacts must also be considered—a business model based around short haul trips has more landing and take-off movements, with associated noise and air-quality consequences. Low fares can also induce additional trips, as discussed in the section on passenger behavior.

Airport choice is significant. Air quality issues are most severe for large airports, where the combination of aircraft and access traffic means that targets can be difficult to meet. Rapid growth in the use of secondary or regional airports can bring noise concerns to previously unaffected communities, which can be a significant cause of local opposition.

ICAO has produced recommendations to reduce fuel use (and hence CO_2 emission) through a range of operational measures.[16] For airlines, these options can be grouped into three broad categories. The first is optimization, with taxiing, flight routes, cruise altitude and operating speed all selected to maximize fuel efficiency. Aircraft options include minimizing the empty take-off weight, improving the selection of aircraft type for route, and improving maintenance. Finally, operating practices like increasing load factor, reducing the carriage of unnecessary fuel, and reducing nonrevenue (positioning) flights offer further scope for improvement. Although measures taken by the airlines can have a significant impact, the largest fuel reductions are expected to result from improved route optimization through improvements in communication, navigation, surveillance, and air traffic management.

The selection of aircraft for a given route is a key factor in optimizing environmental performance. Payload fuel efficiency (payload multiplied by distance per unit of fuel) will typically be maximized at a route length much less than the maximum operating range of the aircraft type.[23] One suggested method to improve fuel efficiency for very-long-haul routes is to undertake the journey in stages using an aircraft optimized for shorter range. This reduces both the mass of fuel carried and the empty mass of the aircraft. Estimates suggest that for a trip of 15,000 km, the fuel to move the same passenger load could be halved by replacing one long stage with three short stages using an appropriately optimized aircraft, offering a 6 to 10 percent reduction in the total global fuel burn.[23] From a business perspective, one obstacle to the matching of aircraft types to routes is the long lifetime of aircraft, given the need to respond quickly to changes in demand to maximize revenues.

Airlines can play a role in reducing engine idle operations to reduce emissions. Options include the use of powered tugs, rather than aircraft engines, to maneuver aircraft. Virgin Atlantic began trials of such a system at several airports in the United Kingdom in 2006. Airlines can also influence noise and air-quality impacts through their selection of aircraft types for purchase or lease, through their scheduling decisions, and through their operational guidance for pilots. One approach to reduce noise on landing is to adopt a low power/low drag approach that reduces thrust and delays the need to deploy flaps. Reduced thrust can also be used during climb-out (above 800 ft.), with additional benefits for engine maintenance. These practices, and measures such as continuous descent approach, are implemented through negotiations between dispatchers, pilots, and air traffic controllers.

3.4 The Passengers: Aviation, Travel Behavior, and Equity

Despite massive growth in air travel in recent decades, aviation remains a minority activity. Of the global population, only around 1 percent have ever flown,[24] largely confined to the affluent, both in terms of the global distribution of traffic and of the demographic patterns of passengers within society.

The availability of cheap flights can induce changes in behavior, changing social expectations and increasing the frequency of both leisure and business trips.

Table 2 Growth in Low-Cost Carriers Source

	Seat Capacity on LCC Flights (% of total)	
	2001	2006
United States	18	27
Europe	4	23
Asia Pacific	1	8

Note: From Ref. 20.

Established initially in the United States, the low-cost carrier business model has spread, with some adaptations, across Europe and is now growing in the Asia Pacific region, establishing a substantial share of the air travel market (Table 2). Alongside this growth, traditional carriers have adopted some characteristics of the low-cost business model in order to reduce fares. It has been argued that this growth has opened up access to air travel, but evidence is limited. A recent study in the United Kingdom revealed that the growth in traffic by low cost carriers over the past decade was associated not with an increase in travel by low income groups, but with an increase in the frequency of travel by those in middle and higher income groups.[25]

The question of induced demand is critical. If the growth in low-cost airlines has simply drawn market share from conventional carriers, so the increase in passenger-kilometer would have been the same without their intervention, then the substitution of low cost for traditional business models can be broadly equated with an improvement in environmental performance. However, if the low fares themselves are inducing additional trips that would not otherwise be made, the benefits of improved efficiency are lost.

Transfer to an alternative transport mode, or alternative destination, offers one option for mitigation. Comparing the carbon intensity of transport modes, passenger-kilometer carbon emissions for short-haul flights can be similar to those for a single occupant in a car or light truck. Rail emissions per passenger-kilometer depend on the power source, but can be half the aviation emissions over a short journey. Journey times can also be competitive once check-in times and city center access are taken into account. Over longer distances, emissions per passenger-kilometer for air reduce as a larger fraction of the flight is in the relatively efficient cruise phase. For very long journeys, emissions per passenger-kilometer rise further as the additional fuel required to sustain long-distance flight reduces the payload that can be carried.

Comparisons between transport mode using only passenger-kilometers emissions are incomplete. Typical trip lengths for aviation are much greater than for other modes, so even with similar emissions per passenger-kilometer, per trip emissions can be much larger. In many cases, destinations can be substitutable, particularly for leisure travel.

3.5 Beyond Civil Passenger Transport—Other Airspace Users

While this chapter focuses on commercial passenger air transport, other airspace users also contribute to environmental impacts and have a role to play in mitigation.

Military

Military aircraft contributed less than a fifth of total global emissions in 1992, with that share forecast to fall to less than 7 percent by 2015.[1] However, military flight also has implications for the environmental impacts of civil commercial air transport, which accounts for the bulk of emissions. The designation of large swathes of air space for military use can force civilian and commercial traffic to take considerable diversions from a direct route, incurring additional emissions and extending journey times. This can be a particular issue in Europe, where air traffic is dense and each country retains sovereignty over its airspace. Improving communication, navigation, and surveillance could reduce the amount of airspace set aside for exclusive military use.

Shared used of airspace presents challenges for security for air traffic management. Some military aircraft can operate at much higher speeds and climb/descent rates than civilian traffic, and the safe management of mixed traffic requires additional skills and techniques.

General Aviation

General aviation encompasses all noncommercial civilian traffic. Most operations are for personal or business transport, recreation, or training. There are about 385,000 general aviation aircraft worldwide, flying a total of 34 million hours in 2005.[3]

In 1992, the general aviation contribution to civil aviation CO_2 was 3.2 percent. For NO_x, the contribution was 3.5 percent. These contributions have been forecast to fall to 2.1 percent (CO_2) and 1.8 percent (NO_x) by 2015.[1]

Air Freight

Air freight moved on scheduled services grew by 79 percent in the 10 years to 2005, compared with a 57 percent increase in passengers over the same period. The increase in freight-kilometers was 70 percent, indicating that the mean journey length has decreased.[3,25] Although much of this freight is carried in the hold of passenger aircraft, there is also a growing provision of air freight services using dedicated aircraft and, in some instances, dedicated freight airports. The cost of transporting goods by air is high compared to other modes, so its use is generally confined to high value or perishable goods. The use of air freight has particularly transformed food supplies—seasonal produce can now be flown in from around the world to secure year-round supplies. Emissions for transport of freight by air per tonne-kilometers far exceed maritime and rail alternatives.

4 BEYOND AIRCRAFT IN FLIGHT—THE WIDER IMPACTS OF AVIATION

4.1 Airports

Measures to constrain aircraft emissions on the ground have already been discussed, but controls on other airport activities can reduce environmental impacts. In freezing conditions, deicing might be needed to avoid compromising the aerodynamic performance of the aircraft. This can be carried out by either heating the ice (using forced air or infrared technology) or by applying fluids.[26] Anti-icing fluids can also be applied as a preventative measure. These chemicals can contaminate rivers and groundwater close to airports.[27,28]

Airport terminals are large public buildings, making significant demands on energy and other resources. Focused efforts in planning can reduce these impacts by incorporating efficiency into power, heating/cooling, and lighting systems, and by installing local power-generation methods. Demand for water can also act as a constraint on airport expansion, and the environmental impact of waste generated by airport shops and services should not be overlooked.

The land required for a major airport is considerable, making construction on new sites a difficult challenge in most urban environments. For expansion of existing airports, the encompassing of existing sites of natural or historical interest can prove an obstacle.

Arrangements for airport access are also important. For a growing airport, passengers and staff can strain existing road infrastructure, increasing congestion. Changes in working patterns, parking pricing, vehicle sharing incentives, and improved provision of public transport alternatives can all contribute to a comprehensive travel plan to reduce vehicle-kilometers traveled. At the same time, a high proportion of airport revenues can come from car parking charges, resulting in a need to balance commercial and environmental priorities.

Intermodal options are being considered in a bid to reduce emissions, particularly by improving interactions between air and rail. At the extreme, one option considered has been to separate air-side and land-side operations to allow one large hub airport to serve a number of cities, facilitating the high occupancy use of large aircraft on routes to maximize efficiency. However, wider security and environmental concerns must also be assessed.

4.2 Aircraft Lifecycle: Manufacture, Maintenance, and Disposal

Even before an aircraft has taken to the skies, it has had an impact on the environment. The manufacturing process is complex and energy intensive, and it can involve hazardous materials. Aircraft assembly often brings together components from widely dispersed manufacturing sites. The raw materials may also have been transported long distances even before manufacture begins.

An integrated approach to reducing impacts is required, taking account of all actors in the manufacture process and the full product lifecycle. This requires distribution of responsibility for the environmental performance of the finished product across the supply chain, to ensure that each contributor is required to properly consider the environmental consequences of their component and their production processes.[29]

From the design stage to the retirement of the last manufactured aircraft, there is typically a life of 45 to 65 years, so environmental considerations in the design of new aircraft today will be slow to propagate through the entire operating fleet.[29]

5 GOVERNING AVIATION: REGULATING GLOBAL AND LOCAL IMPACTS

An extensive web of international, national, and local legislation governs the aviation industry. Many of the controls relate to the safe and efficient operation of traffic, but global and local regulations have also driven environmental improvements through restrictions on aircraft, fuel, airports and on total emissions.

5.1 Aircraft Legislation

Regulation of new aircraft is the main process controlling air-quality-related emissions from aviation. ICAO provides certification standards for a range of species. As these restrictions apply only to newly certified engines, changes associated with long-term use and maintenance practices are not regulated. The emission standards relate to a landing–take-off (LTO) cycle, which describes the time in mode and the engine operation in each of four modes: approach, idle, take-off, and climb-out (Figure 6). Standards relate to emissions divided by take-off thrust, to reflect differences in aircraft size and power. The most recent standards for NO_x emission differentiate by both engine thrust and pressure ratio. Some approaches to improving engine efficiency increase NO_x emission, so the two environmental priorities must be balanced in the engine design process.

Particulate emissions are targeted by controls on smoke number, which is a dimensionless number calculated from the reflectance of a paper filter before and after the exhaust gas passes through it.

These engine certification limits have provided the incentive for emission-reduction technologies to be included in new engine designs and have reduced the emissions from individual engines. In the case of NO_x, reductions in landing–take-off emissions have had a secondary benefit for climate change mitigation by reducing emissions at cruise altitude, which are currently exempt from regulation.

Aircraft noise mitigation has also involved restrictions on permitted aircraft types. A framework for the noise classification of aircraft, reflecting aircraft size and engine type, is set out in Annex 16 to the Convention on International Civil Aviation. These classifications are linked to technological improvements to reduce aircraft noise and are used internationally for regulation at the state

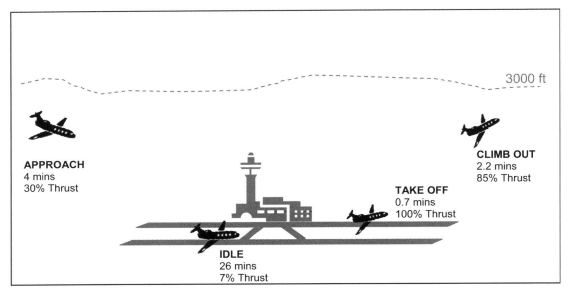

Figure 6 The standard landing–take-off cycle, used for emissions standards, describes an aircraft's movements in the vicinity of an airport below 3000 ft and includes specified time in mode and engine thrust settings.

or airport level. Restrictions can be applied to limit aircraft types (using the Annex 16 classifications), either by banning groups of aircraft completely from particular airports or by imposing restrictions on the number of movements or the time of day for operations.

Regulations have become progressively stricter, first addressing very early (non-noise certificated) jet aircraft, then seeking to reduce Chapter Two aircraft or to fit *hush-kits* to reduce their noise. The ICAO provided a recommended structure to phase out these aircraft in the late 1990s.

To control noise from the loudest of the next generation, Chapter Three aircraft, ICAO is recommending a balanced approach for noise management, including operational procedures and restrictions and improved approaches to land-use management. This balanced approach also recognizes the challenge to operators from developing countries, who may be prevented from maintaining or expanding their route network by the need to invest in fleet modernization to comply with progressively stricter noise regulations. New aircraft for certification after January 2006 are subject to even more stringent Chapter Four conditions.

Restrictions on aircraft and engines typically apply to newly certified aircraft. This reduces the cost to the airline industry and avoids penalties for the early adopters of new technologies, who could otherwise be forced to replace recently purchased aircraft upon the introduction of new legislation. Whilst requiring new equipment across the global fleet would give a quicker return in terms of reduced impacts, it would place impractical demands on manufacturers and airlines.

These controls on aircraft types can have unintended consequences. One is the second-hand market in aircraft, which allows older, more polluting, aircraft to continue operating in countries where environmental legislation is less stringent. This means that environmental problems are transferred, rather than solved. A second consequence is the potential distortion of competition on emerging routes, particularly those linking nations at different stages of development. Before the accession of an additional 10 states to the European Union in 2004, the need to switch to aircraft meeting strict noise criteria forced many of the airlines based in those states to reduce the size of their fleet, so reduced their ability to compete with carriers from existing member states on new routes.[10]

5.2 Controlling Fuel

Aviation fuel is governed by global legislation to ensure consistency, as potential differences in fuel content could jeopardize aircraft performance and ultimately safety. This standardization of fuel for civil air transport also allows regulation on environmental grounds. The scope is, however, limited. Emissions of sulphur dioxide are proportional to the sulphur content of kerosene. These emissions contribute both to the degradation of air quality and to climate impacts. For contrails, fuel sulphur content affects the number of ice crystals formed and has a small impact on the threshold temperature for formation.[30] Current regulations restrict sulphur content to below 0.3 percent, but the average is much lower than that, due to the use of low-sulphur crude oil. Hydroprocessing can also be used to reduce fuel sulphur content, but at the expense of increased carbon emissions in production.[1]

The ratio of hydrogen to carbon in the fuel will also influence emissions, with a higher hydrogen content leading to increased water vapor and less CO_2. However, the energy required to produce hydrogen means that a net increase in the net CO_2 emitted would be likely to occur.[1] Increasing water vapor emissions would also have consequences for contrail and cirrus formation and, in the stratosphere, for direct radiative forcing by water vapor. Both of these mechanisms would offset any benefit gained by reducing CO_2.

In addition to addressing the content of fuel, some countries have applied taxes on the use of aviation fuel. Currently, only six countries tax aviation fuel for domestic flights (Australia, Brazil, Canada, Japan, Norway, and the United States).[31] The existing structure of bilateral agreements, which sets the legal context for international air travel, is a significant obstacle to the extension of fuel taxes to international flights.

5.3 Controlling Airports

Airports are constrained by planning processes, with permission required to expand terminal or runway capacity. Authorities may also impose controls on the number and type of aircraft movements. These issues can be addressed locally or as part of the development of a national strategy for the air transport industry.

Land-use planning is a significant tool in the suite of methods to address aircraft noise impacts. In particularly, sites near to approach or departure routes for a major or growing airport should not normally be allocated for residential development. This land may be more appropriately used for support infrastructure like warehousing and parking, potentially freeing land that is less exposed to noise for residential use. At the extreme, the airport may be relocated away from densely populated areas, as has occurred in Hong Kong.

For development of a new airport or new runway, compulsory purchase of properties likely to be severely affected by noise may be required. Compensation can also be an option, alongside subsidies for mitigation measures such as double-glazing, loft insulation, or cavity wall insulation, all of which reduce noise exposure.

5.4 Controlling Total Emissions

Although many of the measures outlined in this chapter act to constrain the impact of individual flights by reducing emissions or noise per passenger-kilometer or per aircraft movement, the total environmental impact also reflects growth in transport demand.

Near airports, the total emission of pollutants rises wherever traffic growth is faster than emission reductions from technological improvements. Air quality legislation can prove a real constraint to airport expansion. At London's Heathrow airport, the need to mitigate air quality impacts in line with new European Union legislation to become binding in 2010 has contributed to delays in giving consent for a requested third runway.

For climate change, the largest political approach to mitigation to date is the Kyoto protocol, which came into force in 2005, ratified by 168 countries. It provides a framework for emissions reductions for six greenhouse gases, including CO_2, with 35 signatory countries having separately defined targets for emissions in the commitment period 2008 to 2112, relative to emissions in 1990. Countries committed to the strictest (8 percent) reduction include the European Union—15 states that have a collective agreement to redistribute targets among themselves. Developing countries have no restriction on emissions under the protocol, but can be the location for emission reduction or offset schemes through the Clean Development Mechanism, which allows countries to use certain overseas projects towards their own emission reduction targets.

The Kyoto protocol includes domestic aviation in national targets for emission reduction, but excludes international aviation, calling on developed countries to agree strategies to reduce or control emissions through ICAO. No agreement has yet been reached, with the most recent ICAO statement on environmental policy recommending more research and expressing a preference for technological measures over market-based solutions.

Aviation presents special challenges for policy making. The international nature of operations makes regulation difficult. Some policies, including the Kyoto protocol, make a distinction between domestic and international air travel. Although this can be instructive to determine jurisdiction and the allocation of emissions to national targets, it is arbitrary in differentiating between the impacts of flights. Short-haul flights typically have higher fuel burn per passenger-kilometer than long-haul,[20] but not all very-short-haul flights are domestic. Applying the distinction between domestic and international traffic for international policies is discriminatory. If the United States were to ratify the Kyoto protocol, emissions from 72 percent of its air travel (by passenger-kilometer) would be included in national targets for reduction. For the United Kingdom and Germany, who are signatories to the protocol, the figure is close to 5 percent.[3]

Market-based measures, like fuel taxes or emissions trading, are often proposed to encourage modal shift or to ensure that the environmental costs of aviation are fully reflected in the price. There are, however, political and institutional barriers. Fuel taxes are explicitly precluded under the existing networks of bilateral agreements that underpin cooperation in international air transport. In addition, a state acting alone would risk a switch in demand to neighboring states not imposing a task or charge. As a result, global or multinational approaches are required.

At the European level, emissions trading has received most attention, with aviation set to be added to the existing CO_2 permit trading scheme before 2012. In the absence of parallel measures to address the non-CO_2 impacts of aviation, there may be unintended effects if, as expected, aviation becomes a net purchaser of permits from other sectors.

The use of an *uplift* factor to account for the additional impacts of aviation has been considered, but placing a value on it requires a subjective judgment on the relative importance of the different climate impacts, and using radiative forcing for its calculation poorly reflects significant spatial and temporal differences in the impacts due to different mechanisms.[32]

6 FUTURE IMPACTS—GROWTH, TECHNOLOGY, AND THE ENVIRONMENTAL CHALLENGE

Throughout the growth of the aviation industry in recent decades, technological developments have acted to reduce the emissions and impacts per passenger-kilometer. However, rising traffic has swamped the effects of these improvements and the total environmental impacts have tended to rise.

6.1 Market Growth

In the near and medium term, growth in passenger traffic is expected to continue. Table 3 shows annual average growth rates in passenger-km for 1995-2005 along with forecast annual growth rates for 2006 to 2008.[3] Forecast growth is highest

Table 3 Average Annual Growth Rates in Passenger-km Performed for Historic 1995 to 2005 and Forecast 2005 to 2008 Data

Region	Average Annual Growth 1995–2005 (\%)	Forecast Annual Average Growth 2006–2008 (\%)
Africa	5.4	6.3
Asia/Pacific	5.7	6.7
Europe	5.9	6.2
Middle East	9.7	10.7
Latin America/Caribbean	3.9	4.6
North America	4.0	4.4
World	5.2	5.8

Note: From Ref. 3.

for the Middle East and lowest for North America. For all regions, the forecast growth rate is faster than the historic growth. In the longer term, worldwide traffic has been forecast to reach 2.5 times its 1996 level by 2015. Beyond this, uncertainties increase, but travel demand models have suggested a three- to eight-fold increase in the 55 years to 2050.[1]

6.2 New Technologies

Substantial reductions in aircraft CO_2 emission have been achieved through technological advances in engine and airframe design. In the 40 years following the introduction of the Comet 4 aircraft, fuel consumption for new aircraft entering production fell by 40 percent. A further 30 percent reduction per seat-kilometer was achieved through increases in aircraft size and aerodynamic improvements. The proportion of seats occupied (the load factor) has also increased in recent years, further reducing fuel burn (and hence CO_2 emissions) per passenger-kilometer. This reduction in fuel per unit of productivity (averaged over the global fleet) is expected to continue at around one per cent per year.[15] Although high and rising fuel prices and pressure to increase payload and range provide incentives to reduce fuel burn and hence CO_2 emissions, no such incentives exist for the other climate impacts of aviation. Measures to improve engine fuel efficiency can increase NO_x emission and contrail formation, so the trade-offs between these climate impacts will determine the net effect of technology improvements.

Fuel technologies may contribute to reduced emissions for aviation in the future. Synthetic kerosene produced from biomass may offer significant emissions reductions, but for its widespread introduction, the land required for biomass production would be considerable. For the longer term, the most commonly considered alternative to kerosene, liquid hydrogen, is much less dense, requiring four times the volume to deliver the equivalent amount of energy.[33] It would require significant changes in the aircraft design to accommodate the greater volume of fuel, and to fuel storage and delivery infrastructure. There may also be an increased probability of contrail formation and a change in the physical properties of the contrails formed. In addition, any environmental benefit would depend on the hydrogen source.[34]

Research into new composite materials for aircraft manufacture is ongoing and new aircraft, including the Airbus A350 and the Boeing 787, will incorporate some of these advances, replacing aluminium with composite structural materials.[22] These materials offer a range of potential benefits, including reduced aircraft weight and control of the small-scale turbulence at the aircraft's surface, to reduce drag. Current obstacles to their widespread use include the more stringent restrictions placed on maintenance and checking. Conventional construction methods have tolerance levels for cracks, as the development of faults over time has been well characterized so can be predicted. The behavior of composites under such conditions is less predictable. The potential fuel saving offered by these composites has been estimated at 10 to 15 percent for new aircraft entering service in 2020 compared to those in production in 2000.[35] Composite materials may also play a significant role in noise reduction.

The most ambitious attempts to address noise are focused on radical aircraft redesign, with the ultimate target an aircraft inaudible outside the airport boundary, at least in urban areas. This highly ambitious goal is unlikely to be achieved without a move away from the conventional "tube with wings" aircraft design, with the most promising alternative thought to be a configuration based on a blended wing body design with engines embedded into the aircraft in order to shield the ground from noise.[36] This would have added benefits for fuel efficiency, and so would reduce global as well as local impacts.

The very high operating costs of supersonic flight have so far prevented its widespread adoption for commercial use, despite offering considerable reductions in journey time. Should commercial use of supersonic aircraft become widespread, different climate concerns would arise because supersonic aircraft operate at higher altitudes, and the impacts of stratospheric flight on water vapor and ozone are different from those in the upper troposphere, with the net radiative impact estimated to be five times larger.[1]

6.3 Research

ICAO resolution A35-5 states that the organization "will strive to:

1. limit or reduce the number of people affected by significant aircraft noise;
2. limit or reduce the impact of aviation emissions on local air quality; and
3. limit or reduce the impact of aviation greenhouse gas emissions on the global climate."[37]

In the absence of specific, quantitative commitments on these limits or reductions, several targets have been identified by other organizations.

The Advisory Council for Aeronautics Research in Europe (ACARE) sets ambitious, strategic research objectives for European air transport. These targets cover five areas including the environment. The first review called for fuel consumption, CO_2 emission and perceived external noise to be halved for new aircraft in 2020 and NO_x emission reduced by 80 percent, along with reductions in the impacts of manufacture, maintenance and disposal processes.[38] The second review expanded on this, providing specific aims for airlines, airports, aircraft, and air traffic management in order to achieve an Ultra Green Air Transport System.[39]

In the United States, the National Aviation Research Plan places environmental concerns explicitly in the context of increasing capacity. The research plan sets out targets for the near term to 2009. For noise, the objective is a 1 percent per year reduction in the number of people exposed to significant noise (defined as a day–night sound level of 65 decibels or more). The performance target for fuel efficiency is a 1 percent per year improvement per revenue plane mile.[40]

6.4 Challenges

Both technological advances and political measures are likely to play significant roles in governing the future of air transport. Difficult trade-off decisions will be needed, balancing between the social benefits and costs of aviation, between global and local impacts, and between shorts and long-term effects. The priority given to different impacts will affect how policy and development decisions are made.

These decisions are already being taken; the Airbus A380 will emit an extra 1 to 2 percent CO_2 due to airline requests to match the noise performance to that of much smaller aircraft to facilitate night landings at London Heathrow.[41] This decision is in conflict with social and economic assessments of the relative issues that indicate that climate change mitigation should be the priority in aircraft design.[42]

Balancing local and global impacts influences assessment of the relative impact of long-1 and short-haul flights, shaping policies that could influence traveler behavior. In particular, any policy applying market measures to discourage relatively inefficient short-haul flights should guard against reducing the price differential between short- and long-haul trips to avoid encouraging passengers to choose farther destinations.

The great challenge for future development of an air transport system that meets its environmental responsibilities is that the priorities between impacts cannot be defined objectively. There is no single comparison measure to recommend a climate change mitigation measure over one for noise, nor is there an objective way to assess either one against the social benefit of aviation. The decision process should be informed by both global and local priorities.

7 CONCLUSION

The massive growth in air transport since the middle of the twentieth century has transformed the way the world interacts, but not without environmental consequences, both locally and globally. Significant achievements have been made in reducing both emissions and noise, but the environmental impacts of aviation will continue to grow unless technological and operational changes can deliver reductions sufficiently large to counter the effects of industry growth. These technological advances may include changes to air traffic systems, as well as to aircraft and engines. Operational measures will require cooperation between airlines, airports, and air traffic controllers to be fully effective. Even if these advances are achieved, some constraints on demand are likely to be required if total impacts are to be reduced.

REFERENCES

1. J. E. Penner, D. H. Lister, D. J. Griggs, D. J. Dokken, and M. McFarland, eds, *Aviation and the Global Atmosphere: A Special Report of Intergovernmental Panel on Climate Change Working Groups I and III*, Cambridge University Press: Cambridge, 1999, 373 pp.
2. I. C. Prentice, G. D. Farquhar, M. J. R. Fasham, M. L. Goulden, M. Heimann, V. J. Jaramillo, H. S. Kheshgi, C. Le Quéré, R. J. Scholes, and D. W. R. Wallace, "The Carbon Cycle and Atmospheric Carbon Dioxide, in Climate Change 2001: The Scientific Basis," Contribution of Working Group I, in *Third Assessment Report of the Intergovernmental Panel on Climate Change*, 'J.T. Houghton, et al. (eds.), Cambridge University Press: Cambridge, UK, 2001, pp. 183–237.
3. International Civil Aviation Organization, "Annual Review of Civil Aviation 2005", *ICAO Journal*, **61**(5) 6–42 (2006).
4. N. Stuber, P.M.d.F. Forster, G. Rädel, and K. Shine, "The Importance of the Diurnal and Annual Cycle of Air Traffic for Contrail Radiative Forcing", *Nature,* **441**: 864–867 (2006).
5. R. Sausen, I. Isaksen, V. Grewe, D. Hauglustaine, D. S. Lee, G. Myhre, M. O. Kohler, G. Pitari, U. Schumann, F. Stordal, and C. S. Zerefos, "Aviation Radiative Forcing in 2000: An Update on IPCC (1999)." *Meteorologische Zeitschrift*, **14** (4): 555–561 (2005).
6. E.A.M. Franssen, C.M.A.G. van Wiechen, N. J. D. Nagelkerke, and E. Lebret, "Aircraft Noise around a Large International Airport and Its Impact on General Health and Medication Use", *Occupational and Environmental Medicine*, **61**: 405–413 (2004).
7. F. Carlsson, E. Lampi, and P. Martinsson, "The Marginal Values of Noise Disturbance from Air Traffic: Does the Time of the Day Matter?" *Transportation Research Part D—Transport and Environment*, **9**: 373–385 (2004).
8. M. M. Haines, S. A. Stansfeld, R. F. S. Job, B. Berglund, and J. Head, "Chronic Aircraft Noise Exposure, Stress Responses, Mental Health and Cognitive Performance in School Children", *Psychological Medicine*, **31**(2): 265–277 (2001).
9. S. A. Stansfeld, B. Berglund, C. Clark, I. Lopez-Barrio, P. Fischer, E. Öhrström, M. M. Haines, J. Head, S. Hygge, I. van Kamp, and B. F. Berry, "Aircraft and Road Traffic

Noise and Children's Cognition and Health: A Cross-national Study", *The Lancet*, **365**: 1942–1949 (2005).
10. C. Thomas and M. Lever, "Aircraft Noise, Community Relations and Stakeholder Involvement," in *Towards Sustainable Aviation*, P. Upham, et al. (eds.), Earthscan: London, 2003, p. 97–112.
11. Civil Aviation Authority, "Noise Exposure Contours for Heathrow Airport 2004," Environmental Research and Consultancy Department, 2005. Report 0501.
12. Civil Aviation Authority, "Noise Exposure Contours for Stansted Airport 2004," Environmental Research and Consultancy Department, 2005. Report 0503.
13. National Atmospheric Emission Inventory, www.naei.org.uk. Accessed 2006.
14. S. Guibert and L. Guichard, "Paradigm SHIFT—Dual Airspace Concept Assessment," 4th Eurocontrol Innovative Research Workshop, Bretigny, France, 2005.
15. International Civil Aviation Organization, "Operational Opportunities to Minimize Fuel Use and Reduce Emissions," ICAO Circular, 2004. 303-AN/176.
16. R. Ehrmanntraut, "Performance Parameters of Speed Control & the Potential of Lateral Offset," Eurocontrol Experimental Centre, Brétigny, France, 2005. Note 22/05.
17. H. Mannstein, P. Spichtinger, and K. Gierens, "A Note on How to Avoid Contrail Cirrus", *Transportation Research Part D: Transport and Environment*, **10**: 421–426, (2005).
18. P. Spichtinger, K. Gierens, U. Leiterer, and H. Dier, "Ice Supersaturation in the Tropopause Region over Lindenberg, Germany", *Meteorologische Zeitschrift*, **12** (3): 143–156 (2003).
19. OAG, "European Low Cost Carriers," White Paper. 2006.
20. R. Babikian, S. P. Lukachko, and I. A. Waitz, "The Historical Fuel Efficiency Characteristics of Regional Aircraft from Technological, Operational, and Cost Perspectives", *Journal of Air Transport Management*, **8** (6): 389–400 (2002).
21. V. Williams and R. B. Noland, "Comparing the CO_2 Emissions and Contrail Formation from Short and Long Haul Air Traffic Routes from London Heathrow", *Environmental Science & Policy*, **9**: 487–495 (2006).
22. J. E. Green, "Civil Aviation and the Environment—the Next Frontier for the Aerodynamicist", *The Aeronautical Journal*, **110** (2006): 469–486.
23. I. Humphreys, "Organizational and Growth Trends in Air Transport," in *Towards Sustainable Aviation*, P. Upham, et al, (eds.) Earthscan: London, 2003, 19–35.
24. Civil Aviation Authority, "No-Frills Carriers: Revolution or Evolution? A Study by the Civil Aviation Authority," http://www.caa.co.uk/docs/33/CAP770.pdf, 2006. Accessed
25. International Civil Aviation Organization, "Annual Civil Aviation Report: Year-In-Review", *ICAO Journal*, **51** (6): 5–22 (1996).
26. Association of European Airlines, *Recommendations for De-icing/anti-icing of Aircraft on the Ground*, 21st ed., 2006.
27. D. A. Turnbull and J. R. Bevan, "The Impact of Airport De-icing on a River: The Case of the Ouseburn, Newcastle upon Tyne", *Environmental Pollution*, **88** (3): 321–332 (1995).
28. Y. Jia, L. R. Bakken, G. D. Breedveld, P. Aagaard, and A. Frostegard, "Organic Compounds that Reach Subsoil May Threaten Groundwater Quality; Effect of Benzotriazole on Degradation Kinetics and Microbial Community Composition", *Soil Biology and Biochemistry*, **38** (9): 2543–2556 (2006).

29. J. J. Lee, "Greener Manufacturing, Maintenance and Disposal—Towards the ACARE Targets", *The Aeronautical Journal*, **110** (1110): 567–571 (2006).
30. U. Schumann, F. Arnold, R. Busen, J. Curtius, B. Karcher, A. Kiendler, A. Petzold, H. Schlager, F. Schroder, and K. H. Wohlfrom, "Influence of Fuel Sulfur on the Composition of Aircraft Exhaust Plumes: The Experiments SULFUR 1-7", *Journal of Geophysical Research-Atmospheres*, **107** (D15) (2002).
31. OECD, The Political Economy of the Norwegian Aviation Fuel Tax, 2005, COM/ENV/EPOC/CTPA/CFA(2005)18/FINAL.
32. P. M. d. F. Forster, K. P. Shine, and N. Stuber, "It Is Premature to Include Non-CO_2 Effects of Aviation in Emission Trading Schemes", *Atmospheric Environment*, **40**: 1117–1121 (2006).
33. F. Haglind, A. Hasselrot, and R. Singh, "Potential of Reducing the Environmental Impact of Aviation by Using Hydrogen Part I: Background, Prospects and Challenges", *The Aeronautical Journal*, **110** (1110): 533–565 (2006).
34. M. Ponater, S. Pechtl, R. Sausen, U. Schumann, and G. Hüttig, "Potential of the Cryoplane Technology to Reduce Aircraft Climate Impact: A State-of-the-art Assessment", *Atmospheric Environment*, **40** (36): 6928–6944 (2006).
35. Royal Aeronautical Society, "Air Travel—Greener By Design: Report of the Science and Technology Sub Group," 2005.
36. A. P. Dowling and T. Hynes, "Towards a silent aircraft", *The Aeronautical Journal*, **110** (1110) (2006) 487–494.
37. International Civil Aviation Organization, "Assembly Resolutions in Force," Doc 9848, 2004.
38. Advisory Council for Aeronautics Research in Europe (ACARE), "Strategic Research Agenda Volume 2: The Challenge of the Environment," http://www.acare4europe.org, 2002.
39. Advisory Council for Aeronautics Research in Europe (ACARE), "Strategic Research Agenda 2," http://www.acare4europe.org/html/sra_sec.shtml, 2004.
40. Federal Aviation Administration, "National Aviation Research Plan. An Investment in Aviation's Future," Report to the United States Congress pursuant to 49 United States Code 44501(c), 2005.
41. J. E. Green, "Greener by Design", Proceedings of the AAC-Conference, June 30 to July 3, 2003, Friedrichshafen, Germany, 2003: pp. 334–342.
42. P. Brooker, "Civil Aircraft Design Priorities: Air Quality? Climate Change? Noise?" *The Aeronautical Journal*, **110** (1110): 517–532 (2006).

Index

A

AASHTO. *See* Association of State Highway and Transportation Officials
aaSIDRA, 122
ABS. *See* Antilock braking system
Accelerator position, 192
Accumulators, 200–201
    example, 200f
    gas bladder design, 200–201
    storage device, 193–194
Acid precipitation, 49
Acid rain potential (AP), 272
Active components valves, usage, 201–202
Activities. *See* Transportation environmental impacts, 276t
Advanced technology vehicles, introduction, 187
Advanced traveler information systems (ATIS), 107
Advanced vehicle power transmission control, 208
Advisor (vehicle simulation computer program), 151–152, 166
Aesthetics, impact, 36. *See also* Transit facilities
AFC. *See* Alkaline fuel cell
Airborne lead (Pb), 48
Airbus A350, 325
Aircraft
    air quality, 310
    assembly, 318
    CO_2, reductions, 324–325
    emissions, quantification, 310
    engines, combustion products, 310
    environmental challenge, 323–326
    future impacts, 323–326
    impact
        mitigation, 313–314
        operational measures, 313
        quantification, 309–310
    relationship. *See* Stakeholders
    landing, impact, 307–310
    legislation, 319–321
    lifecycle, 318–319
    low-cost operations, 314
    manufacture, composite materials, 325
    market-based measures, 323
    market growth, 323–324
    noise impacts, 308f
    operators, 313–314
    radiative forcing, 307
    research, 325–326
    take-off, local impact, 307–310
    transport, technological advances/political measures, 326
Aircraft in flight
    global impacts, 303–307
    impacts, 302–310
Air freight, 317
Airlines, operating practices/environmental impact, 314–315
Air migration, environmental hazard, 23
Air pollutants
    consideration, 47
    emissions. *See* Vehicles
    factors, estimation, 54
    reduction, 50–51
        gasoline reformulation, impact, 51
Air pollution, environmental costs, 80–81
Airports
    control, 321–322
    impact, 318
    total emissions, control, 322–323
Air quality. *See* Aircraft
    designations, 51, 53
    modeling, TDM integration, 54f
    monitoring, 52–53
    planning. *See* Transportation transportation activity, impact, 47–48
Airspace users, 317
Air supply, subsystems, 157
Air toxics, 50
Alkaline fuel cell (AFC), characteristics, 156t
Alkyl esters (biodiesel), 225
All-electric range, zero level, 167
Alternative investment options (net present value evaluation), annualization method (comparison), 65
American Society of Testing and Materials (ATSM)
    approved quality standards. *See* Biodiesel
    D03 subcommittee, 237
    standards, 225
Anhydrous ammonia (NH_3), 247
Animal fats, usage, 233
Annualization method, comparison. *See* Alternative investment options
Annual operating/maintenance cost, 78
Antilock braking system (ABS) function, 205
AP. *See* Acid rain potential
APUs. *See* Auxillary power units
Arterials, buses (traffic signal control priority), 112–113
ASCE, natural resource demand statement, 261
Asclepiadaceae, 251
Asset deterioration, modeling, 268

331

Asset management, 258–259
  components. *See* Generic asset management system components
  databases, 291
  defining, 284–286
  emphasis, 262
  FHWA promotion, 285
  framework, 259f, 286–287
  investment level, 292–293
  LCA, 294
  mapping, 292
  performance measurement, 294
  reliance, 259
  simulation models, 291–292
  strategies, 294
  sustainability, relationship, 294–296
  systems, 291
  tools
    types, 291–292
    usage, 291–293
  trade-off analysis tools, 292–293
  TRB task force, 285
  visualization tools, 292
AssetManager NT, 293
Association of State Highway and Transportation Officials (AASHTO), 288
  Task Force on Asset Management, formation, 285
ASTM D 5798 standard specification. *See* Denatured fuel ethanol
ATIS. *See* Advanced traveler information systems
Atkinson cycle gasoline engine (Prius), 152–153
  map, 153f
Atmospheric ozone, 49
ATSM. *See* American Society of Testing and Materials
Automatic vehicle location (AVL) system, 102–103
Automotive fuel economy, NRC review/analysis, 64
Automotive spark-ignition engines, denature fuel ethanol E75-E85 (ASTM D 5798 standard specification), 218t
Auxiliary power units (APUs), 140

  usage, 153–154
Average link speed, variation, 103
Aviation
  challenges, 323
  climate impact, 306
  global/local impacts, regulation, 319–323
  governing, 319–323
  impacts, 318–319
  travel behavior/equity, comparison, 315–316
Aviation sector impacts. *See* Environment
AVL. *See* Automatic vehicle location
A-weighted sound level, 29
  example, 30f
Axial piston pump/motor, noise pressure spectrum, 210f

B

B100. *See* Biodiesel
Backward-facing driveline, torque/speed request, 208–209
Bagasse, theoretical yield, 216
Balance of plant (BOP), subsystems, 157–158
Ball valves, necessity, 202
Baseline ICE vehicles
  characteristics, 166t
  mileage, 178
Baseline PFI ICE vehicles, comparison, 175
Baseline vehicles, characteristics, 185t
Battery
  capacity, fraction, 180
  charging, 136
  cycle life correlations, 146f
  development, 144–150
  energy storage capacity, 135
    increase, 179
  OEM costs, 178
  performance characteristics. *See* High-power battery
  relative costs, 150t
  SOC, 165
  technologies, 144–147
    characteristics, 145t

  type/chemistry, defining, 144–145
  types, characteristics, 145t
  unit cost, 179
  usage. *See* Energy storage
Battery-electric vehicles
  characteristics, 131t
  driveline schematic, 134f
Battery management system (BMS), 146–147
Battery-powered electric vehicles (BEVs), 130
  characteristics, 135t
  control strategy, 163
  cost characteristics, 178t
  economic considerations, 177–179
  energy, 169, 171
    consumption, 177
  initial price differential, recovery, 178
  lithium batteries, usage, 147
  market penetration, 177–179
  operation/performance, simulations, 166
  transmission, requirement (absence), 154–155
  usage, 134–136
Bay Area Rapid Transit (BART), environmental impacts (identification), 34, 36
Benefit/cost ratios, 262
Benefits, costs (contrast). *See* Motor vehicle use
Bent axis hydraulic pump/motor, 199f
BEVs. *See* Battery-powered electric vehicles
Bicycle park-and-ride facilities. *See* Train stations
Biodiesel, 224–234
  ASTM approved quality standards, 227–228
  availability, 231–232
  average emissions impacts. *See* Heavy-duty highway engines
  B20, sale, 231–232
  B100
    power torque curve, 229f
    usage, advantages/disadvantages, 228t

Index

biomass source, 225
blends, 231–232
  percentage, 229
cloud point, 229
emissions, 229–231
  reduction, 233
fuel transportation issues, 231
introduction, 224–225
manufacturing methods, 225–227
performance, 229
production capacity, 232–233
quality standards, 227–228
research/future, 233–234
safety, 233
specifications, 227t
storage issues, 231
subsidy, 233
transport, 231
vehicle modifications, 228–229
Biofuels, 250–251
  closing remarks, 251
  gas station, example, 232f
  introduction, 213–215
  renewable materials, usage, 214
  selection, considerations, 214f
  usage. *See* Petroleumlike biofuels
Biological impacts, 36
Biomass
  digestibility, 215–216
  sources. *See* Biodiesel; Ethanol; Vegetable oil
Biomethane, reforming, 236
BMS. *See* Battery management system; Bridge Management system
Boeing 787, 325
Boil-off, problem, 247
Boolean logic, usage, 208–209
BOP. *See* Balance of plant
Borohydride (NaBH$_4$), 247
Bottleneck mitigation, 109–110
  plus-lane concept, impact, 110f
BQ-9000 (quality assurance program), 228
Braking system, components (right-sizing), 192
Breakeven fuel cost, estimate, 185–186

Breakeven fuel prices, range, 186–187
Breakeven gasoline prices. *See* Plug-in hybrid electric vehicle; Series hybrid vehicles
  sensitivity, 179–180
Bridge
  management/maintenance environmental issues, 289–291
  historical background, 287–289
  implementation, challenges, 296
  introduction, 283–291
Bridge Management System (BMS), 293
Bronze, usage, 201
Brownfields
  environmental impact, 22–23
  mitigation, 23
  factors, categories, 23
Building. *See* Green building/design
  maintenance activities, 38–39
Bundled costs, 78
Buses
  fuels, CO$_2$ emissions, 27t
  signal priority, 112–113
  study. *See* Volvo Environmental Concept Bus
Business activities, 39
Butanol, usage, 250

C

CAFE. *See* Corporate Average Fuel Economy
California Air Resources Board, carcinogen identification. *See* Diesel particulate
California bus engine standards (g/bhp-hr), 27t
California emissions standards, 173t
Capital, replacement value, 65
Carbon dioxide (CO$_2$)
  formation, 50
  fossil fuel emissions, 304f
  increase, international aviation (impact), 304t
  reduction, 321

Carbon dioxide (CO$_2$)
  emissions, 242, 310. *See also* Buses
  achievement, 175
  external cost, 91
  factors. *See* Fuels
  impact, 175
  reduction, 174
Carbon emission, fossil fuels (usage), 214
Carbon monoxide (CO), 48
  ambient levels, 83
  CO/CO$_2$ ratio, determination, 117–118
  driving mode, impact, 101t
  emission factors
    example, 102
    VSP bins, contrast, 103f
  emissions, 114, 230
  motor vehicle emissions, 83
  NAAQS, 52t
  percentage, 98
Carbon monoxide (CO) emissions
  decrease, 112
  hot spot, 111
  vehicle speed, impact, 100t
Carbon sequestration methods, 242–243
Carboxy-hemoglobin, formation, 48
Catalyst layer, assistance, 155
Cavitation, impact, 207
Cells. *See* Fuel cells
  voltage/efficiency/power, 159f
Center valves. *See* Closed-center valves; Critically center valves
CFIs. *See* Continuous flow intersections
Changeable message signs (CMS), 107
Charge depleting hybrid, 136–137
Charge sustaining (CS) full hybrid, 139
  cost characteristics, 181–182
  fuel savings, 182
  total CO$_2$ emissions, considerations, 182–183
Charge sustaining (CS) hybrid, 136–137
  design, 163
  fuel economy, 171

Charge sustaining (CS) hybrid vehicles
  economic considerations, 177, 181–183
  market penetration, 177, 181–183
  on-off engine operation, 164f
Charge-sustaining hybrid electric vehicle (CSHEV), 131–132
Check valve design, 204f
Chemical cycling, usage. *See* Hydrogen
Chemical releases, OD potential, 290
Chlorofluorocarbons (CFCs), 50
CIA. *See* Community impact analysis
Cirrus. *See* Contrail-cirrus
City bus traffic, hydraulic hybridization candidate, 204–205
Civil passenger transport, 317
Clean Air Act of 1990, 50
  Amendments (1990), 53
  Section 211(b), 229–230
Clean-up costs, 81–82
Climate change, impact, 38
Closed-center valves, 202–203
Clostridium acetobutylicum, 250
Cloud point. *See* Biodiesel
  variation, biodiesel concentration, 230f
CMS. *See* Changeable message signs
CNS. *See* Communication, navigation, and surveillance
Collector support transit systems, 41
Combustion sources, 38
Communication, navigation, and surveillance (CNS), 312
Community annoyance. *See* Sound levels
Community impact analysis (CIA), 24
COMMUTER2.0 model, 120
  run, travel/emission impacts (example), 121f
Composite materials. *See* Aircraft
Composite tanks, pressurized hydrogen (usage), 245

Compression-ignition (diesel) engines, 225
Computer-based tools, usage, 267
Confirmed Climate Change Policy and Package, 295
Conformity Rule, 53
Congestion
  mitigation measures, impact, 108
  TDM strategies, 105
Construction-related traffic congestion, 290
Consumers, ignorance/irrationality, 76
Continuous descent approach, 312–313
Continuous flow intersections (CFIs), 112
Continuously variable transmission (CVT). *See* Hybrid driveline
  pulleys/belts, usage, 155
Contrail-cirrus, 305–306
Contrails, 305–306
  dissipation, 306f
  reduction, 312
Control
  algorithms, 204
  strategies
    development, 118
    usage. *See* Energy storage systems, bang-bang, 192
Control valves, selection, 202
Corn
  impact, 215
  usage, 223f
  utilization. *See* Ethanol
Corn-derived ethanol, impact. *See* Food shortage
Corporate Average Fuel Economy (CAFE), 175
  fuel economy values, 176
  impact, 175t
CORSIM model, 123
Critical-center valves, selection, 202–203
Critically center valves, 202–203
Cryogenic liquid tasks, hydrogen (usage), 247
CS. *See* Charge sustaining
CSHEV. *See* Charge-sustaining hybrid electric vehicle
Current vehicular travel modal shift, 105

spatial shift, 105
temporal shift, 105
CVT. *See* Continuously variable transmission
Cycle life
  correlations. *See* Battery issue, 146

D

Damages, nonsynonymity. *See* Externalities
Day-night average sound level (Ldn // DNL), 29, 309
DC/AC inverter
  schematic, 143f
  switching devices, inclusion, 142
  usage, 139
DC/DC converters, 158
DC motors, brushes (usage), 140
Dead band parallel regenerative braking system, usage, 205
Deaths, 9
Denatured fuel ethanol, ASTM D 5798 standard specification, 217. *See also* Automotive spark-ignition engines
Department of Energy (DOE)
  alternative fueling station locator, 248
  analysis period, 265–266
  description. *See* Proton exchange membrane fuel cell
  goals, 160t
  hydrogen storage system goals, comparison, 162
Department of Transportation (DOT), asset inventory assessment, 286–287
Depth-of-discharge, 146
Design-optimization process, 192
Deterministic LCCA, 267
Diesel, Rudolf, 224
Diesel engines
  durability/efficiency, 51
  usage, 169
Diesel fuel
  chemical/physical properties, 174t
  linkage (U.S. EPA). *See* Fine particulate matter

Diesel-fueled internal combustion engine, inefficiency, 192
Diesel particulate, California Air Resources Board (carcinogen identification), 26
Diesel-powered light-duty vehicles, 173
Direct costs, indirect costs (contrast). *See* Motor vehicle use
Direct effects, contrast. *See* Transportation
Direct emissions. *See* Surface transportation
Direct-injection gasoline engine, 152
Direct methanol fuel cell (DMFC), characteristics, 156t
DMFC. *See* Direct methanol fuel cell
DNL. *See* Day-night average sound level
DOE. *See* Department of Energy
Domestic producer surplus, subtraction. *See* Price-times-quantity
DOT. *See* Department of Transportation
Drag coefficient, 135
Driveline characteristics. *See* Plug-in hybrid electric vehicle
Driveline component technologies, 140–155
Driveline schematics. *See* Battery-electric vehicles; Parallel hybrid vehicles; Series hybrid vehicles
Driver pedal positions, 208
Driving cycle, differences, 141
Driving mode. *See* Emission factors
  approach, application, 101
  impact. *See* Carbon monoxide
Dynamometer testing, 123

E

Eaton Corporation Hydraulic Launch Assist (HLA) system, 195
ECE-EUDC schedule, 173
Ecological risk (ER), 272
Economic activity, 2–3
Economic considerations. *See* Technology
Economic impacts. *See* Transit projects
Economic theory, mechanisms, 86
ECU. *See* Electronic control unit
EEI. *See* Environmental Energy, Inc.
E85 (ethanol, 85 percent), 215
EF. *See* Emission factors
Efficiency of use, 69–70
Efficient marginal-cost pricing, factors, 74t
EIS. *See* Environmental impact study
Electrical energy storage, 144–150
Electric drivelines, peak power ratings, 141–142
Electricity
  demand, 36
  substitution, 165
Electric motors
  battery, usage, 165
  simulation results, 143t
  usage, 140–141
Electric vehicle (EV)
  battery pack, 134
  characteristics, 167t
  long-term commercialization, 135
  motors, classification, 141f
  operation/performance, simulations, 166–172
  range, 135
Electric vehicle (EV) design/performance
  computer simulation results, 130
  introduction, 130
  summary/conclusions, 187
Electrolysis. *See* Hydrogen
Electronic control unit (ECU), 219
Emission factors (EF), 99
  driving mode, 101f
  example. *See* Carbon monoxide; Modal emission factors
EMission FACtors (EMFAC), 54

Emissions. *See* Pollution emissions; Transportation
  changes, summary, 118
  comparison, 306
  considerations, 172–174
  control strategies, 50–51
  gasoline/ethanol blend, impact, 221
  impacts, 124
  inventories, development, 118
  reductions, 121f
    fuel technologies, contribution, 324–325
  RSD, operation (schematic), 118f
  trading, 323
  VSP, correlation, 102
EMME2, 122
Employer-based programs, objective, 105
EMS. *See* Environmental management systems
Endangered species, transit facility environmental impacts, 21
  mitigation, 21
Energy
  cane, 216
  consumption, 28–29, 290
    mitigation, 29
  density, statement, 144
  efficiency, 38
  policy, 35t
  recovery, regenerative braking, 134–135
  use, 3–4. *See also* Transportation; U.S. energy use
Energy Information Agency (EIA), diesel fuel usage, 232
Energy Policy Act of 2005, Section 1344 (tax credit extension), 233
Energy-security costs, tax, 81
Energy storage
  battery, usage, 131–132
  capacity, 130. *See also* Battery
  maximization, control strategy (usage), 204
  unit characteristics, 149t

Engineering management system, 259
  evolution, 258f
Engine/generator, necessity, 153–154
Engines, 150–153
  characteristics, 150
  data scanner, 114
Environment
  aviation sector impacts
    conclusion, 327
    introduction, 301–302
    relationship. See Public transportation
Environmental costs, 80–81
Environmental Energy, Inc. (EEI), butanol/hydrogen production process, 250
Environmental factors, integration, 289
Environmental hazards, identification, 23
Environmental impacts. See Operating transit systems; Ridership-based environmental impacts; Technology-based environmental impacts; Transit facilities; Transit operations
  avoidance/minimization/mitigation, 21
  mitigation, 34–36
Environmental impact study (EIS), impacts (evaluation), 22
Environmental justice, 21
  issues. See Transit facilities
Environmental linkages. See Public transportation
Environmentally preferable purchasing, 38
Environmental management systems (EMS), usage, 36–37
Environmental policy, 35t
Environmental projects. See Transportation
Environmental Protection Agency (EPA)
  emssions reductions regulations, 50–51
  fuel economy ratings. See Honda FCX vehicle
  indicators, development. See Transportation

Office of Air Quality Planning and Standards (OAQPS), 52
Office of Transportation and Air Quality (OTAQ), 120
  report, 274, 277
Environmental stewardship. See Transit environmental stewardship
  factors, transit systems (impact). See Longer-term environmental stewardship factors
  sustainable development, relationship, 36–42
EPA. See Environmental Protection Agency
Equity, issues. See Transit facilities
ER. See Ecological risk
Ethanol, 215–224
  availability, 223
  biomass sources, 215–216
  blends, incompatible/compatible materials. See High ethanol blends
  cold-start properties, 217, 219
  comparison. See Gasoline
  corn utilization, 217
  E100, usage, 221
  E10 blend, 223
  E20 blend, 219
  emissions, 221
  energy, 219
  fuel storage, 222
  fuel transportation/distribution, 221–222
  introduction, 215
  links, 224
  manufacturing methods, 215–216
  manufacturing research and development, 217
  85 percent. See E85
  performance, 219–221
  production, methods (schematic), 216f
  quality standards, 217
  safety, 222
  subsidy, 222–223

tax money, waste (criticisms), 222
undigested residue, gasification, 217
usage, 174
vehicle modifications/use, 217–219
Ethanol-fueled vehicles, 217, 219
Euphorbiaceae, 251
Euphorbiaceae lathyris (E. lathyris), 251
European emissions standards, 173t
EV. See Electric vehicle
Externalities
  damage, nonsynonymity, 80
  definition, usage, 82–83
  distinction. See Personal nonmonetary costs
  economic problems, 81
  unaccounted for cost, synonymity, 80

F
Facility life cycle, 284f
Fallow farm land, usage, 233
Fare-free rides, inclusion, 106
Fatalities, 9. See also Passenger-miles; Ton-miles
  sector analysis, 10t
Fault diagnosis, 208
FC. See Fuel cells
Federal Clean Air Act (1990), 118–119
Federal emissions standards, 173t
Federal Highway Administration (FHWA)
  encouragement. See Life-cycle cost analysis
  highway-cost allocation study, 64–65
  Life-Cycle Cost Analysis Primer, 265
  Office of Asset Management, formation, 285
  promotion. See Asset management
Federal Urban Driving Schedule (FUDS), 164
  driving cycle, 173

Index   337

simulation results. *See*
Plug-in hybrid
electric vehicle
Highway cycles, combination (improvement factors), 169
Feedstock, usage, 215
Fees, estimated net, 73–76
Fermentation. *See* Hydrogen
FFVs. *See* Flexible fuel vehicles
Financial criteria, 262
Fine particulate matter ($PM_{2.5}$), 48
   diesel fuel, linkage (U.S. EPA), 26
   NAAQS, 52t
   production, 26
Fine particulate matter (PM_{10}), 48, 98
   NAAQS, 52t
Flexible fuel vehicles (FFVs), 216
   availability, 219
Flexible work hours, TDM strategies, 105
Food shortage, corn-derived ethanol (impact), 222–223
Ford, Henry, 215
Fossil fuel emissions. *See* Carbon dioxide
Four-mode modal emission models, 122
Four-wheel drive (4WD) SUV, fuel economy improvement, 191
Free parking, unpricing, 76
Freeway
   MOBILE6 emission factors, sensitivity, 119f
   operational strategies, 109–111
Freeway bottlenecks, 109
Front motors, regenerative brake force changes, compensation, 208
FUDS. *See* Federal Urban Driving Schedule
Fuel-based approach, Synchro (usage), 122
Fuel cell (FC) electric vehicles, 165
Fuel cell (FC) vehicles. *See* Hydrogen fuel cell vehicles
   breakeven fuel prices, 186t
   characteristics, 133t, 185t
   comparison. *See* Gasoline
   component/driveline costs, 186t
   fuel savings, 186t
   improvement factor, assumption, 171–172
   operation/performance, simulations, 171–172
   specifications, comparison, 238f
   transmission, requirement (absence), 154–155
Fuel cells (FC), 155–162
   advantages, 240
   balance/control, 158–159
   characteristics, 156t
   disadvantages, 241
   driveline schematic, 139f
   efficiency, 172
   operating characteristics, 159
   output, supplementation, 185
   Plug Power manufacture, 241f
   system, operation, 157–160
   technology, 155–157
   listing, 239t
   usage, 134
Fuel cycle emissions. *See* Upstream full fuel cycle emissions
Fuel economy, control (optimization), 207–209
Fuel economy improvement, 197
   factors, simulation results (basis), 172t
   ratios. *See* Port Fuel Injection engine-powered vehicles
Fuel efficiency map. *See* Vehicles
Fuels
   CO_2 emission factors, 176t
   controlling, 321
   flash point comparison, 233f
   supply, subsystem, 157
   taxes, 323
   technologies, contribution. *See* Emissions
   use, sector analysis, 4t
Full cycle fuel-related emissions (upstream emissions), 172–173
Full-function BEVs, 177–178
Full hybrids
   approach, 138–139
   engine/vehicle classes, cost results, 182t
   simulation results, 170t
   vehicle, 179
Fuzzy controller, proposal, 209

G

Gas bladder accumulator, 200f
Gas bladder design. *See* Accumulators
Gas-guzzler tax, 81
Gasification. *See* Hydrogen
Gasoline
   consumption, 91–92
   cost savings, 121f
   ethanol
      blend, impact. *See* Emissions
      cold-start properties, comparison, 221
   fuel, chemical/physical properties, 174t
   ICE vehicles, fuel cell vehicles (comparison), 140
   reformulation, impact, 51
Gasoline direct injection (GDI)
   engine, 152
   map, 153f
Gasoline-fueled internal combustion engine, inefficiency, 192
Gasoline/water/ethanol tenary phase diagram, 220
Gas piston accumulator, 200f
   fluid form, separation, 201
Gas turbine engine, 154
   schematic, recuperator (inclusion), 154f
GDI. *See* Gasoline direct injection
General Dynamics Corporation, 195
Generic asset management system components, 285f
Geographic Information Systems (GIS), 292
Geological fault, impact, 36
GIS. *See* Geographic Information Systems
Glass spheres, hydrogen (usage). *See* Microscale glass spheres

Global warming
  environmental costs, 80–81
  environmental externalities, 72
Global warming potential impact (GWPI), 272
GNP. See Gross national product
Green building/design, 37
Greenhouse emissions, 172–173
Greenhouse gas emissions, 174–175
  hybrid vehicle technologies, implementation (impact), 175
Gross national product (GNP)
  GNP-type accounts, inclusion/exclusion, 85t
  loss, 72
  motor vehicle use, total social cost (comparison), 92
Ground-level ozone (O_3), 48, 49
Ground water migration, environmental hazard, 23
GWPI. See Global warming potential impact

H

Habitats, transit facility environmental impacts. See Sensitive habitats
Hard braking cycle, 193
Hazardous air pollutants (HAPs), 50
Hazardous materials
  environmental impact, 22–23
  mitigation, 23
  factors, categories, 23
Hazardous substances, release (likelihood), 23
HC. See Hydrocarbon
HCM. See U.S. Highway Capacity Manual
Heavy-duty highway engines, biodiesel (average emissions impacts), 231f
HERS. See Highway Economic Requirements System
HERS/ST, 269
  model, 292
HEVs. See Hybrid electric vehicles

HHR. See Human health risk
HHVs. See Hydraulic hybrid vehicles
High-acceleration events, 117–118
High emissions, episodes (minimization), 113
High-energy density batteries, 135
High ethanol blends, incompatible/compatible materials, 220t
High occupancy toll (HOT) facilities, TDM strategies, 106
  lanes, 105, 122
  usage, 110–111
High-power battery, performance characteristics, 145
High-pressure accumulator, 208
  pressure, sufficiency, 196–197
High-pressure hydraulic fluid flows, 203
High-pressure port, 199
High-speed rail (HSR), comparison, 64
High-speed tolling, 111
Highway Economic Requirements System (HERS)
  usage. See Oregon Department of Transportation
Highway Economic Requirements System (HERS) simulation tool, 292
Highway information services, TDM strategies, 107
Highway life-cycle cost analysis software, 269
HLA. See Hydraulic Launch Assist
Honda FCX vehicle, 171–172
  EPA fuel economy ratings, 172t
  storage system, usage, 245
  test data, 172
Honda Insight VTEC engine, map, 152f
Honda VTEC gasoline engine, map, 151f
HOT. See High occupancy toll

Hot spot. See Carbon monoxide emissions
Hourly equivalent sound level (Leq)(h), 29
H_2SO_4. See Sulfuric acid
HSR. See High-speed rail
Hub-and-spoke model, 314
Human health/ecosystems, toxicological stress, 270
Human health impacts, avoidance/minimization/mitigation, 21
Human health risk (HHR), 272
Hush-kits, usage, 320
Hybrid driveline, CVT schematic, 138f
Hybrid electric vehicles (HEVs), 130–132
  California standards, meeting, 173
  CS mode, 179
  liquid hydrocarbon fuels, usage, 174
  sale, 132
Hybrid fuel cell vehicle, 134
Hybridization
  pump/monitor technologies, usage, 198
  reasons, 192
Hybrid vehicle
  design/performance
  computer simulation results, 130
  introduction, 130
  summary/conclusions, 187
Hybrid vehicles
  characteristics, 169t
  definition, 192
  energy storage, ultracapacitors/batteries (comparison), 147–150
  operation/performance, simulations, 166–172
  technologies, implementation (impact). See Greenhouse gas emissions
Hybrid VSP-S approach, 99, 103–104
Hydraulic actuator, model, 208
Hydraulic hybrid components, 198–206
Hydraulic hybridization, 197
  city buses, candidates, 204–205

Index

Hydraulic hybrid system, 192–197
  benefits, 197
Hydraulic hybrid vehicles (HHVs)
  control, 203
    performance, 208
  introduction, 191–192
  off-the-shelf components, 191–192
  research areas, 206–210
  summary, 210–211
Hydraulic Launch Assist (HLA)
  module, 195f
  system. *See* Eaton Corporation Hydraulic Launch Assist system
Hydraulic pump/motor
  feedback, 205
  optimal power distribution, calculation, 208
  tolerances, 207
Hydraulic ram position (control), spool valves (usage), 202f
Hydraulic systems, reliability, 207
Hydraulic vehicles
  control systems, 204–206
  storage capacity, increase, 206–207
Hydrocarbon (HC)
  emissions, 114, 230, 310
  usage. *See* Synthesis hydrocarbons
Hydrogen, 234–250
  availability, 248
  commercialization, 243
  economy, realization, 234
  electrolysis, 235–236
  emissions, 242–243
  FC vehicle prototypes, development/demonstration, 240
  fermentation, 236
  fueling stations, network (distribution cost), 244f
  fuel transmission/distribution, 243–245
  gasification/pyrolysis, 236–237
  introduction, 234
  links, 250
  manufacturing methods, 234–237
  on-board reforming, 247

  photolytic/photoelectrolysis, 235
  photosynthesis, 236
  production, 242
    cost, 237
    green routes, 235f
  quality standards, 237
  research, 249
  safety, 248–249
    advantages/disadvantages, 249
  solid-state storage, 247
  storage
    chemical cycling, usage, 247
    costs, 246f
    issues, 245–247
    systems. *See* On-board hydrogen storage systems
    technologies, volumetric/gravimetric energy densities, 246f
  stored energy, 234
  thermochemical cycles, 236
  transmission costs, 243f
  usage. *See* Composite tanks; Cryogenic liquid tanks; Microscale glass spheres
  use, potential. *See* Internal combustion engine vehicles, 237–242
  well to wheel, 249
Hydrogen fuel cell vehicles, 132–134, 139–140
  economic considerations, 177, 184–187
  market penetration, 177, 184–187
  simulation results, 172t
Hydrogen-powered ICE, 241
  advantages, 241–242
  disadvantages, 242
Hydrogen storage
  characteristics. *See* Vehicles
  maximum volume, availability, 161
  requirements, estimates, 161
  subsystem, 157
  system, 160–161
    goals. *See* Department of Energy
Hydrolyzed sugars, fermentation, 217
Hydrous E100, 219

I

ICAO. *See* International Civil Aviation Organization
ICE. *See* Internal combustion engine
IDAS, 293
Incident management, 109
Incremental costs, reduction, 197
Indirect costs, contrast. *See* Motor vehicle use
Induction motor, efficiency map, 142f
Inefficiency sources, externalities (relationship), 91
Infrastructure assets
  changes, 260
  system framework, 260f
  types, 259–260
Infrastructure cost
  approximation, 75
  consideration. *See* Motor vehicle infrastructure
Infrastructure management, 258
Injuries, 9. *See also* Passenger-miles; Ton-miles
  sector analysis, 10t
In-line sensors, development, 207
Institute of Transportation Studies, 166
INTEGRATION model, 123
Intelligent Transportation Systems Benefits Page. *See* U.S. Department of Transportation
Intermodal Surface Transportation Efficiency Act (ISTEA), 288
  mandates, response, 291
Internal combustion engine (ICE). *See* Hydrogen-powered ICE
  California standards, 173
  combustion, mechanical energy, 157
  driveline components, direct link (removal), 195–196
  E100, usage, 221
  fuel economy, 171
  hydrogen use, potential, 241–242
  inefficiency. *See* Diesel-fueled internal

Internal combustion engine (ICE). *See* Hydrogen-powered ICE *(Continued)* combustion engine; Gasoline-fueled internal combustion engine
  powertrain, 175
  usage, 237
Internal combustion engine (ICE) vehicle, 132, 160
  characteristics. *See* Baseline ICE vehicles
  curb weight, 161
  fuel cell vehicles, comparison. *See* Gasoline
  weight/road load, comparison, 135
International aviation, impact. *See* Carbon dioxide
International Civil Aviation Organization (ICAO), 312
  recommendations, 313–314. *See also* Noise resolution A35-5, 325
International Organization for Standardization (ISO)
  ISO 14040, indication, 271
  ISO 14041, 271
  ISO 14042, 271
  ISO 14043, 271
  ISO 14000 standards, 36
  quality assurance program, analogy, 228
Intersection. *See* Continuous flow intersections
  design, innovation, 112
In-vehicle navigation/routing systems, 107
ISTEA. *See* Intermodal Surface Transportation Efficiency Act

K

Kinematic viscosity, 228
Kyoto Protocol, 303
  usage, 322, 323

L

Landing-take-off (LTO) cycle, 319
  usage, 320f
Land-use influence, 17
Land use policy, 35t
Land-use strategies, TDM strategies, 106–107
Lane control, 110–111
LCA. *See* Life-cycle assessment
LCCA. *See* Life-cycle cost analysis
LCI. *See* Life-cycle interpretation
LCIA. *See* Life-cycle impact assessment
Ldn. *See* Day-night average sound level
Lead-acid battery, 144
Lead-based paints, usage, 283
Lead (Pb), 49
  NAAQS, 52t
Lean burn engine, 152
  usage, 169
Left-turn movements, separation/servicing, 112
Leq(h). *See* Hourly equivalent sound level
Level of service (LOS), 120
LID. *See* Low-impact development
Life cycle. *See* Facility life cycle
Life-cycle assessment (LCA), 270–277. *See also* Asset management; Sustainable transportation infrastructure management
  approach. *See* Transportation
  framework/application, 271f
  methodology, 271–274
Life-cycle cost analysis (LCCA), 261. *See also* Deterministic LCCA; Probabilistic LCCA
  concept, 264–265
  deterministic approach, 266
  FHWA encouragement, 265
  probabilistic approach, 266
  procedures, 266
  software tools, 269
  tools, 268–270. *See also* PMS
Life-cycle costing, 264–270
Life-cycle costs, 149
  quantification, 262
Life-cycle effects, direct effects (contrast). *See* Transportation
Life-cycle impact assessment (LCIA), 271, 272
  results, 273
Life-cycle interpretation (LCI) approach. *See* Transportation
  occurrence, 272
Light-duty vehicles stoichiometric port-fuel-injected engines, usage, 150–151
  trade-offs, 135
Lignocellulose, 215
  conversion technologies, 216
  gasification, 217
  percentage, 216
Lignocellulose-to-ethanol processes, 217
Link traffic flow, multiplication. *See* Per-vehicle emission rate
Lithium batteries, usage. *See* Battery-powered electric vehicles
Lithium-ion batteries, 135, 144
  BMS, necessity, 146–147
  ultracapacitors, usage, 147–148
  usage, 177–178
Lithium polymer battery, 144
Lmax. *See* Maximum sound level
Local roads, bundled costs, 78
Long-distance flight, 316
Longer-term environmental stewardship factors, transit systems (impact), 19
Long haul routes, 311f
Long-run annualized capital cost, 78
LOS. *See* Level of service
Low-cost carriers source, growth, 316t
Low-impact development (LID), 38
Low-income populations
  benefits receipt, denial/reduction/delay (prevention), 21
  identification, 22
Low-pressure port, 199
LTO. *See* Landing-take-off

M

Macroscopic traffic operations models, 122
Maintenance, rehabilitation and replacement (MR&R), 287

Index 341

Major arterials, MOBILE6 emission factors sensitivity, 119f
Marginal social cost (MSC), 70, 75
Marginal social value (MSV), marginal social cost (MSC) (equivalence), 70–71, 75
Market penetration, 177–187
Maximum sound level (Lmax), 29
MCFC. See Molten carbonate fuel cell
MDT. See Montana Department of Transportation
Mechanical components, 154–155
Mechanical subsystem, model, 208
Metal conductor, conduction, 155
Methane, usage, 251
Methanol, usage, 250
Metropolitan transit system, example, 17f
MicroBENCOST, 293
MicroBencost, 269
Microcomputer-based controller, necessity, 140
Microscale glass spheres, hydrogen (usage), 247
Microscopic traffic simulation models, 122–124
Mid-size car, simulation results, 169
  Advisor, usage, 170t
Mild hybrids
  approach, 138–139
  engine/vehicle classes, cost results, 182t
  simulation results, 170t
Minority populations
  benefits receipt, denial/reduction/delay (prevention), 21
  identification, 22
MixAlco process, 217, 251
Mixed alcohols, usage, 251
MKP Company, 195
MOBILE6
  emission factors, sensitivity, 119f
  factors, application, 120
  input, 122
MOBILE6.2, 118–121

Mobile emission estimation models, 118–121
MOBILE inputs, 119
MOBILE5 (vehicle emission modeling software), 54
MOBILE6 (vehicle emission modeling software), usage, 54
Modal activity model, application (difficulty), 102
Modal emission factors, example, 100–101
MODE. See Travel time distribution
Mode
  choice, policies (impact), 35t
  definition/characteristics. See Transportation
Molten carbonate fuel cell (MCFC), characteristics, 156t
Monash University, 208–209
Monetary costs, classification, 71
Monetary external cost, 80
Monetary externalities, 72. See also Motor vehicle use
Montana Department of Transportation (MDT) mission statement, 290
Monte Carlo simulation, 267–268
Morey, Samuel, 215
Motor control (development), speed sensor (absence), 208
Motorized vehicle-miles of travel, 105
Motors
  classification. See Electric vehicle
  usage, 140–141. See also Electric motors
Motor vehicle categories, costs (allocation), 88–90
Motor vehicle emissions, regional air quality measures (linkages), 98–99
Motor vehicle goods/services
  efficient pricing, 74t
  private-sector bundling, 76–78
  ranking, 86t
  private-sector pricing, 85t

private-sector production/pricing, 73–76
Motor vehicle infrastructure cost, consideration, 66
  nonmonetary costs, 82–83
  nonmonetary environmental costs, 90t
  public sector input, 78–80
  ranking, 87t
  social costs, 90t
Motor vehicle noise, environmental costs, 80–81
Motor vehicle-related infrastructure/service, 78
Motor-vehicle-related services, addition, 64–65
Motor Vehicles Emissions Simulator (MOVES), 54
Motor vehicle services, public sector input, 78–80
  ranking, 87t
Motor vehicle use
  analysis
    quality/complexity, ratings (description), 84t
    results, 88–93
    results, nonusage, 90–93
  annualized social cost, 65
  benefits, costs (contrast), 67–69
  conceptual framework, 65–72
  cost
    analysis, 68–69
    consideration, 66
    interpretation process, 66–67
    observations, 83–86
    price-times-revenue calculation, 75
    summary, 91t
  direct costs, indirect costs (contrast), 83, 86
  efficiency condition, classification, 70–71
  estimates, quality, 86–88
  external costs, 59t–61t
  inefficiency, 71
  long run, short run (contrast), 83, 86
  methodological organizing criterion, 71–72
  monetary externalities, 80
  ranking, 88t

Motor vehicle use (Continued)
  nonmonetary costs, 59t
  nonmonetary externalities,
    80–82
    ranking, 89t
  personal cost, 59t–61t
  personal nonmonetary costs,
    72–73
    ranking, 84t
  priced costs, distinction, 79
  price/MSC, divergence
    (increase), 83
  private-sector costs, 59t–61t
  public-sector costs, 59t–61t
  social cost, components,
    72–88
  social-cost analysis,
    conceptual issues, 69
  social cost-benefit analysis,
    68f
  standards/regulations,
    economic inefficiency,
    76
  total social cost
    analysis, 92–93
    comparison. See Gross
      national product
    components, classification,
      69–72
  U.S. annualized cost, 65
Motor vehicle use, social cost
  analysis
    information, usage, 62
    purpose, 58–64
    background, 58
    classification, 59t–61t
    summary, 93
MOVES model, 125
MR&R. See Maintenance,
  rehabilitation and
  replacement
MSC. See Marginal social cost
MSV. See Marginal social
  value
Multicriteria multimodal
  tradeoff analysis tools, 269

N

NAAQS. See National Ambient
  Air Quality Standard
NaBH$_4$. See Borohydride
NaOH. See Sodium hydroxide
National Air Monitoring
  Stations (NAMS), 52

National Ambient Air Quality
  Standard (NAAQS), 51
  list, 52t
  ozone limits, 98
  usage, 52–53
National average conditions,
  118–119
National Aviation Research
  Plan, environmental plans,
  326
National Biodiesel Board,
  report, 232–233
National Bridge Investment
  Analysis System (NBIAS),
  292
National Cooperative Highway
  Research Project (NCHRP)
  20–57, 293
  report, 109
  Report 456, 277
  Report 545, 270
  Report 551 definition. See
    Performance
    measurement
National default daily speeds,
  119
National Energy Efficiency and
  Conservation Strategy, 295
National Environmental Policy
  Act (NEPA), 261
National Ethanol Vehicle
  Coalition, flexible fuel
  vehicle guide, 223
National Renewable Energy
  Laboratory (NREL), 166
  report, 230
National Research Council
  (NRC) review/analysis.
  See Automotive fuel
  economy
National Transportation
  Statistics, vehicle
  emissions, 99
Natural
  environment/community,
  transit system
  infrastructure (physical
  design/construction
  impact), 19
Natural resource depletion
  (RD), 272
NBIAS. See National Bridge
  Investment Analysis
  System

NCHRP. See National
  Cooperative Highway
  Research Project
Near-uniform speed profiles,
  maintenance, 113
NEPA. See National
  Environmental Policy Act
Net social benefit, 67–68
Network transportation
  infrastructure management
  decisions, 260
Networkwide emissions,
  generation, 104
New Zealand Biodiversity
  Strategy, 295
New Zealand Waste Strategy,
  295
NH$_3$. See Anhydrous ammonia
Nickel metal hydride battery,
  144
  technology, usage, 181
Nitrogen dioxide, NAAQS, 52t
Nitrogen oxide (NO$_x$)
  annual emissions, 98
  emissions, 50, 114
    percentage, 112
  unsaturated fatty acids,
    impact, 230
  production, 305
  reductions, 319–320
Noise, 29–31. See also Pump
  noise
  assessment, procedure, 31f
  impact, 307–308. See also
    Aircraft
  management, ICAO
    recommendations, 320
  measurement. See Transit
    noise
  mitigation, 31, 313
    measures. See Transit
      noise
  pressure spectrum. See Axial
    piston pump/motor
  reduction, 209–210
Non-CO$_2$ impacts, 303
Nonmonetary costs. See Motor
  vehicle infrastructure
  distinctions, 81
Nonmonetary externalities, 72.
  See also Motor vehicle use
Nonmotorized alternatives,
  TDM strategies, 106
Nonrecurring congestion,
  impacts (mitigation), 109

Nontransportation goods/services, bundling, 58
NO$_x$. *See* Nitrogen oxide; Oxides of nitrogen
NREL. *See* National Renewable Energy Laboratory
Numeric analysis, 22

O

O$_3$. *See* Ground-level ozone
Oak Ridge National Laboratory, 123
OAQPS. *See* Environmental Protection Agency
OD. *See* Ozone depletion
OEM. *See* Original Equipment Manufacturer
Office of Management and Budget (OMB), 266
Office of Transportation and Air Quality (OTAQ). *See* Environmental Protection Agency
Offstreet nonresidential parking, 77t
Off-street parking, 77
Offstreet residential parking, 77t
Off-the-shelf components. *See* Hydraulic hybrid vehicles
Oil-producing plants, production yields, 226t
OMB. *See* Office of Management and Budget
On-board hydrogen storage systems, requirements, 245
On-board reforming. *See* Hydrogen
On-board vehicle measurements, 100–101
On-off engine operation. *See* Charge sustaining hybrid vehicles
On-ramp congestion, 109–110
On-street parking, 77
Operating costs, sum, 65
Operating transit systems, environmental impacts, 19
Operational parallel hybrid vehicles, 207–208
Operations/maintenance costs, 65
Opportunity cost, 66

Optimal power distribution, calculation. *See* Hydraulic pump/motor
Oregon Department of Transportation (ODOT) customized HERS, usage, 293
  definition (adoption). *See* Sustainable transportation system
Organization of Petroleum Exporting Countries (OPEC), oil output restriction, 75
Original Equipment Manufacturer (OEM)
  cost, relationship, 179
  fuel cell system cost, assumption, 184–185
  motor cost, 184
OTAQ. *See* Environmental Protection Agency
Outcomes, reference, 262
Output-based performance measures, 294
Outputs
  reference, 262
  variables, statistical characteristics, 267–268
Oxides of nitrogen (NO$_x$), 48–49
Ozone. *See* Atmospheric ozone; Ground-level ozone
  formation, 98
  NAAQS, 52t
Ozone-depleting chemicals, tax, 81–82
Ozone depletion (OD), 272
  potential. *See* Chemical releases

P

PAFC. *See* Phosphoric acid fuel cell
PAMS. *See* Photochemical Assessment Monitoring Stations
Parallel hybrid architecture, 194f
Parallel hybrid vehicles, 137–139
  control design methodology, development, 208
  design, 163–165
  driveline schematics, 131t

engine/electric motor, power (providing), 164–165
engine output, 137–138
on-off mode, 167
operation/performance, simulations, 167–169
planetary gear torque coupling, 137f
Parallel hydraulic hybrid, 193–195
  method, 194–195
Parallel hydraulic truck, example, 196f
PARAMICS model, 123
Park-and-ride facilities, TDM strategies, 106
Parking
  classification/analysis, 77–78
  lots, classification, 77t
  pricing, TDM strategies, 105
  reduction/management, 41
  social cost-benefit analysis, 78
  unpricing. *See* Free parking
Particulate matter (PM), 49
  emissions, 230
  NAAQS, 52t
Passenger-kilometer carbon emissions. *See* Short-haul flights
Passenger-kilometer performance, average annual growth rates, 324t
Passenger-miles, fatalities/injuries, 10t
Passengers, impact, 315–316
Passenger traffic, increase, 323–324
Pavement
  management/maintenance environmental issues, 289–291
  historical background, 287–289
  implementation, challenges, 296
  introduction, 283–291
Pavement Management System (PMS), 293
  LCCA tools (comparison), 269t
Payload fuel efficiency, 315
Pb. *See* Airborne lead; Lead
PDF. *See* Probability density function
Pedestrian, priority, 40

PEMFC. *See* Proton exchange membrane fuel cell
PEMFCV. *See* Proton exchange membrane fuel cell vehicle
PEMS. *See* Portable emission measurement systems
Percent swing, 146
Performance measures. *See* Sustainable transportation infrastructure management
  examples, 263t–264t
  NCHRP Report 555
  definition, 262
  usage, 262
Performance modeling, data (support), 287
Periodic costs, sum, 65
Periodic operations/maintenance costs, determination, 65
Permanent magnet (PM) motors, 141
  efficiency map, 142f
  usage, 140
Personal nonmonetary costs. *See* Motor vehicle use extremities, distinction, 73
Person-miles of travel, 105
Per-vehicle emission rate, link traffic flow (multiplication), 104
Petroleumlike biofuels, usage, 251
PFI. *See* Port Fuel Injection
Phase separation problems, avoidance, 219
PHEV. *See* Plug-in hybrid electric vehicle
Phosphoric acid fuel cell (PAFC), characteristics, 156t
Photochemical Assessment Monitoring Stations (PAMS), 53
Photochemical oxidant impact (POI), 272
Photolytic/photoelectrolysis. *See* Hydrogen
Photosynthesis. *See* Hydrogen
Physical design/construction, impact. *See* Natural environment/community
Physical highway assets, 284–285
PID. *See* Proportional integrative derivative

Pigou, A.C., 67
Pigovian taxes, 67, 81. *See also* Quasi-Pigouvian taxes
  construing, 82
  expense, 76
  externalities, 73
  levy, 81
  usage, 91–92
Piston accumulator. *See* Gas piston accumulator
Planetary gear torque coupling. *See* Parallel hybrid vehicles
Plug-in hybrid electric vehicle (PHEV), 132
  breakeven gasoline prices, 181t
  compact car, simulation results, 168f
  control strategy, design, 164–165
  cost characteristics, 180t
  design, 163–165
  driveline characteristics, 180t
  economic considerations, 177, 179–181
  FUDS driving cycle, simulation results, 168t
  fuel savings, 182
  full hybrid design, 139
  market penetration, 177, 179–181
  total CO_2 emissions, considerations, 182–183
Plug Power. *See* Fuel cells
PM. *See* Particulate matter; Permanent magnet
$PM_{2.5}$. *See* Fine particulate matter
PM_{10}. *See* Fine particulate matter
PMS. *See* Pavement Management System
POI. *See* Photochemical oxidant impact
Pollutants emissions. *See* Vehicle air pollutant emissions
  reduction, 51
  transportation sector. *See* Priority pollutants
Pollution
  emissions, 5–6
  reduction, 197
Pollution-control equipment, economic cost, 82

Poppet valves, 203
  sequence, usage, 201–202
Population densities, increase (occurrence), 36
Portable emission measurement systems (PEMS), 114
  example, 115f–116f
  usage, 103, 108
  summary, 118t
Port fuel-injected gasoline engine
  efficiency map, 151f
  usage. *See* Light-duty vehicles
Port Fuel Injection (PFI) engine-powered vehicles, fuel economy improvement ratios, 170t
Port Fuel Injection (PFI) gasoline-powered vehicles, hybridization, 169
Port Fuel Injection (PFI) ICE vehicles, comparison. *See* Baseline PFI ICE vehicles
Port-injected gasoline engine, 152–153
Power
  density, trade-off, 144
  electronics, 141–144
  usage, 139
  management, subsystems, 158
  pulse, efficiency, 145
  requirement, 144–145
  storage, increase, 206
Power electronics/controller
  calculation, 184
  electronic transmission role, 154–155
Powertrain
  components, right-sizing, 192
  control strategies, 162–165
Power transmission control. *See* Advanced vehicle power transmission control
Preventive maintenance, effectiveness, 268
Price-times-quantity payments, domestic producer surplus (subtraction), 75
  usage, 75
Price-times-revenue calculation. *See* Motor vehicle use
Price values, impact, 175t
Primary standards, 52
Priority pollutants

Index **345**

emissions, transportation sector, 7t
transportation direct emissions (2000), 6t
transportation life-cycle emissions (2000), 6t
Prius (Toyota)
  electric machine design, 137
  engine. *See* Atkinson cycle gasoline engine
Private-sector production/pricing. *See* Motor vehicle goods/services
Probabilistic LCCA, 267–268
Probability density function (PDF), 267
Producer surplus, estimated net, 73–76
Project development process, TOD (relationship), 41f
Project-level transportation infrastructure management decisions, 260–261
Proof-of-concept prototypes, improvements (acceleration), 192
Proof-of-concept vehicles, design process, 192
Proportional integrative derivative (PID) techniques, 192
Proton exchange membrane fuel cell (PEMFC), 132–134
  characteristics, 156t
  development, 134, 139–140
  DOE description, 238, 240
  fuel economy/equivalents, 238t
  hydrogen/air operation, 159
  internals, 238, 240
  processes
    illustration, 240f
    schematic, 156f
  system diagram, 158f
Proton exchange membrane fuel cell vehicle (PEMFCV), 237
PSAT (vehicle simulation computer program), 151–152, 166
ptBiodiesel, 229–230
Public motor vehicle goods/services, estimates (comparison), 92

Public participation strategies, 22
Public-sector infrastructure costs, monetary externalities (consideration), 72
Public transit projects, impact. *See* Water resources
Public transportation
  conclusion, 43–44
  definition, 16–18
  environment, relationship, 15
  introduction, 15–16
  references, 44–46
  systems, environmental linkages, 18–42
Pump/monitor technologies, usage. *See* Hybridization
Pump/motor
  technologies, consideration, 199–200
  types, flow rate, 199
  volumetric displacement, 199
Pump noise, 209–210
Pure gasoline, blend, 220
Pure public goods, resources usage (inefficiency), 78–79
Pyrolysis. *See* Hydrogen

Q

Quasi-Pigouvian taxes, 81–82

R

Ramp metering, 109
  usage, 122
Raw material construction, 290
RD. *See* Natural resource depletion
Rear-end accidents, 109
Rear motors, regenerative brake force changes, compensation, 208
Rechargeable organic liquids, 247
Recuperator, inclusion. *See* Gas turbine engine
Reduced-fare rides, inclusion, 106
Refinery emissions, Pigovian charge assessment, 82
Regenerative brake force, changes (compensation). *See* Front motors; Rear motors

Regenerative braking. *See* Energy smoothing, 208
  system, usage. *See* Dead band parallel regenerative braking system
Regional node, uses (mixture), 40
Regional Transit District (RTD), TOD policy (adoption), 40–41
Regulatory speed limit signals (erection), 110
Rehabilitation plan, proposal, 268
Remote sensing device (RSD), 114, 117–118
  schematic. *See* Emissions
Residential garages, prices (inclusion), 76
Resource Management Act of 1991, 295
Resources
  consideration, 67
  depletion, 270
  usage, inefficiency, 78–79
Resource-use decisions, 71
Reversible flow lanes, 112
Ridership-based environmental impacts, 33–34
Ridesharing, TDM strategies, 106
Roundabouts, 112
  control, unit emissions, 124t
RSD. *See* Remote sensing device

S

Saccharomyces cerevisiae, 216
Safe Accountable Flexible Efficient Transportation Equity Act: A Legacy for Users (SAFETEA-LU), 288
Safety risk (SR), 272
Secondary standards, 52
SEL. *See* Sound exposure level
Sensitive habitats, transit facility environmental impacts, 21
Series hybrid vehicles, 136–137
  battery
    energy storage capacity, 136

Series hybrid vehicles, (Continued)
  usage, 183
  breakeven gasoline price, 183–184
  comparison, 184t
  characteristics, 171t
  control strategies, 163
  cost characteristics, 183t
  design, 136
  driveline schematics, 131f
  economic considerations, 177–187
  market penetration, 177, 183–184
  operation/performance, simulations, 169–171
Series hydraulic hybrids, 195–197
  architecture, 196f
Service costs, monetary externalities (consideration), 72
SETAC. See Society of Environmental Toxicology and Chemistry
Short-haul flights, passenger-kilometer carbon emissions, 316
Short haul routes, 311f
SIGNAL2000, 122
Signal coordination, impact. See Speed profiles
Signalization, 122
Signal priority. See Buses
Signal retiming programs, 111–112
Signal system coordination project, implementation, 118t
Signal timing/coordination, 111–112
Simple Electric Vehicle Simulation (SIMPLEV) program, usage, 166
Single occupancy vehicle (SOV) drivers, shift, 106
SIPs. See State Implementation Plans
Sketch planning tools, 292
SLAMS. See State and Local Air Monitoring Stations
Slip velocity, algorithm (proposal), 208
SMD. See Solid mass disposal
Smoke control, 209

SO_2. See Sulfur dioxide
SOC. See State of charge
Social cost, estimation, 82
Social-cost accounting, 92
Social-cost analysis
  conceptual issues. See Motor vehicle use
  context, 64–65
  detail/quality, 62
Social cost-benefit analysis. See Motor vehicle use; Parking
  performing, 64
Social impacts. See Transit projects
Society of Environmental Toxicology and Chemistry (SETAC), 270
  definition. See Life-cycle assessment
Sodium hydroxide (NaOH), catalyst, 225
SOFC. See Solid oxide fuel cell
Soil exposure, environmental hazard, 23
Solenoid valves, control system regulation, 205
Solid mass disposal (SMD), 272
Solid oxide fuel cell (SOFC), characteristics, 156t
Solid waste generation, 290
Sound exposure level (SEL), 29
Sound levels, community annoyance, 30f
Sound Transit, environmental policy, 37f
  implementation, 37–38
SOV. See Single occupancy vehicle
Special Purpose Monitoring Stations (SPAMS), 53
Speed-acceleration look-up table, 123
Speed change cycles, number/severity, 111
Speed profiles. See VISSIM-generated speed profiles
  capture/geo-referencing, 102–103
  maintenance. See Near-uniform speed profiles
  representations, signal coordination (impact), 117f
  trips, example, 108f
  VSP distribution, 104f
Speed reduction, 110
Spool valves
  sequence, usage, 201–202
  usage. See Hydraulic ram position
Spring piston accumulator, 200f
SR. See Safety risk
St. Johns Schools, 229
Stakeholders, aircraft impacts (relationship), 311–317
Starch-bearing crops, 216
Starches
  fermentation, 215
  saccharification, 216
State and Local Air Monitoring Stations (SLAMS), 52
State Implementation Plans (SIPs), 53, 118–119
State of charge (SOC). See Battery
  maintenance, 163
  range, 165, 167
  specification, 164–165
STEAM, 293
Stochastic Dynamical Programming, 207
Stoichiometric
  port-fuel-injected engines, usage. See Light-duty vehicles
Stop-and-go cycles, reduction, 113
Storage capacity
  increase. See Hydraulic vehicles
  reversible mechanism, 206–207
Storage tank, pressure regulation, 157
StratBENCOST, 293
StratBencost, 269
Strategic decision-support tools, 259–260
Sucrose-bearing crops, 215
Sugar
  extraction, 216
  impact, 215
Sugar-bearing crops, 216
SULEV. See Super ultra low-emission vehicle
Sulfur dioxide (SO_2), 48, 49
Sulfuric acid (H_2SO_4), 49
Sulfur oxides, NAAQS, 52t
Sulphur dioxide, emissions, 321

Supersonic flight, operating costs, 325
Super ultra low-emission vehicle (SULEV), 173
  standard, 174
Surface speed operational strategies, 111–113
Surface transportation, direct emissions, 98
Surface water migration, environmental hazard, 23
Sustainability, relationship. See Asset management
Sustainable development, transit (linkage), 39–42
Sustainable transportation infrastructure management, 261. See also Transportation life-cycle assessment
  conclusions/recommendations, 278
  introduction, 257–258
  performance measures, 262–264
  problems/solutions, perceptions, 268–269
Sustainable transportation system, ODOT definition (adoption), 19
Swash plate hydraulic pump/motor, 198f
Swash plate pump/motor, consideration, 198–199
Swash plate surface, pistons (riding), 198–199
Switch reluctance AC motors, usage, 140
Syngas, fermentation, 217
Synthesis hydrocarbons, usage, 251
System metrics, comparisons, 160t

T

TAMT. See Transportation Asset Management Today
Tank-to-wheels emissions, 175
Taxation, distortion, 91
Taxes, estimated net, 73–76
Tax policy, 35t
TCMs. See Transportation control measures
TDM. See Travel demand management

TEA-21. See Transportation Equity Act of the 21st Century
Technology, economic considerations, 177–187
Technology-based environmental impacts, 26–33
Thermal management, subsystems, 158
Thermochemical cycles. See Hydrogen
Three-mode modal emission models, 122
Tidal flow lanes, 112
Time horizon, TDM strategies (mapping), 107t
Time-varying power demands, 140
Tire-rolling resistance, 135
TOD. See Transit-oriented development
Ton-miles, fatalities/injuries, 10t
TOS. See Traffic operational strategies
Total CO_2 emissions, considerations. See Charge sustaining full hybrid; Plug-in hybrid electric vehicle
Total emissions, control. See Airports
Total full-cycle emissions, 176
Total government expenditure, portion, 79
Total highway transportation costs, 266
Total social cost, components (classification). See Motor vehicle use
TRADEOFF, 306
Traffic congestion management emission impacts,
  assessment, 113–124
  introduction/background, 97–98
  methods, inventory, 105–113
  scope, 98–99
  summary, 124–125
Traffic management, 311
Traffic operational strategies (TOS), 98
  comparison. See Travel demand management
  usage, 108–113

Traffic performance, 122
  summary, 118t
Traffic signal
  control priority. See Arterials
  unit emissions, 124t
Train stations, bicycle park-and-ride facilities, 106
TRANPLAN, 122
TransCad, 122
Transesterification, 225
Transit alternatives, TDM strategies, 106
Transit environmental stewardship, 36–38
Transit facilities
  aesthetics, impact, 24–25
    mitigation, 25
  design/construction, environmental impacts, 20–25
  environmental impacts, 20f
  environmental justice issues, 21–22
    mitigation, 22
  equity issues, 21–22
    mitigation, 22
  visual impacts, 24–25
    mitigation, 25
Transit linkage. See Sustainable development
Transit New Zealand, 296
Transit noise
  measurement, 29
  mitigation measures, 32t
Transit operations,
  environmental impacts, 20f
  overview, 25–26
Transit-oriented development (TOD), 17
  characteristics, 40–41
  guidelines, 42
  importance, 41–42
  policies, 40
  process, 41
  relationship. See Project development process
Transit projects,
  social/economic impacts, 23–24
  mitigation, 24
Transit stations, town center prominence, 40

348   Index

Transit system
  environmental impacts. *See* Operating transit systems
  impact. *See* Longer-term environmental stewardship factors
  infrastructure, physical design/construction (impact). *See* Natural environment/community
Transportation
  activity
    causes/effects, 274f
    impact. *See* Air quality
  air quality, relationship, 47
    references, 55
  air quality planning, 54–55
  conformity, 53
  decision-making process, participation fairness (ensuring), 21
  direct emissions. *See* Priority pollutants
  economic/environmental footprints, 1
  introduction, 1–2
  references, 12–13
  summary/conclusion, 11–12
  emissions, 11t
  energy
    consumption. *See* U.S. passenger transportation energy consumption
    use (2000), 4t
  environment, relationship. *See* Public transportation
  environmental impacts, USEPA indicators (development), 274
  external costs, 11t
  externalities, 11
  infrastructure
    application, 273–274
    LCI/LCA approaches, 273
    management, sustaining, 261
  investment policy, 35t
  investments, social/economic effects, 277f
  life cycle effects, direct effects (contrast), 9

life-cycle emissions. *See* Priority pollutants
long-range scenarios, costs (evaluation), 62–63, 69–70
mode
  definition/characteristics, 18t
  environmental impacts, examples, 275t
models, 122–124
policies, costs (evaluation), 62–63, 69–70
pricing policy, 35t
projects
  environmental impacts, 274–277
  evaluation, 62–63, 69–70
  impact, 277–278
research/funding, prioritization, 63–64
sectors, emissions, 8t
services
  prices, establishment, 63
  usage, ensuring, 63
statistics, 3t
system
  demand/travel intensity, restriction, 998
  ODOT definition, adoption. *See* Sustainable transportation system
  operational performance, improvement, 98
Transportation Asset Management Today (TAMT), 295
Transportation control measures (TCMs), 53
Transportation Equity Act of the 21st Century (TEA-21), 288
Transportation Research Board (TRB) task force. *See* Asset management
Transport modes, carbon intensity, 316
TRANUS, 122
Travel, mode, 16–17
Travel demand management (TDM), 105–107
  strategies, 98, 105–107
  integration. *See* Air quality

mapping. *See* Time horizon; Vehicle miles of travel
TOS strategies, comparison, 113
Travel demand models, 122
Travel-related inputs, 54
Travel time distribution
  driving mode (MODE) usage, 99, 100–101
  vehicle activity descriptor, 101–102
  hybrid VDP-S approach, usage, 99
  by VSP mode, 99
TRC function, 205
Triglycerides, 225
  transesterification, alcohol (inclusion), 225f
Trip emission rates, measurement, 109t
Trip making behavior, impact, 107
Tropospheric ozone, 98
Turbo-charged diesel engine, 152–153

U

ULEV. *See* Ultra low-emission vehicle
Ultracapacitors, 144–150
  batteries, comparison. *See* Hybrid vehicles
  calculations, 184
  characteristics. *See* Vehicles
  cost (cents/Farad), 149
  high-power devices, 147
  relative costs, 150t
  technologies, 147
Ultra low-emission vehicle (ULEV), 173. *See also* Super ultra low-emission vehicle
Unaccounted for cost, synonymity. *See* Externalities
Unimproved land, parking, 77t
United Nations Framework Convention on Climate Change (UNFCCC), 305
United States Advanced Battery Consortium (USABC), 135
  goal, battery cost (assumption), 178

Index 349

Unit emissions, reduction, 108–109
University of California-Davis, 166
University of Toledo, 195
Unpriced costs, addition (avoidance), 90–91
Unsaturated fatty acids, impact. *See* Nitrogen oxide
UPS package delivery truck
  development, 197
  example, 197f
Upstream emissions. *See* Full cycle fuel-related emissions
  expression, 175
Upstream full fuel cycle emissions, 175–177
Urban daily VMT, 97
Urban Mobility Report of 2005 (Texas Transportation Institute), 97–98
U.S. Army hydraulic hybrid shuttle bus, 194f
U.S. Department of Transportation (USDOT), Intelligent Transportation Systems Benefit Page, 109
U.S. energy use, 3–4
U.S. Highway Capacity Manual (HCM), 122
U.S. passenger transportation energy consumption, 28t
USABC. *See* United States Advanced Battery Consortium
Use, efficiency. *See* Efficiency of use
Utah Transit
  cost assessment, 39
  EMS adoption, 38–39

V

Valves, 201–204
  design. *See* Check valve design
  necessity. *See* Ball valves
  selection. *See* Control valves; Critical-center valves
  sequence, usage. *See* Poppet valves; Spool valves
  usage. *See* Active components valves
Vanpool programs, TDM strategies, 106
Variable Valve Timing and Lift Electronic Control System (VTEC), 151, 152. *See also* Honda Insight VTEC engine
  gasoline engine, 152
VEETC. *See* Volumetric Ethanol Excise Tax Credit
Vegetable oil (VO), 224–234
  availability, 231–232
  biomass source, 225
  emissions, 229–231
  fuel transportation issues, 231
  introduction, 224–225
  manufacturing methods, 225–227
  performance, 229
  production capacity, 232–233
  quality standards, 227–228
  safety, 233
  storage issues, 231
  subsidy, 233
  vehicle modifications, 228–229
  viscosity, 228
Vehicle emissions
  direct measurement methods, 114–118
  estimation, fundamentals, 99–104
Vehicle-kilometers travel, reduction, 318
Vehicle miles of travel at specified speeds/speed ranges (VMT-S), 99, 100
Vehicle miles of travel (VMT), 99–100. *See also* Urban daily VMT
  direct reduction, 106
  distribution, 118–119
  usage, 119
  reduction, 66
  TDM strategies, mapping, 107t
Vehicle-miles traveled, servicing, 111–112
Vehicle powertrain
  component selection, 134–140
  configurations, 134–140
  control options, 134–140
Vehicles
  activity, nature, 100
  air pollutant emissions, 26–27
  mitigation, 27
  applications, hydrogen storage, 160–162
  applications, ultracapacitors (characteristics), 148
  attribute requirements, 162t
  characteristics, 161t
  design options, 130–134
  exhaust emissions, regulation, 173–174
  frontal area, 135
  hydrogen storage characteristics, 162t
  maintenance, reduction, 197
  modifications, 219t
  operation, ranges (fuel efficiency map), 193f
  performance, improvement, 197
  simulation computer programs, 151–152
  speed, impact. *See* Carbon monoxide
Vehicle specific power-speed (VSP-S) approach, 99. *See also* Hybrid VSP-S approach
Vehicle specific power-speed (VSP-S) emission factor models, 120
Vehicle specific power (VSP) bins, 124
  contrast. *See* Carbon monoxide
  correlation. *See* Emissions
  distribution. *See* Speed profiles
  emission factor models, 120
  modal definitions, 102t
  mode, 99, 101–103
  values, 124
Vehicle Stability Control (VSC) functions, 205
Vehicular travel activities, demand (reduction), 105
Vibration, 31–33
  impact, 36
  mitigation, 33
Video detection technology, usage, 117
Virgin Atlantic, system trials, 315
VISSIM-generated speed profiles, 123f
VISSIM model, 123

Visual impacts. *See* Transit facilities
VMT. *See* Vehicle miles of travel
VMT-S. *See* Vehicle miles of travel at specified speeds/speed ranges
VO. *See* Vegetable oil
Volatile organic compounds (VOCs), 48–50
percentage, 98
Volumetric Ethanol Excise Tax Credit (VEETC), 222
Volvo Environmental Concept Bus, study, 205

VSC. *See* Vehicle Stability Control
VSP. *See* Vehicle specific power
VSP-S. *See* Vehicle specific power-speed
VTEC. *See* Variable Valve Timing and Lift Electronic Control System

W

Wankel engine, 154
Waste, characteristics, 23
Waste cooking oil, 233
Water management subsystems, 157–158

Water pollution, environmental costs, 80–81
Water-reactive metal hydrides, 247
Water resources, public transit projects (impact), 25
Water vapor emissions, increase, 321
Well-to-tank emissions, studies, 175
Well-to-wheels emissions, studies, 175
What-if tools, 292
Wheel cylinder, pressure, 208
Work-related activities, 39